A NEW SWING-CONTRACT DESIGN FOR WHOLESALE POWER MARKETS

IEEE Press
445 Hoes Lane
Piscataway, NJ 08854

A NEW SWING-CONTRACT DESIGN FOR WHOLESALE POWER MARKETS

Leigh Tesfatsion

IEEE PRESS SERIES ON POWER ENGINEERING

IEEE PRESS

WILEY

Published by John Wiley & Sons, Inc., Hoboken, New Jersey.

Published simultaneously in Canada.

Library of Congress Cataloging-in-Publication Data

Names: Tesfatsion, Leigh, author.
Title: A new swing-contract design for wholesale power markets / Leigh Tesfatsion.
Description: Hoboken, New Jersey : Wiley-IEEE Press, [2021]. | Series: IEEE press series on power engineering | Includes index.
Identifiers: LCCN 2020027003 (print) | LCCN 2020027004 (ebook) | ISBN 9781119670124 (hardback) | ISBN 9781119670131 (adobe pdf) | ISBN 9781119670148 (epub)
Subjects: LCSH: Electric utilities. | Electric utilities–Costs. | Options (Finance)
Classification: LCC HD9685.A2 T466 2021 (print) | LCC HD9685.A2 (ebook) | DDC 333.793/23–dc23
LC record available at https://lccn.loc.gov/2020027003
LC ebook record available at https://lccn.loc.gov/2020027004

Cover design by Wiley
Cover images: (Top image) © Sam Robinson/Gettyimages, (Center image) © dowell/Getty Images

Set in 10/12pt TimesLTStd by SPi Global, Chennai, India

MIX
Paper from responsible sources
FSC www.fsc.org FSC® C013604

This book is dedicated to Herman C. Quirmbach for his steadfast support of this challenging book project and for his insistence, by word and example over many years, that reasoning be both precise and clear.

CONTENTS

PREFACE

THE NEED FOR FLEXIBLE DEPENDABLE RESERVE provision in electric power systems has dramatically increased in recent years. Growing reliance on variable energy resources and greater encouragement of demand-side participation have led to greater uncertainty and volatility of net load. Consequently, system operators are finding it harder to secure reserve with sufficient flexibility to permit the continual balancing of net load, a basic requirement for power system reliability. This study reconsiders the design of wholesale power markets in light of these concerns. Four design principles are stressed: (i) Wholesale power markets must necessarily be forward markets due to the speed of real-time operations; (ii) Only one type of product can effectively be offered in a wholesale power market: namely, reserve, an insurance product offering availability of net load balancing services for future real-time operations; (iii) Net load balancing services offered into wholesale power markets primarily take the form of power *paths* that can be dispatched at specific grid locations over time; (iv) All dispatchable resources should be permitted to compete for the provision of power-paths in wholesale power markets without regard for irrelevant underlying technological differences. If these four principles are accepted, current trade and settlement arrangements for wholesale power markets need to be fundamentally altered. This study proposes a new linked swing-contract market design, consistent with principles (i)–(iv), that could meet the needs of centrally managed wholesale power markets better than currently implemented designs.

AUTHOR BIOGRAPHY

LEIGH TESFATSION, PhD, is a Research Professor, and Professor Emerita of Economics, Mathematics, and Electrical and Computer Engineering, at Iowa State University. She is a Senior Member of the IEEE whose work has been supported by grants and contracts from the US Department of Energy, Pacific Northwest National Laboratory, Sandia National Laboratories, Los Alamos National Laboratory, and NSF. She has served as guest editor and associate editor for a number of journals, including the IEEE Transactions on Power Systems, the IEEE Transactions on Evolutionary Computation, the Journal of Energy Markets, and the Journal of Economic Dynamics and Control.

ACKNOWLEDGMENTS

THIS STUDY HAS BEEN SUPPORTED by US Department of Energy (DOE) grants DE-AR0000214 and DE-OE0000839, by Contract No. 339051 with the Pacific Northwest National Laboratory (PNNL) operated by Battelle for the US DOE under Contract DE-AC05-76RL01830, by Contract No. 1163155 with Sandia National Laboratories, by Project Award No. M-40 from the Power Systems Engineering Research Center (PSERC), and by various grants from the Iowa State University Electric Power Research Center (ISU EPRC).

I am especially grateful to Joe Dygert, Joe Eto, Marija Ilić, and Lorenzo Kristov for early support and encouragement of the swing-contract ideas expressed in this study, and to Swathi Battula, Raymon H. Bryne, Ryan T. Elliott, James F. Ellison, Ross Guttromson, Deung-Yong Heo, Wanning Li, Verne Loose, Shanshan Ma, César A. Silva-Monroy, and Zhaoyu Wang who have collaborated with me on previous swing-contract research. I am indebted to Ross Baldick, Ramón Alberto León Candela, Kwok Cheung, Paul Gribig, Bill Hogan, Akshay Korad, Dick O'Neill, Jim Price, Fereidoon Sioshansi, Di Wu, Nanpeng Yu, and Tongxin Zheng for thoughtful comments on earlier draft versions. I thank Yonghong Chen, Hailong Hui, Daniel Kirschen, and Eugene Litvinov for patiently responding over the years to my many questions about current US RTO/ISO-managed wholesale power market operations. Finally, I thank five anonymous referees for helpful constructive comments on preliminary drafts of this monograph.

CHAPTER 1

INTRODUCTION

"Design to the mission, design as a system, keep it simple."

—[109, p. 20]

CENTRALLY MANAGED WHOLESALE power markets operating over high-voltage transmission grids support the steady flow of electric power from bulk power sellers to bulk power buyers, for ultimate resale and distribution to retail customers. This mission has been complicated in recent years by a dramatic surge in the availability and use of *variable energy resources (VERs)*.

A VER is a power source whose power injections into a transmission grid cannot be fully dispatched in a controlled manner to balance changes in power withdrawals or to meet other system requirements. Examples include solar panels and wind turbines that are not fully firmed by storage. The increased participation of VERs in wholesale power markets, together with the increased encouragement of active demand-side participation, increases the uncertainty and volatility of grid *net load*, i.e. power withdrawal net of non-dispatched power injection.

In consequence, as discussed more fully in Chapters 2–3, US RTO/ISO-managed wholesale power markets[1] are finding it harder to secure dependable reserve with sufficient flexibility to permit the continual balancing of net load, a basic requirement for power system reliability. Trade and settlement arrangements in these markets are still largely based on rigid reserve definitions, eligibility requirements, and settlement processes that make it difficult to ensure adequate provision and appropriate compensation of needed reserve from multiple types of resources. Emphasis is placed on the designation and compensation of artificially separated product concepts such as energy, ramping, and capacity, whereas value

[1]The US Federal Energy Regulatory Commission [58] defines an *RTO/ISO-managed wholesale power market* to be the collection of all capacity, energy, and/or ancillary service markets operated by a *regional transmission organization (RTO)* or an *independent system operator (ISO)*. The key distinction between an RTO and an ISO is that RTOs have larger regional scope.

A New Swing-Contract Design for Wholesale Power Markets, First Edition, Leigh Tesfatsion.

in power markets in fact principally arises from the dispatchable availability and delivery of *power-paths*, i.e. flows of power into and out of a grid at specific grid locations during designated operating periods.

This study reconsiders the design of US RTO/ISO-managed wholesale power markets in light of these concerns. Four market design principles are stressed:

[MD1:] All wholesale power markets must necessarily be forward markets[2] due to the speed of real-time operations.

[MD2:] Only one type of product can effectively be offered in a wholesale power market: namely, reserve, an insurance product offering availability of net load balancing services for future real-time operations.

[MD3:] Net load balancing services offered into wholesale power markets generally take the form of power-paths that can be dispatched at specific grid locations over time.[3]

[MD4:] All dispatchable resources should be permitted to compete for the provision of power-paths in wholesale power markets without regard for irrelevant underlying technological differences.

A *swing-contract market design* is proposed that is in accordance with principles MD1–MD4. This design envisions an ISO-managed wholesale power market M(T) organized as a reserve market for some designated future operating period T. Reserve consists of dispatchable power-paths for period T. As illustrated in Figure 1.1, a *power-path* for period T refers to a sequence of power injections and/or withdrawals at a single designated grid location during period T.[4] Dispatchable resources offer reserve (dispatchable power-paths) into M(T) by means of "swing contracts."

More precisely, as carefully explained in Chapter 4, a *swing contract* SC_m issued by a dispatchable resource m is a reserve contract that m can offer into a swing-contract market M(T) in either firm or option form.[5] SC_m consists of four components, each specified by m: (i) an offer price α_m; (ii) an exercise set $\mathbb{T}_m^{\mathrm{ex}}$; (iii) a

---

[2] A *forward market* is a market involving the purchase and sale of a product for which the payment method for the product is contractually determined in advance of its delivery date. In contrast, in a *spot market* the delivery and payment for a product are determined at the same time.

[3] As discussed in [46, 59], *primary frequency response* is synchronized reserve capacity that autonomously responds to changes in system frequency; consequently, it is not dispatched. The provision and compensation of primary frequency response is not considered in the current study.

[4] Since a power-path refers to the injection and/or withdrawal of power at a single grid location over time, a power-path is characterized without reference to spatial transmission. As illustrated in Figure 1.1, power-paths can be depicted in a time-power plane.

[5] As explained more fully in Chapter 4, a *firm contract* is a noncontingent contract that imposes obligations on both the issuer and the holder. An *option contract* is a contingent contract that gives the holder the right, but not the obligation, to exercise the contract at one or more contractually specified exercise times. The exercise of an option contract converts it into a firm contract.

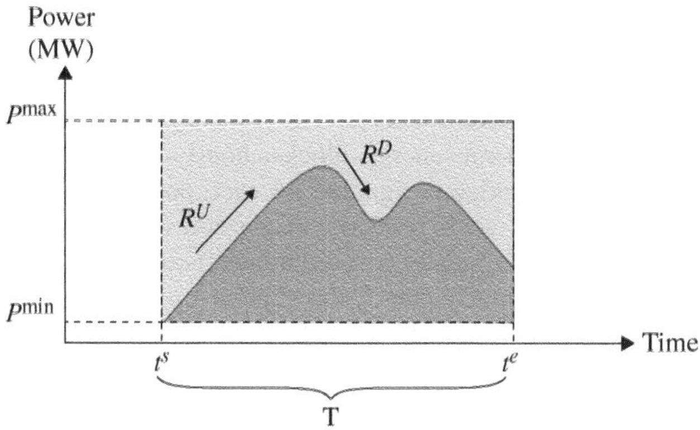

Figure 1.1 One of many possible power-paths that a dispatchable resource with swing (flexibility) in down/up ramping and power amplitude could be signaled to deliver at its grid location during operating period $T = [t^s, t^e)$.

physically characterized set $\mathbb{PP}_m$ of power *paths* for period T, each of which m could feasibly deliver at a designated grid location during T in response to dispatch signals; and (iv) a performance payment method ϕ_m.

If SC_m is cleared, the offer price α_m (if positive) is paid to m either directly or in amortized payment-schedule form. The offer price thus permits m to *cover ex ante* any cost that m would have to incur to ensure the *availability* of the power-paths in $\mathbb{PP}_m$. This availability cost could include capital investment cost, start-up cost, no-load cost, and opportunity cost. The exercise set $\mathbb{T}_m^{ex}$ consists of designated times between the close of M(T) and the start of T at which the ISO can exercise SC_m, assuming SC_m has been cleared. The form of this exercise set determines whether SC_m is a firm contract or a type of option contract.[6]

The dispatchable power-paths in $\mathbb{PP}_m$ are characterized in terms of attributes such as delivery location, start-time, minimum down/up time, active and reactive power limits, ramp-rate limits, duration limits, and energy capacity. The precise specification of these attributes determines the degree of swing (flexibility) in m's offered reserve. Finally, the performance payment method ϕ_m permits resource m to *recover ex post* any cost that m incurs for verified period-T service *performance*, i.e. for the verified period-T delivery of a power-path in $\mathbb{PP}_m$ in response to dispatch signals. This performance cost could include fuel cost, labor cost, transmission service charges, and machinery wear and tear caused by fast ramping.

Reserve offers submitted into M(T) take the form of portfolios of swing contracts offered by dispatchable resources for operating period T. These dispatchable resources can include generators, distributed-resource aggregators, and storage facilities. Reserve offers in firm form effectively constitute regulation reserve,

[6] As will be clarified in Section 4.3, standard types of option contracts are distinguished by the number and positioning of their exercise times.

whereas reserve offers in option form effectively constitute contingency or planning reserve.

As demonstrated in Chapter 5, these reserve offers can take the standard supply-offer forms required by current US RTO/ISO-managed wholesale power markets. Examples include: must-run energy blocks; hourly step-function power supply schedules with a separate price designated for each power-step; and power self-scheduled by power traders to secure needed transmission support for the power outcomes of privately negotiated physically covered bilateral contracts.

However, as is also demonstrated in Chapter 5, the general formulation of a swing contract can accommodate reserve offers with a much broader range of offered attributes than envisioned in these standard supply offer forms. Moreover, the issuer m of a swing contract SC_m can use the performance payment method ϕ_m included in SC_m to specify m's required compensation ex post for dynamic aspects of a delivered power-path, such as ramping, duration, and reactive power support, as well as static aspects such as total delivered energy.

Reserve bids submitted into a swing-contract market M(T) take the form of price-sensitive and/or fixed demands for power-path delivery during operating period T. Reserve bids can be submitted by load-serving entities to service the forecasted loads of their customers during T, and by power traders who need to self-schedule the power outcomes of privately negotiated physically covered bilateral contracts in order to secure needed transmission support.

As detailed in Chapters 6–9, an ISO managing a swing-contract market M(T) solves a contract-clearing optimization problem to determine which reserve offers and price-sensitive reserve bids to clear for operating period T. The objective of the ISO is to maximize the expected total net benefit of the market participants, conditional on initial state conditions and subject to system constraints.

Total net benefit consists of total benefit net of total avoidable cost. The system constraints include power balance, transmission line, and reserve constraints. These constraints incorporate, as exogenous inputs: (i) all fixed demands; (ii) all forecasts for non-dispatched power injection; (iii) all of the power-path attributes included by dispatchable resources in their reserve offers; and (iv) system-wide and zonal reserve requirements set by the ISO to ensure coverage of net load uncertainty sets as a robust means of protection against net load forecast errors.

The ISO functions as a clearing house for M(T), collecting payments and overseeing payouts to market participants. However, the ISO does not have any financial stake in market operations. To maintain this independent status, all net reserve cost[7] and transmission service cost incurred through market operations are passed through to market participants. Net reserve cost is allocated across market participants based on the relative volatility and size of their net must-service load.[8] Transmission service

[7]*Net reserve cost* is reserve procurement cost net of any price payments for cleared price-sensitive reserve bids and net of any penalty payments for real-time deviations from dispatch signals.

[8]The *net must-service load* of a market participant at a particular grid location is the amount of its non-dispatched power withdrawal at that location, if any, minus the amount of its non-dispatched power injection at that location, if any.

cost is allocated across market participants based on the power imbalance[9] at their grid locations.

More generally, Chapter 10 proposes a *linked* collection of swing-contract markets whose look-ahead horizons for designated future operating periods can range in duration from multiple years to minutes. The linkage among these markets is achieved by having the reserve offers and price-sensitive reserve bids cleared in earlier markets be carried forward on the books of the ISO as a portfolio of contracts that can be adaptively updated in subsequent markets. This linkage facilitates reserve procurement by permitting a successively refined understanding of resource availability and system conditions for future operating periods.

The key features of this *Linked Swing-Contract Market Design* in comparison with current US RTO/ISO-managed wholesale power market designs, elaborated in Chapters 10–15, are summarized below:

- permits the robust-control management of uncertain net load
- handles uncertain net load by ensuring flexible dependable reserve supply
- eliminates the need for detailed net load scenario specifications
- facilitates a level playing field for resource participation
- recognizes the forward nature of wholesale power markets
- recognizes all offered product in these forward markets is a form of reserve
- identifies reserve as dispatchable power-paths available for future operations
- requires resources to internally manage commitment and capacity constraints
- permits co-optimization across a wide range of reserve attributes
- ensures settlement time-consistency through two-part pricing
- compensates reserve availability ex ante and reserve deployment ex post
- permits resource owners to cover ex ante their full costs of availability
- permits resource owners to recover ex post their full real-time performance costs
- eliminates the need for out-of-market payment adjustments
- provides system operators with real-time flexibility for net load balancing
- encourages close following of dispatch signals through performance incentives
- reduces the complexity of market rules.

Chapter 16 considers how current US RTO/ISO-managed day-ahead markets could gradually transition to a swing-contract market design. As shown, a swing contract submitted by a dispatchable resource into a day-ahead market can in principle

[9] *Power imbalance* is said to occur at a particular bus in a transmission grid if there is a nonzero net power injection at this bus that requires the transmission of power to or from other buses in order to ensure power balance across the transmission grid as a whole.

be incorporated as follows. First, the swing-contract's offer price and performance payment method can be incorporated into the objective function for the optimization used by the RTO/ISO to solve for generator unit commitments and scheduled dispatch levels for next-day operations. Second, the power-path attributes designated by this swing contract can be incorporated into the system constraints for this optimization.

However, in order for this incorporation to result in accurate merit-order dispatch for next-day operations, the optimization would have to account *fully* for the expected total net benefit associated with each possible configuration of generator unit commitments and scheduled dispatch levels. At present this is not the case. For example, the unit commitment costs appearing in the objective function typically cover (at most) the start-up, no-load, and minimum-run costs of generators, not their full availability costs. Also, voltage limits are typically not included among the system constraints, thus preventing consideration of the benefits provided by offered voltage-support services. In consequence, swing contracts offering diverse dispatchable power-paths, with explicit offer prices and performance payment methods ensuring full coverage of availability and performance costs, could be incorrectly omitted from the merit-order dispatch stack on the grounds they are too costly.

To illustrate what might be done to address this issue, Chapter 16 presents an extended day-ahead market optimization in complete analytical form that permits a fuller range of costs to be incorporated in the objective function. It is shown, explicitly, how swing contracts offering dispatchable power-paths with swing (flexibility) in power amplitude and ramp rate can be incorporated into this extended optimization

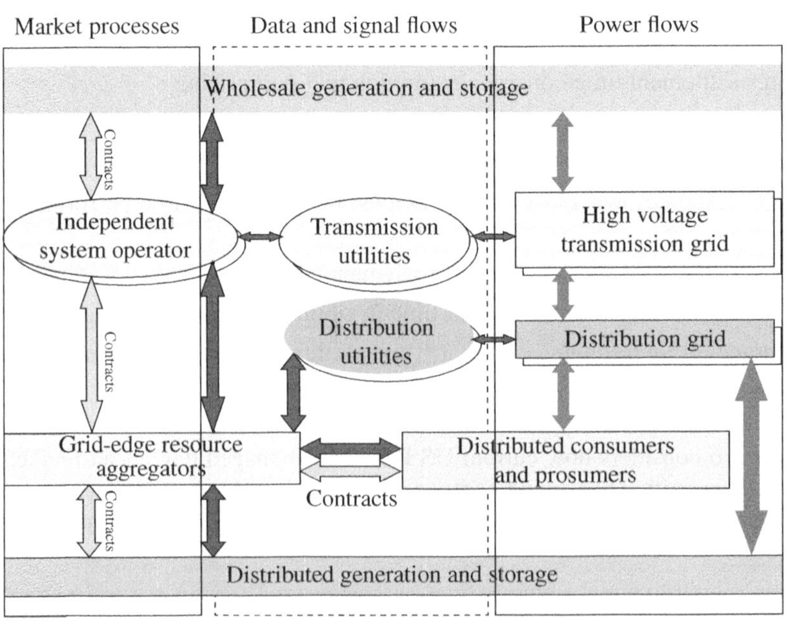

Figure 1.2 Illustration of an integrated transmission and distribution system.

along with standard types of supply offers while still retaining a mixed-integer linear programming formulation. The solution of this extended optimization results in an accurate merit-order dispatch stack.

Chapters 17–18 explore swing-contract support for integrated transmission and distribution system operations; see Figure 1.2. Special attention is focused on the possibility that *independent distribution system operators (IDSOs)*, functioning in distribution systems as grid-edge resource aggregators,[10] could use swing contracts to facilitate their participation in transmission systems as reserve providers as well as load-serving entities. The reserve provision of an IDSO could take the form of swing contracts offering the availability of dispatchable power-paths harnessed from grid-edge resources in return for appropriate compensation. This IDSO participation would permit retail customer interests to be more directly and completely represented at the wholesale power market table.

Potential future research directions are outlined in Chapter 19, and concluding remarks are given in Chapter 20. Glossaries and nomenclature tables for terms used to describe market operations in both standard and swing-contract forms are provided in Appendix A.

[10]In this study, a grid-edge resource is defined to be any entity capable of power usage and/or power output that is directly connected to a distribution grid. A grid-edge resource aggregator is any entity that manages power usage, power supply, and/or ancillary service provision for a collection of grid-edge resources.

CHAPTER 2

US RTO/ISO-MANAGED WHOLESALE POWER MARKETS: OVERVIEW

2.1 CHAPTER PREVIEW

The decision to propose a major reformulation of current US RTO/ISO-managed wholesale power markets in this study has been prompted by a careful consideration of eight broadly accepted goals for wholesale power market design. A summary statement of these eight design goals is provided in Section 2.2.

A brief overview of daily operations in current US RTO/ISO-managed wholesale power markets is provided in Section 2.3. Increasing stresses faced by these markets are discussed in Section 2.4.

As will be elaborated below in Chapter 3, problematic aspects of current US RTO/ISO-managed wholesale power market operations are hindering the ability of these markets to deal with the stresses outlined in Section 2.4, hence also their ability to adhere to the eight design goals outlined in Section 2.2.

2.2 GENERAL GOALS FOR WHOLESALE POWER MARKET DESIGN

Eight broadly accepted goals for wholesale power market design have motivated this study. These goals are based in large part on the goals set forth by Oren [141, Section II.A] and Tesfatsion et al. [188, Section 2]. Briefly summarized, these goals are as follows:

Goal-1. Incentive alignment: The market design should be well-aligned with the private objectives and constraints of the market participants, thus ensuring their voluntary participation.

A New Swing-Contract Design for Wholesale Power Markets, First Edition, Leigh Tesfatsion.

Goal-2. Resource adequacy: The market design should provide incentives for new resources to enter in sufficient quantity to accommodate retirements, de-ratings, and increases in power demand over time while maintaining adequate reserve to address uncertainty and volatility in net load.

Goal-3. Efficiency: The market design should be *efficient*, i.e. it should not waste resources. To achieve *short-run efficiency*, i.e. the non-wastage of *existing* resources, the design should require the bid/offer-based participation of both buyers and sellers. To achieve *long-run efficiency*, the design should encourage the development and adoption of new technologies permitting increased benefits from power usage as well as lower costs for power production and transmission.

Goal-4. Reliability and resiliency: The market design should ensure the continual balancing of net load during normal power system operations, despite weather events and other types of anticipated disturbances. The design should also support rapid recovery and return to net load balancing following sudden major disruptions, such as the loss of a line or a generation unit.

Goal-5. Open access: The market design should be fair, providing an even playing field for all potential and actual market participants. It should permit and encourage resources to compete for the provision of reserve and the procurement of power-path deliveries, and it should discourage the exercise of market power.

Goal-6. Conceptual coherency and transparency: The market design should be conceptually coherent, and market rules and operations under the design should be as transparent as possible.

Goal-7. Minimum administrative intervention: The market design should discourage ad-hoc rule-making and decision-making by administrators. To further this goal, market rules and operations should be based on service requirements rather than on the physical and operational attributes of resources, to an extent compatible with the attainment of other design goals. Wherever possible, mechanisms should be instituted to permit and encourage transition to a design with limited administrative control.

Goal-8. Supportive of previous reform efforts: The market design should be in accordance with recent FERC and RTO/ISO initiatives promoting increased market access, pay for performance, demand-side participation, and encouragement of private initiative.

2.3 US RTO/ISO-MANAGED MARKET OPERATIONS

Bulk power trading in all US electric power regions includes bilateral transactions managed by exchanges (centralized trading platforms), over-the-counter trading managed by dealers, and privately negotiated bilateral contracting managed by

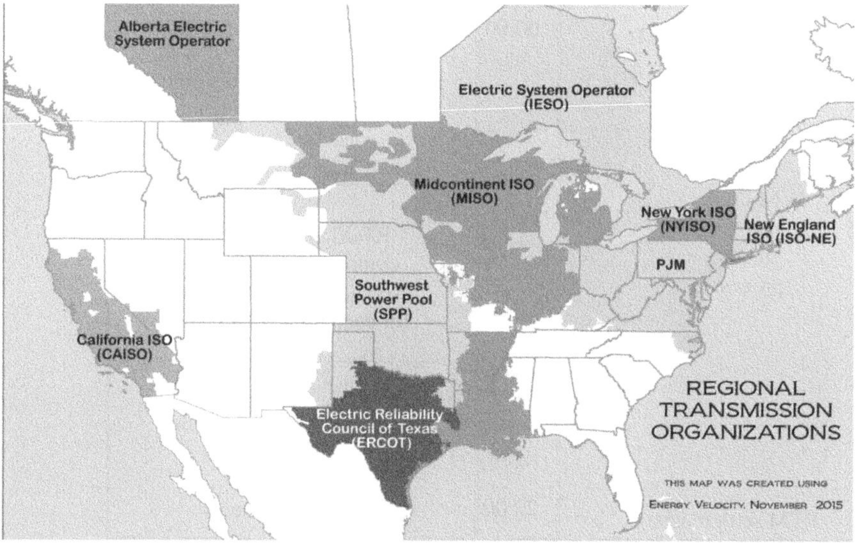

Figure 2.1 North American RTO/ISO-managed wholesale power markets. Public domain: [61].

brokers [55, Chapters 3]. Traditionally, the transmission services supporting this bulk power trading were provided by regulated high-voltage transmission grids owned and operated by vertically integrated utilities or other types of entities.

However, in a series of notices culminating in a 2003 White Paper [52],[1] the US Federal Energy Regulatory Commission (FERC) recommended that US electric power regions be re-organized as RTO/ISO-managed wholesale power markets operating over RTO/ISO-managed transmission grids. Power traders would thus be provided an additional means for conducting their bulk power trades: namely, submission of bids (to buy) and offers (to sell) into a wholesale power market. Cleared bids and offers would automatically be scheduled to ensure needed transmission support. For all other forms of physically covered bulk power trades, transmission support would have to be obtained by the self-scheduling of these trades in the wholesale power market.

Seven US electric power regions are currently operating in accordance with this FERC design: California Independent System Operator (CAISO), Electric Reliability Council of Texas (ERCOT), Independent System Operator for New England (ISO-NE), Midcontinent Independent System Operator (MISO), Independent System Operator for New York (NYISO), PJM Interconnection (PJM), and Southwest Power Pool (SPP); see Figure 2.1. Nearly two thirds of US electric power usage is currently serviced by these seven regions.

---

[1]This white paper is based on seminal work by Schweppe et al. [28, 159, 160] and Hogan [79, 80].

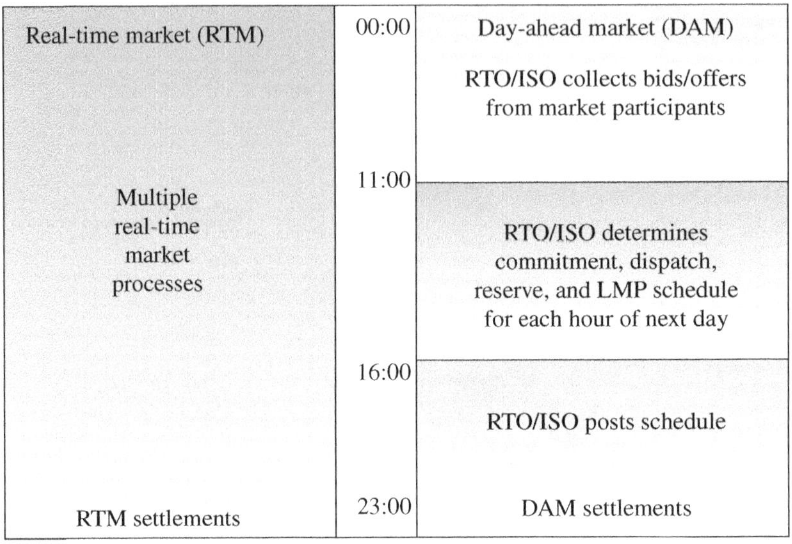

Figure 2.2 Simplified depiction of daily operations for an RTO/ISO-managed two-settlement system.

The FERC design envisions electric energy delivered at particular grid locations during standardized time-steps (e.g. hourly) to be the basic transacted commodity.[2] At the core of the FERC design is a *two-settlement system* consisting of a *day-ahead market (DAM)* operating in tandem with a *real-time market (RTM)* [135]; see Figure 2.2. The purpose of these DAM/RTM processes is to secure sufficient generation availability in advance of real-time operations to permit the continuous real-time balancing of net load.

DAMs rely on SCUC/SCED optimizations,[3] either in combined or separate form, to determine binary (0/1) unit commitments for dispatchable generators, scheduled generation dispatch levels, and reserve[4] for next-day operations. More precisely, the DAM SCUC/SCED optimization held on each day D determines unit commitments and scheduled dispatch levels for dispatchable generators for each hour H of day D+1. In addition, this optimization determines scheduled levels of reserve for each hour H of day D+1 to hedge against the possibility of net load forecast errors.

[2]In modern economic and financial usage, a *commodity* is an asset with a standard unit of measurement that permits one unit to be substituted for another unit with no change in valuation. Hence, commodities can be valued and traded in bulk multi-unit form.

[3]SCUC is an acronym for *Security-Constrained Unit Commitment*, and SCED is an acronym for *Security-Constrained Economic Dispatch*.

[4]In US DAM/RTM operations, reserve essentially refers to the unencumbered generation capacity of committed dispatchable generators, i.e. to the portion of their capacity that is not encumbered by scheduled dispatch responsibilities.

RTMs are balancing mechanisms consisting of one or more sub-markets with shorter look-ahead horizons whose purpose is to reduce discrepancies between DAM generation dispatch schedules and actual real-time net loads.[5] RTMs rely on SCED optimizations to determine scheduled generation dispatch levels and reserve for near-term operating periods, taking unit commitments as given.

The goal of the DAM/RTM two-settlement system is to permit the commitment, dispatch, and pricing of energy at particular grid locations during standardized time-steps to be determined by the demands and supplies of energy buyers and sellers. This determination process is meant to be in accordance with the determination of market-clearing quantities and prices in ordinary competitive commodity markets, to an extent consistent with the reliable operation of the transmission grid.

In pursuit of this goal, DAM/RTM settlements for scheduled generation dispatch levels are determined by locational marginal pricing, i.e. by the pricing of energy in accordance with the location and timing of its injection into, or withdrawal from, a physical grid. *Locational marginal prices (LMPs)*, expressed as US dollars ($) per megawatt hour (MWh), are derived from the solutions for DAM/RTM SCED optimizations, conditional on unit commitments.[6]

The generation dispatch levels scheduled in a day-D DAM at a grid location b for some hour H during day D+1 are settled in accordance with the LMP determined in the day-D DAM at b for H. Any adjustments needed in these dispatch levels, as determined in RTMs held after the close of the day-D DAM but prior to H, are settled using the LMPs determined in these RTMs. Finally, generators whose DAM/RTM revenues do not fully cover their reported unit commitment costs receive additional *out-of-market (OOM)* make-whole payments to ensure this coverage.

However, the FERC design does not provide any specific guidance for the provision and settlement of ancillary services.[7] Rather, these aspects of power system management are explicitly left to the individual states participating in each electric power region [52, p. 11]. In consequence, as detailed in [46, Tables 1 and 2], [133, Figure 4], and [146, Table 1], ancillary service procurement and settlement processes differ widely across the seven US energy regions currently organized as RTO/ISO-managed wholesale power markets.

[5]For example, the CAISO RTM is composed of a series of sub-markets run at different times with look-ahead horizons ranging from hours to minutes in advance of a designated future operating period, called the "binding interval" [26].

[6]As elaborated more carefully in Section 3.2.4, the LMP at a grid location b for a time-step k is the (right-hand) dual variable of the DAM/RTM SCED power balance constraint at b for k, calculated at the DAM/RTM SCED optimal solution point. See [5, 120] for detailed discussions of the mathematics of LMP determination in US RTO/ISO-managed wholesale power markets.

[7]*Ancillary services* are support services for the maintenance of net load balancing in accordance with required grid reliability standards, as determined by the North American Electric Reliability Corporation (NERC). Examples include: black-start capability, reserve, and reactive power production and absorption for voltage control.

2.4 STRESSES FACED BY CURRENT US RTO/ISO-MANAGED MARKETS

Since the official release of the FERC design in 2003, the conditions facing US RTO/ISO-managed wholesale power markets have dramatically changed. The increased participation of variable energy resources (VERs), together with the increased encouragement of active demand-side participation, have substantially increased the uncertainty and volatility of net load as well as the frequency of strong ramp events.

In consequence, flexibility in net load balancing services has become increasingly important to maintain the efficiency and reliability of power system operations. To meet this challenge, US RTO/ISOs have introduced major changes in their market rules and operational procedures. As reviewed in [44, 58, 71, 84], these changes have included: new demand response initiatives; new flexibility products (e.g. ramping products); revised market eligibility requirements (e.g. for storage); and a restructuring of capacity market payments in an attempt to ensure the continued availability of thermal generation as a backstop reserve.

For example, as discussed by [27, 136], CAISO and MISO have each proposed the introduction of "flexible ramping" products. Also, as discussed by [161], ISO-NE has introduced a major rule change called "Energy Market Offer Flexibility." In addition, some RTO/ISOs are exploring innovative ways to incorporate VERs more fully into DAM/RTM operations. Along these lines, MISO has introduced a new resource category called *dispatchable intermittent resource (DIR)*, designed primarily for its wind resources [128].

Nevertheless, several major issues remain unresolved for US RTO/ISO-managed wholesale power markets. As previously noted, one key issue is that reserve products are variously defined and compensated across these markets. This lack of uniformity makes it difficult to evaluate and compare their market operations.

A second key issue is ensuring appropriate compensation for flexibility in reserve provision. Reserves are specified in broad rigid terms (e.g. regulation, spinning reserve, non-spinning reserve) that do not permit resources to be further differentiated and compensated on the basis of additional types of valued flexibility, such as an ability to ramp up and down between minimum and maximum values over very short-time intervals. FERC Order 755 [39, 54, 78], issued in 2011, requires pay-for-actual-performance for frequency regulation (secondary frequency control) in US RTO/ISO-managed markets in an attempt to ensure higher compensation for higher quality performance.[8] To date, however, this order has been implemented differently across these markets, hindering evaluations and comparisons, and raising concerns regarding uncertain possible effects of these implementations [21]. Moreover, other types of reserve are not covered by this order.

[8]The types of performance stressed in FERC Order 755 [54] are ramping ability and an ability to respond accurately to a system operator's dispatch signals in real-time operations.

A third key issue is that attempts to introduce new flexibility products (e.g. ramping products) have led to the introduction of additional OOM compensation processes. As stressed by Bushnell [20], the increased complexity resulting from OOM compensation processes provides increased opportunities for market participants to gain unfair profit advantages through strategic behaviors.

A fourth key issue is that these new flexibility products are conceptually problematic. For example, how can "ramping" be evaluated apart from other critical characteristics of the power-paths that deliver this ramping, such as start-time, power amplitude, and power duration?

CHAPTER 3

MOTIVATION FOR CURRENT STUDY

3.1 CHAPTER PREVIEW

As discussed in the previous chapter, current US RTO/ISO-managed wholesale power markets face increasing stresses due to rapidly changing economic and technological conditions. This chapter highlights problematic design aspects that are hindering the ability of these markets to respond to these stresses.

Section 3.2 carefully explains eight conceptual concerns regarding the design of US RTO/ISO-managed wholesale power markets. Section 3.3 discusses past work on swing-contract market design undertaken in response to some of these concerns and clarifies the relationship of the current study to this past work.

3.2 PROBLEMATIC DESIGN ASPECTS OF US RTO/ISO-MANAGED WHOLESALE POWER MARKETS

3.2.1 Artificial Distinction Between Energy and Reserve

First and foremost, US RTO/ISO-managed markets maintain an artificial distinction between "energy" and "reserve." More precisely, they distinguish between: (i) the optimal determination of a dispatch schedule for the delivery of "energy" during some future operating period T; and (ii) the provision of "reserve" to handle any needed real-time deviations from this dispatch schedule.

In actuality, however, nothing is dispatched in these forward markets. Each of these markets is simply a planning device intended to ensure sufficient *availability* of dispatchable power to balance uncertain net load during a future operating period.

Consider, for simplicity, the case in which all load for a future operating period T takes a must-service (non-price-sensitive) form. Let $\mathbb{Z}$ denote a partitioning of the collection of grid buses into reserve zones z, and let $\mathrm{nl(T)} = \{\mathrm{nl}(z, \mathrm{T}) \mid z \in \mathbb{Z}\}$ denote the forecasted system-wide net load profile for period T.

A New Swing-Contract Design for Wholesale Power Markets, First Edition, Leigh Tesfatsion.

A forecasted zonal net load profile nl(z,T) has zero probability of being exactly realized during period T. Rather, it functions as a centroid; it anchors the determination of a zonal net load uncertainty set U(z,T) taking the form of a corridor in the time-power plane whose range above and below nl(z,T) is determined by ISO-configured zone-z reserve requirements. The RTO/ISO-determined "optimal dispatch schedule" for period T ensures that nl(T) can be covered by committed dispatchable generation, and the ISO-determined "optimal reserve" for period T ensures more broadly that each zonal net load uncertainty set U(z,T) can be covered by committed dispatchable generation.

Viewed in this light, there is no fundamental conceptual distinction between an optimal dispatch schedule and optimally determined reserve; both take the form of available dispatchable power-paths whose purpose is to ensure coverage of uncertain net load for a future operating period.

3.2.2 Problematic use of Hedonic Pricing

In economics, *hedonic pricing* is the pricing of a product on the basis of prices separately assigned to its intrinsic physical attributes as well as to its external circumstances. A second conceptual concern with current US RTO/ISO-managed markets is their attempt to use hedonic pricing for the attributes of power-paths *treated as independent products*.

More precisely, SCED optimization routines are currently used in these markets to determine a power-path dispatch schedule at designated grid locations for a designated future operating period T, where T is represented as a sequence of standardized time-steps (e.g. hours). Dual-variable solutions for SCED system constraints are interpreted as "market clearing" prices for derived products (energy, ramp, capacity, reserve, …) at each of these grid locations for each of these time-steps, where each product is assumed to be a separate well-defined commodity with a standard unit of measurement (MWh, MW/min, MW, MW, …).

These dual-variable prices are then subsequently used to determine settlement payments for market participants. For example, the (right-hand) dual variable solutions for SCED power balance constraints, referred to as locational marginal prices (LMPs), are used to determine settlements for scheduled generation levels at each grid location for each time-step.

One major problem with this hedonic pricing approach to settlement is that it ignores the fact it is flexibility in the attributes of dispatchable power-paths that provides value both to RTO/ISOs and to customers. This flexibility is needed and desired because of the uncertainty and volatility of net load realizations during future operating periods. Single power-paths scheduled for dispatch at specific grid locations during an operating period T do not provide this flexibility. Rather, flexibility is obtained by ensuring the availability of *diverse collections* of dispatchable power-paths at each grid location for period T.

A second major problem with this approach is that it presumes the value of a power-path can be appropriately measured by means of separate prices assigned to its attributes, treated as independent products. It thus ignores the fact that power-path attributes such as start-time, duration, ramp-rate, maximum down/up power capacity, and delivered energy are correlated "joint products."[1] A change in any one attribute of a power-path can necessitate changes in its other attributes.

For example, the ramp-rate r_m (MW/min) of a dispatchable resource m limits m's maximum down/up power capacity (MW) during an operating period $\mathrm{T} = [t^{\mathrm{s}}, t^{\mathrm{e}})$ as well as m's maximum deliverable energy (MWh) for T. Suppose m is injecting power at level $p_m^{\mathrm{s}} \geq 6\,\mathrm{MW}$ at the start-time t^{s} for T, the duration of T is six minutes (0.1 hours), and $r_m = 1$ MW/min. Then m's maximum down/up power capacity during T is $p_m^{\mathrm{s}} \pm 6$ MW, and m's maximum deliverable energy for T is $[p_m^{\mathrm{s}} \cdot 0.1$ hours $+ 0.3$ MWh].

As will be illustrated in Chapter 7, in some circumstances it might be desirable to use a hedonic-price approximation for the performance cost $\phi_m(\mathbf{p})$ associated with any given power-path $\mathbf{p}$ that a dispatchable resource m could deliver during an operating period T. For example, this performance cost might be approximated as a linear combination of metric functions that separately assign costs for delivered energy, ramping, and maximum down/up power capacity:

$$\phi_m(\mathbf{p}) = \mathrm{C}^{\mathrm{E}}(\mathbf{p}) + \mathrm{C}^{\mathrm{R}}(\mathbf{p}) + \mathrm{C}^{\mathrm{MaxCap}}(\mathbf{p}) \tag{3.1}$$

However, it would be conceptually incorrect and highly problematic in practice to ignore that the energy, ramping, and capacity "products" whose costs are evaluated in (3.1) are in fact highly correlated attributes of a single underlying power-path **p**.

3.2.3 Revenue Insufficiency and Incentive Problems

Revenue insufficiency is said to exist in a market if seller revenues are not sufficient to cover seller costs [127, Sec. 3.2]. Revenue insufficiency results in incentive problems to the extent that it discourages sellers from participating in transactions that could, in fact, result in increased net benefit. A third conceptual concern regarding current US RTO/ISO-managed wholesale power markets is that these markets have experienced revenue insufficiency and incentive problems since their inception, despite being designed to mimic the operations of a competitive market.

Roughly stated, a *competitive market* is a two-sided commodity market with numerous buyers j and sellers i who are unable to affect the commodity price π by their own individual actions. Consequently, they act as price takers, determining their quantity demands $q_j^{\mathrm{d}}(\pi)$ and quantity supplies $q_i^{\mathrm{s}}(\pi)$ as functions of the commodity price π. Competitive market theory then predicts that the commodity price π will converge to a single commodity price π^* that brings aggregate buyer demand into equality with aggregate seller supply: $\sum_j q_j^{\mathrm{d}}(\pi^*) = \sum_i q_i^{\mathrm{s}}(\pi^*)$.

[1] In economics, a collection of products are said to be *joint products* if their production results jointly from the use of common inputs in a production process. The use of a single fuel source to co-generate electricity and steam is a joint-product example well known to power engineers.

Stoft [173, Ch. 2-2, p. 123] argues that revenues from buyer payments in a competitive market should be sufficient *on average* to cover all seller costs, assuming buyers and sellers can freely enter the market in pursuit of profits and exit the market to avoid losses. Stoft concludes (p. 132): "The (revenue insufficiency) problems of power markets do not arise from fixed costs but from the difficulty of achieving competitive prices in the face of demand-side flaws."

However, the studies by Chao [31], Eldridge et al. [45], and Gribik et al. [65] demonstrate that the persistent revenue insufficiency problems in current US RTO/ISO-managed wholesale power markets do arise in part from fundamental cost-formulation issues.

In the shorter run, the difficulty is that DAM/RTM-determined energy prices do not ensure that dispatched generators are able to cover their unit commitment costs. Specifically, there may be no set of linear prices[2] that supports both an optimal unit commitment solution and an optimal economic dispatch solution. As a result, in order to retain a linear LMP pricing mechanism for DAM/RTM SCED optimizations, RTO/ISOs rely on out-of-market (OOM) payments to ensure that the generators included in the dispatch schedules resulting from these SCED optimizations receive full compensation for their unit commitment costs as well as their variable costs.

In the longer run, the difficulty is that entities contemplating investments in new generation capacity cannot be sure their subsequent DAM/RTM revenues will permit them to recoup their investment costs. To the extent that DAM/RTM revenues do not signal the full benefits of these investments, this revenue insufficiency problem becomes an incentive problem. As will be explained with care in Section 15.4, US RTO/ISO-managed wholesale power markets are currently exploring three different ways to ensure adequate long-run generation capacity: namely, scarcity pricing (ERCOT); forward capacity markets (ISO-NE, MISO, NYISO, PJM); and load-serving entity contractual obligations (CAISO, SPP).

In contrast, as will be elaborated in subsequent chapters, the reserve offers proposed in this study for swing-contract markets are *two-part* pricing contracts specifically designed to guarantee that cleared reserve providers are fully able to recoup their costs. This guarantee holds regardless of the length of a swing-contract market's look-ahead horizon because compensation for performance is paid ex post based on verified actual performance.

3.2.4 Computational Fragility of LMP Derivations

A fourth conceptual concern is the computational fragility of LMP derivations in US RTO/ISO-managed DAMs and RTMs. Seemingly inconsequential differences in computational procedures can produce large consequential changes in LMPs.

As discussed in Section 2.3, US DAMs rely on SCUC/SCED optimizations, either in combined or separate form, to determine binary (0/1) unit commitments

[2]A pricing mechanism for a commodity Q is said to be *linear* if the same price π is charged for each unit of this commodity. Thus, given any amount q of Q, the total charge for q is $\pi \cdot q$.

for dispatchable generators, scheduled generation dispatch levels, and reserve for next-day operations. US RTMs rely on SCED optimizations to determine scheduled generation dispatch levels and reserve for near-term operating periods, taking unit commitments as given. DAM/RTM settlements for scheduled generation dispatch levels are based on LMPs, which are determined from DAM/RTM SCED optimal solutions taking unit commitments as given.

More precisely, LMPs are derived as dual variable solutions for power balance constraints in SCED optimizations, taking unit commitments as given. The power balance constraints appearing in a SCED optimization for an operating period T partitioned into standardized time-steps k impose Kirchhoff's Current Law[3] at each grid location b for each time-step k. Given regularity conditions, the dual variable for a SCED power balance constraint PBC(b,k) at b for k measures the change in the optimized value of the SCED objective function with respect to a change in the constraint constant cc(b,k) for PBC(b, k), calculated at a SCED optimal solution point. This constraint constant cc(b, k) is typically taken to be the forecasted amount of must-service load at b for k that appears in PBC(b, k).

A *unique* dual variable solution exists for PBC(b, k) if the optimized SCED objective function is a differentiable function of the constraint constant cc(b, k). A *range* of dual variable solutions exists if the optimized SCED objective function is right and left differentiable with respect to cc(b, k) but not differentiable with respect to cc(b, k). As carefully explained in [120], the LMP(b, k) at b for k is conventionally defined to be the *right-hand dual variable* for PBC(b, k), i.e. the change in the optimized SCED objective function with respect to an *increase* in cc(b, k).

As shown by Eldridge et al. [45, Sec. 3.2] and Sioshansi et al. [166], LMPs can vary erratically in response to seemingly small changes in unit commitments arising, for example, from the use of different stopping rules in the SCUC optimizations used to determine approximately optimal 0/1 solutions for these unit commitments.[4] These LMP variations in turn induce variations in OOM make-whole payments; together, these variations can result in substantial systematic (non-random) wealth transfers across market participants.

In addition, LMPs are also extremely sensitive to seemingly small changes in the solution algorithms used for SCED optimizations because these changes can substantially affect which SCED inequality constraints are binding in the resulting solution. Vendor software used in US RTO/ISO-managed markets is proprietary; hence, it is not possible to determine which precise computational procedure is used in each of these markets to compute SCED solutions. It is highly likely, however, that these computational procedures differ in a number of regards [62].

[3]Roughly stated, Kirchhoff's Current Law applied to an electrical network, such as a transmission grid, asserts that the sum of all currents entering into a node of the network must equal the sum of all currents exiting from this node.

[4]As noted in [45, Sec. 3, p. 10], current US RTO/ISO-managed wholesale power markets typically solve SCUC optimizations to within a predetermined "Mixed Integer Programming (MIP) gap" and not to full optimality.

For example, is a simplex, interior-point, or active-set method employed? How are initial points selected for the start of computations? What steps are taken to protect against round-off error, truncation error, and error accumulation? What stopping-rule metric is used to decide that "convergence" to a solution has occurred, or that computation should be halted because convergence has failed to occur?

Another concern is that differences between DAM and RTM SCED formulations can induce systematic deviations between DAM and RTM LMP solutions for a given operating period. *Load-serving entities (LSEs)* participating in DAMs are permitted to submit demand bids that include price-sensitive demand schedules as well as fixed demands.[5] However, LSEs are not permitted to participate in RTMs except as some form of demand-response resource.[6] RTO/ISO forecasts are used in RTMs to represent the fixed (non-price-sensitive) power demands of customers.

In consequence, even if DAM and RTM SCED optimizations for a time-step k are based on the same generation supply offers and have the same forecasted net load at each bus b, their LMP solutions can differ. Specifically, a solution $\mathrm{LMP}^{\mathrm{DA}}(b,k)$ resulting from a DAM SCED optimization for which at least some demand is expressed in the form of LSE price-sensitive demand schedules can differ from a solution $\mathrm{LMP}^{\mathrm{RT}}(b,k)$ resulting from an RTM SCED optimization for which all demand is expressed as ISO fixed-demand forecasts.

This follows because LMPs are dual variable solutions for power balance constraints. For a DAM SCED optimization that incorporates price-sensitive demand schedules, an incremental increase in the must-service load appearing within a power balance constraint $\mathrm{PBC}(b,k)$ can be balanced either by an increase in scheduled generation dispatch or by a decrease in the amount of cleared price-sensitive demand. The latter option is not available for an RTM SCED optimization in which all demand takes a fixed-demand form.

3.2.5 Performance Payment in Advance of Performance Delivery

A fifth conceptual concern is payment for performance in advance of actual performance. Current US RTO/ISO-managed DAM SCED optimizations rely on dual variable solutions for system constraints to determine DAM LMPs with energy, transmission, and loss components as well as prices for variously defined reserve products. The DAM LMP solutions are then used to settle DAM-scheduled generation dispatch levels for next-day operations in advance of actual next-day operations.

---

[5]A *fixed demand* is a demand submitted into a US RTO/ISO-managed wholesale power market for a specified future amount of *must-service load*, i.e. non-price-sensitive load that must be serviced (balanced) by the RTO/ISO under normal operating conditions.

[6]For example, ERCOT [50] permits *qualified scheduling entities (QSEs)* to submit RTM bids on behalf of LSEs who represent *controllable load resources (CLRs)* with installed real-time telemetry that enables them to incrementally increase or decrease their power usage every five minutes in response to dispatch instructions received from their controlling QSEs.

However, DAM LMPs are *forecasted system marginal costs* in the following sense: They measure the rate of change of the optimized DAM SCED objective function with respect to incremental changes in power balance constraint constants, conditional on forecasted conditions for next-day operations. The key issue here can be succinctly stated as follows: Is it necessary, or desirable, to use forecasted system marginal costs for future operations to calculate and implement performance *settlements* for market participants in advance of actual performance?

The computational fragility of DAM LMP valuations was previously noted in Section 3.2.4. These valuations are also conceptually fragile because they depend on the following five types of forecasts as DAM SCED inputs: (i) LSE-forecasted loads for next-day operations, as indicated by LSE demand bids; (ii) ISO-forecasted non-dispatched power injection during next-day operations; (iii) forecasted marginal costs and power capacity limits for next-day operations, as indicated by generator supply offers; (iv) ISO-forecasted transmission line limits needed to ensure grid integrity during next-day operations; and (v) ISO-forecasted reserve requirements needed to ensure net-load balancing during next-day operations. DAM LMPs are thus subject to forecast errors arising from five different sources.

FERC Order 755 [54] concludes that performance payment in advance of actual performance is neither necessary nor desirable for frequency regulation (secondary frequency control). The present study suggests the same is true for all reserve, broadly interpreted to mean dispatchable power-paths, with the important proviso that the need for *availability* cost coverage must not be overlooked.

3.2.6 Minimal Direct Representation of Retail Customer Interests

The US power industry was originally based on regulated vertically integrated utilities [182, Sec. 1]. The servicing of retail customers was viewed as *load following*, i.e. as the need to meet retail customer power-usage requirements whatever form these requirements took, regardless of cost.

A sixth conceptual concern is that this passive conception of retail customer servicing has been carried forward into restructured US RTO/ISO-managed markets. The wholesale power requirements of most retail customers[7] are serviced by private or publicly owned for-profit companies, generally referred to as LSEs. These intermediaries are charged wholesale prices for wholesale power procurement. In return, they receive payments at administratively set flat rates for the delivery of wholesale power to their retail customers. These flat rates are designed to ensure that the revenues earned by LSEs exceed their charged costs by a normal profit margin.

Power economists have stressed for many years the need for more active demand-side participation in US RTO/ISO-managed wholesale power markets to balance the traditionally dominant role of suppliers. To paraphrase the famous

[7] Exceptions include retail customers participating in municipally owned utilities or electric power cooperatives that are managed on behalf of their owners/members.

economist Alfred Marshall [125, Sec. 7], operating a market without consideration of price-sensitive demand is like trying to operate a pair of scissors with a missing blade. In response to these criticisms, LSEs participating in day-ahead markets have been permitted to submit hourly demand bids that include a price-sensitive demand schedule as well as fixed demand. However, LSE demand bids still predominately take a fixed-demand form, presumably reflecting the continued predominance of flat-rate pricing at the retail level.

As discussed at greater length in ([171], [182, Sec. 5.1], [189, Sec. II]), recent breakthroughs in technology (e.g. smart metering with advanced communication capabilities) have radically improved the potential for more active retail customer participation in wholesale power market operations. For example, these breakthroughs have encouraged increased interest in various types of RTO/ISO-managed initiatives permitting retail customers to participate in day-ahead markets as dispatchable suppliers of power absorption as an ancillary service.

Nevertheless, a major concern remains. Most retail customers in the footprint of a US RTO/ISO-managed wholesale power market still do not have a direct seat at the wholesale power market table.[8] Instead, they must rely on for-profit LSEs to manage their wholesale power market participation. These for-profit LSEs do not have a fiduciary responsibility to represent the preferences and interests of retail customers. Rather, their responsibility is to secure profits for their owners/shareholders.

For example, if an ISO raises equipment maintenance standards to increase grid reliability, generators can raise their reported marginal costs and LSEs can petition their local state boards to raise retail flat rates to cover the resulting increase in wholesale power prices. The cost of this increased reliability is thus ultimately borne by retail customers who have no means to express to the ISO their true willingness to pay for this increased reliability. Instead, RTO/ISOs rely on problematic metrics such as *value of lost load (VOLL)* to estimate the cost of power interruptions to retail customers [133, Sec. 5.3].

3.2.7 Reliance on Overly Simplistic Cost Conceptions

A seventh conceptual concern is that US RTO/ISO-managed markets rely on microeconomic cost concepts that are too simplistic for their intended purpose. Specifically, they rely on the traditional division of total cost into "fixed cost" (not dependent on production level) and "variable cost" (dependent on production level).

[8] The Electric Reliability Council of Texas (ERCOT) is an exception [169]. The US Federal Energy Regulatory Commission (FERC) has authority in the US over rates and conditions for the *interstate* sale and transmission of wholesale electric power. ERCOT is not under FERC jurisdiction because the ERCOT transmission grid has been deliberately designed to ensure the flow of electric power within the ERCOT system is contained within the single state of Texas. Authorized by Texas state statute and overseen by the Power Utility Commission of Texas (PUCT), ERCOT has gradually assumed fiduciary responsibility for electric power operations within its footprint at both the wholesale (transmission) and retail (distribution) levels. For example, ERCOT explicitly includes encouragement of retail choice as part of its stated mission.

A more economically meaningful division of total cost is between cost that is non-avoidable and cost that is avoidable [9]. An *avoidable cost* associated with a voluntary future activity at a given present time is a cost that can be prevented if the future activity is not undertaken but that is incurred if the future activity is undertaken. A *non-avoidable cost* associated with a voluntary future activity at a given present time is a cost that is incurred whether or not the future activity is undertaken.

The *total cost* associated with a voluntary future activity at a given present time can thus be divided into sunk cost and avoidable cost as follows:

Sunk cost: Non-avoidable fixed cost

Avoidable cost: Avoidable fixed cost + Variable cost

The *net earnings* associated with the voluntary future activity are then defined to be the earnings resulting from the undertaking of this activity net of avoidable cost. The *profit* associated with the voluntary future activity is defined to be the earnings resulting from the undertaking of this activity net of total cost.

For example, consider a currently off-line dispatchable generator m at the start of some day D, immediately prior to the start of a day-ahead wholesale power market. An example of a sunk cost for m would be an amount of money that m has previously spent to purchase a piece of generation equipment that now has no resale value. An example of an avoidable fixed cost for m would be the start-up cost that m would have to incur to transit from an off-line state to a synchronized state[9] at the start of day D+1 if m submits an offer into the day-ahead market and the ISO clears this offer.[10] An example of a variable cost for m would be the fuel cost that m would have to incur in order to maintain a positive power injection level p during some hour H of day D+1.

For product producers, a production activity should be undertaken only if the revenues earned over time from this activity are sufficient to cover the avoidable cost associated with this activity. Stated in more standard economic terms, the net earnings resulting over time from this activity should be non-negative. If this is not the case, the avoidance of the activity results in avoidance of losses. Conversely, for product buyers, the payments they offer to sellers for product purchases will typically have to be sufficient over time to cover seller avoidable cost in order to induce production. However, sunk cost is sunk, whether or not a future production activity is undertaken. Consequently, sunk cost should not play any role in determining whether or not a future activity is undertaken, or what compensation should be provided if it is undertaken.

The settlement processes used in current US RTO/ISO-managed wholesale power markets are not fully in accordance with these basic economic principles. For example, the OOM "make-whole" payments that ERCOT awards to power sources

[9]A generator is said to be in a *synchronized state* if it is an operating state that permits it to inject power into the grid, even if no such power injection is currently being undertaken.

[10]Kirschen and Strbac [97] refer to start-up cost as a "quasi-fixed cost."

when their DAM revenues are insufficient to cover their start-up and minimum-energy costs do not ensure coverage of their full avoidable fixed costs.

Also, FERC [57] has proposed that the unit commitment costs of fast-start resources be incorporated into DAM/RTM energy and ancillary service prices. As reviewed in Hartman [73, pp. 6-7], this FERC proposal has led subsequent commentators to suggest that the unit commitment costs for other types of generation should be added into these prices as well. However, any such incorporation would result in a blurring of the operationally critical distinctions among sunk cost, avoidable fixed cost, and variable cost.

3.2.8 Use of Spot-Market Pricing for Forward Markets

Finally, an eighth conceptual concern regarding current US RTO/ISO-managed markets is their singular focus on spot-market pricing for all resource transactions. Prices are expressed as the demand-equal-supply outcomes of competitive market processes occurring at a given point in time at a given location.

This singular focus misses three critical points. First, as shown in Chapter 12, spot-market pricing presumes all fixed costs are sunk, hence irrelevant for current price determination. Second, avoidable fixed costs often take a discrete ("lumpy") form, e.g. make a new capital expenditure or not. Third, purchase and sale decisions for future operating periods, determined at points in time for which at least some portion of fixed costs are avoidable, not sunk, are forward decisions best handled by means of appropriately designed two-part pricing contracts[11] to ensure coverage of avoidable fixed cost as well as variable cost.

3.3 RELATION OF CURRENT STUDY TO PREVIOUS SWING-CONTRACT WORK

In response to some of the concerns expressed in Section 3.2, a group of researchers from Iowa State University and Sandia National Laboratories began working in 2011 on a new approach to the design of wholesale power markets.

In 2013, these researchers released a conceptual report Tesfatsion et al. [188] proposing the use of energy and reserve contracts in standardized swing-contract forms permitting separate pricing for service availability and real-time service performance. Firm swing contracts are recommended for meeting normal net load balancing requirements, whereas option swing contracts are recommended for the handling of contingencies.

[11] As is well understood in economics, two-part pricing contracts can be used by monopolistic suppliers in spot-market settings as price-discrimination instruments permitting extraction of net surplus from buyers. The context here is quite different; forward-market suppliers have avoidable fixed costs that must be covered in order to stay in business. Potential market power and strategic gaming concerns within this latter context are addressed in Section 15.8.

The trading of these swing contracts is envisioned to take place within a linked collection of RTO/ISO-managed forward markets whose look-ahead horizons range from multiple years to minutes ahead of a targeted real-time operating period. The linking is to be achieved by having the swing contracts cleared in earlier markets be carried forward on the books of the RTO/ISO as a portfolio of contracts that can be adaptively updated in subsequent markets.

The authors of [188] were strongly influenced by previous electric power research stressing the relevance of options and two-part pricing contracts for electric power trading as well as the importance of risk hedging in advance of real-time operations. Key works include the "safe passage" writings by Oren and his collaborators [42, 141, 142]. Other major influences include work by Bidwell [13], Bunn [18], Chao and Wilson [32, 33], and Hogan [80].

In addition, the authors of [188] were motivated by the use in practice of relatively simple forms of "electricity swing options" in over-the-counter trades. Typically these contracts specify that the contract holder must purchase energy over a contractual period, where the amount of this energy is no less than a specified minimum amount and no more than a specified maximum amount. However, the contract holder is provided some degree of flexibility ("swing") with regard to the amount of power purchased (or sold) at contractually specified "strike prices" at contractually specified times within the contractual period. See, for example, Haarbrücker and Kuhn [70] and Ross and Zhu [153] for further discussion of such contracts.

Building on the concepts expressed in [188], Heo and Tesfatsion [76, 77] focus on the design and implementation of linked day-ahead and real-time markets based on portfolios of firm contracts with swing (flexibility) in their offered service attributes. Numerical examples are used to demonstrate how these two linked markets can be successively cleared in a manner permitting real-time balancing of net load.

Subsequently, Hansen et al. [72] and Li and Tesfatsion [113, 115] significantly extended Heo and Tesfatsion [76, 77] by developing complete analytical *mixed integer linear programming (MILP)* optimization formulations for a swing-contract day-ahead market subject to standard types of system constraints. Simple test cases are used to illustrate the implementation of these formulations.

In 2019, Ma et al. [122] extended the swing-contract day-ahead market optimization formulation developed by Li and Tesfatsion [115] to help ensure real-time reserve deliverability. The optimization formulation is augmented with a method permitting the adaptive updating of reserve zones based on past realized grid congestion conditions. Five-bus and thirty-bus test cases are used to demonstrate the practicality and effectiveness of this extended optimization formulation.

This previous work on swing-contract market design stresses the use of swing contracts for the joint procurement of energy and reserve in wholesale power markets. However, as stressed in Chapter 1 of the current study, wholesale power markets are necessarily forward markets; and the only product effectively offered in these forward markets is reserve (dispatchable power-paths). This recognition greatly simplifies the manner in which swing-contract market designs can be developed, motivated, implemented, and compared to alternative market designs. The current study incorporates this recognition.

Moreover, this previous work is primarily focused on the use of swing contracts by generation resources participating in wholesale power markets. The current study explores this usage as well. However, the current study also considers with care the possible use of swing contracts by ancillary service providers. Particular attention is focused on the possible use of swing contracts by entities operating at distribution-transmission interfaces that function in distribution systems as grid-edge resource aggregators and in transmission systems as prosumers able to offer dispatchable down/up power-paths in flexible dependable forms.

Finally, the swing-contract market design developed in the current study distinguishes carefully between sunk cost and avoidable cost. This is seen both in the two-part pricing formulation for swing-contract reserve offers and in the proposed ISO allocation method for net reserve cost and transmission service cost.

CHAPTER 4

SWING CONTRACTS FOR ISO-MANAGED WHOLESALE POWER MARKETS

4.1 SWING CONTRACT OVERVIEW

To provide a viable alternative to contractual arrangements in current US RTO/ISO-managed wholesale power markets, swing contracts must fulfill two basic roles. First, swing contracts must permit the efficient pricing and procurement of reserve (dispatchable power-paths). Second, swing contracts must provide blueprints for the delivery of power-paths in real-time operations to ensure net load balancing.

Key properties of swing contracts formulated specifically for the support of centrally managed wholesale power market operations are explained and illustrated in this chapter. As will be seen, these swing contracts share some of the characteristics of standard forward contracts. However, they also have distinct characteristics tailored to the special purposes and requirements of power systems.

4.2 SWING CONTRACTS: GENERAL FORMULATION

As will be explained with care in Chapter 6, a swing-contract market M(T) for a future operating period T is an ISO-managed market whose goal is the efficient procurement of reserve (dispatchable power-paths) to meet net load balancing needs in period T. For simplicity of exposition, it is assumed in this study that the contracts cleared for M(T) are announced at the close of M(T). This section explains the general swing-contract formulation envisioned for reserve offers in such a market.

Let SC_m denote a swing contract submitted by a dispatchable resource m into a swing-contract market M(T). As defined in this study, SC_m takes the following general form:

$$\mathrm{SC}_m = (\alpha_m, \mathbb{T}_m^{\mathrm{ex}}, \mathbb{PP}_m, \phi_m) \tag{4.1}$$

where each component of SC_m is designated by m.

A New Swing-Contract Design for Wholesale Power Markets, First Edition, Leigh Tesfatsion.

The first component, α_m, is the *offer price* of SC_m. A non-negative[1] offer price designates the amount that must be paid to m if SC_m is cleared, either directly or in amortized payment-schedule form. More precisely, it is the amount that must be paid to m to procure from m the availability of reserve during period T under the particular contractual terms stated in SC_m.

The offer price α_m permits m to ensure recovery of all avoidable fixed cost[2] associated with guaranteeing this availability. This avoidable fixed cost could include: *capital investment cost*, e.g. land acquisition, building construction, and equipment purchase or rental cost; *transaction cost*, e.g. insurance, code compliance, licensing, and employee search cost; *unit commitment cost*, e.g. start-up, no-load, minimum-run, and shut-down cost; and *opportunity cost*, e.g. foregone opportunity to use generated power directly (behind-the-meter) for a local purpose.[3]

The second component, $\mathbb{T}_m^{\text{ex}}$, is a collection of possible *exercise times* t_m^{ex} for SC_m occurring between the close of M(T) and the start-time for the future operating period T. As discussed more fully in Section 4.3, the number and positioning of these exercise times determine whether SC_m is a firm contract or some type of option contract.

The third component, $\mathbb{PP}_m$, is a set of power-paths for period T, each of which m could feasibly deliver at a designated grid location during period T in response to dispatch signals. Feasible deliverability of the power-paths in $\mathbb{PP}_m$ requires these power-paths to be feasible continuations of the dispatchable power-paths offered by m in previously cleared swing contracts (if any). In addition, the signals communicated to m during period T to dispatch a particular power-path in $\mathbb{PP}_m$ must be conditioned on m's operating state at the start of T so that this particular power-path is a feasible continuation of m's previously dispatched power-paths (if any).[4]

The power-paths in $\mathbb{PP}_m$ are characterized in terms of attributes such as delivery location, start-time, minimum down/up time, power limits, ramp-rate limits, duration limits, and energy capacity. The precise specification of these attributes determines the degree of swing (flexibility) in m's offered reserve.

---

[1]Swing-contract offer prices are not restricted in sign. For example, a generator m in a must-run operating state could designate an offer price $\alpha_m < 0$ for a swing contract SC_m offering power-paths with a minimum power level $P_m^{\text{min}} > 0$ in an attempt to ensure this swing contract is cleared. In this case, generator m is committing to pay $-\alpha_m$ should SC_m be cleared.

[2]See Section 3.2.7 for a definition and discussion of avoidable cost for electric power systems.

[3]This study uses the following definitions for costs related to power usage; see Chapter 5 for illustrative examples. *Start-up cost* is the cost a power source incurs in order to change from an off-line state to a synchronized state. *No-load cost* is the behind-the-meter cost a power source incurs in order to maintain itself in a synchronized state during a designated time-step. *Minimum-run cost* is the before-the-meter cost a power source incurs to support a minimum sustainable power injection level during a designated time-step. *Shut-down cost* is the cost a power source incurs in order to change from a synchronized state to an off-line state. Finally, the standard economic definition of *opportunity cost* for any resource is the net earnings the resource could have obtained in its next-best alternative use.

[4]These important linkage issues will be addressed with care in Chapter 10 as part of a broader discussion of a proposed linked swing-contract market design.

The final component, ϕ_m, is a performance payment method designating how m is to be compensated ex post for any cost incurred for verified period-T service performance, i.e. for the verified period-T delivery of a power-path in $\mathbb{PP}_m$ in response to dispatch signals. This performance cost could include payments for inputs (fuel, labor, ...) needed for power production, charges allocated for transmission services, and the wear-and-tear depreciation of physical equipment.

The performance payment method ϕ_m can take a wide variety of forms for multiple different purposes. However, a critical requirement is that ϕ_m be explicitly formulated in terms of metrics that permit the ISO and m to agree ex ante on the intended meaning of m's offered period-T service performance and to verify ex post the extent to which this period-T service performance has actually been delivered.

For example, m could be solely focused on ensuring appropriate compensation ex post for total energy delivered during period T in response to dispatch signals. In this case, one possibility would be for ϕ_m to designate a pre-specified energy price ($/MWh) to be used for this ex post compensation, which would eliminate all price risk for m.[5] Alternatively, ϕ_m could require this ex post compensation to be based on ex-post locational marginal prices (LMPs) calculated for period T at m's particular grid location subsequent to actual net load realizations.

More broadly, ϕ_m could require compensation ex post for various dynamic attributes of any power-path that resource m delivers during period T in response to dispatch signals, such as ramping ("power-path mileage")[6] and duration. In addition, ϕ_m could require compensation ex post for the correlated provision of reactive as well as active power, where reactive power production and absorption are used for voltage support. Another possibility is that ϕ_m could specify penalties that m agrees to pay for uninstructed deviations from dispatch signals, and/or additional compensation that m requests for exceptionally close following of dispatch signals.[7]

4.3 SWING CONTRACTS IN FIRM OR OPTION FORM

Let SC_m denote a swing contract (4.1) offered by a dispatchable resource m into a swing-contract market M(T) for some future operating period T. A decision by the ISO to clear SC_m at the close of M(T) immediately triggers an offer-price payment obligation. If the offer price α_m is positive, resource m must be paid this offer price; otherwise, m is obligated to pay $-\alpha_m$.

[5]Roughly defined, the *risk* associated with an action is the possibility that the action will result in adverse unexpected consequences [199]. *Price risk* for a buyer (seller) is the risk that prices might be higher (lower) than expected.

[6]For example, CAISO defines the *mileage* of an automatic generation control (AGC)-instructed power-path to be the summation of the absolute changes in the consecutive AGC set points that are communicated to a dispatchable resource as power-path dispatch instructions; see [78].

[7]See Section 6.6 for a more extended discussion of performance penalties and incentives.

However, any further obligations entailed by the clearing of SC_m depend on whether SC_m is in firm or option form. This, in turn, depends on the specification of the exercise set $\mathbb{T}_m^{\text{ex}}$ included within SC_m.

Suppose, first, that the exercise set $\mathbb{T}_m^{\text{ex}}$ for SC_m consists of a single exercise time t_m^{ex} that coincides with the closing time for M(T). In this case, a decision by the ISO to clear SC_m at the close of M(T) triggers the immediate exercise of SC_m as a *firm contract*. That is, the following three obligations are firmly entailed. First, the ISO is obligated to dispatch a power-path from $\mathbb{PP}_m$ during period T. Second, resource m is obligated to deliver this power-path during period T in response to the ISO's dispatch signals. Third, the ISO is obligated to ensure that m is compensated ex post for this delivered power-path in accordance with the performance payment method ϕ_m included within SC_m.

Alternatively, SC_m could be some type of *option contract*. Specifically, SC_m could be a European, American, or Bermudan option.[8]

SC_m is a European option if its exercise set $\mathbb{T}_m^{\text{ex}}$ consists of a single exercise time t_m^{ex} occurring strictly after the close of M(T) but prior to the start of the operating period T. SC_m is an American option if its exercise set $\mathbb{T}_m^{\text{ex}}$ consists of all times occurring strictly after the close of M(T) and no later than a designated expiration date occurring prior to the start of T. Finally, SC_m is a Bermudan option if its exercise set $\mathbb{T}_m^{\text{ex}}$ consists of multiple designated exercise times occurring after the close of M(T) and no later than a designated expiration date occurring prior to the start of T.

In each of these option cases, if the ISO clears SC_m at the close of M(T), the ISO then has the right *but not the obligation* to exercise SC_m at any exercise time designated within the exercise set $\mathbb{T}_m^{\text{ex}}$. Further obligations between the ISO and m are only entailed if this exercise takes place. Specifically, if the exercise takes place, the swing contract reduces to a firm contract with delivery and performance payment obligations as described above.

Swing contracts in firm form are suitable for routine short-term net load balancing purposes. Swing contracts in option form are suitable for protection against low-probability high-impact contingencies such as line or generator outages, or for longer-term planning purposes for which uncertainty is relatively high. Alternatively expressed, firm swing contracts effectively constitute regulation reserve, whereas option swing contracts effectively constitute contingency and planning reserve.

As will be demonstrated in Chapter 5, firm swing contracts include, as special cases, standard types of supply offers submitted into current US RTO/ISO-managed wholesale power markets. In addition, firm swing contracts can accommodate offers

[8] A *European option* can only be exercised at a single designated exercise time, called the *expiration date* of the option. An *American option* can be exercised at a designated expiration date, or at any time prior to this expiration date. Finally, a *Bermudan option* is a blend of the European and American option forms; it can be exercised at a designated expiration date, or at designated exercise times occurring prior to this expiration date. Electricity swing options used in over-the-counter energy trading typically take a Bermudan form; see [153].

of reserve (dispatchable power-paths) with a much broader range of offered attributes than envisioned in these standard types of supply offers.

Swing contracts in European option form provide flexibility to the ISO with regard to the entailment of dispatch obligations. Swing contracts in American and Bermudan option form provide further flexibility to the ISO with regard to the exercise time itself. Exercise-time flexibility could be particularly important for ensuring the availability of spinning reserve with multiple feasible ramp rates and the availability of non-spinning reserve with multiple feasible start-up times.

CHAPTER 5

ILLUSTRATIVE SWING-CONTRACT RESERVE OFFERS

5.1 CHAPTER PREVIEW

As noted in Chapter 4, the practical usefulness of the general swing-contract formulation $SC = (\alpha, \mathbb{T}^{ex}, \mathbb{PP}, \phi)$ depicted in (4.1) poses two key requirements. First, the power-path set $\mathbb{PP}$ must provide a blueprint, in advance, for a collection of dispatchable power-paths $\mathbf{p}$ that could be made available to facilitate net load balancing in future real-time operations through the exercise of SC at some time t^{ex} in $\mathbb{T}^{ex}$. Second, the offer price α and performance payment method ϕ must provide a way to assess, in advance, the availability cost and expected ex post performance cost for SC, thus permitting SC to be cleared in proper merit order.

How these two goals are met in practice is a function of current capabilities. One can imagine, for example, that future capabilities might permit a power-path $\mathbf{p}$ to be represented as a linear combination of basis functions, as a calculus-of-variation perturbation, in pictorial or bit form rather than through analytical expression, or in ways that today cannot even be conceived. Similarly, one can imagine future optimization capabilities permitting a great range of expression for availability and ex post performance costs.

The purpose of this chapter is modest in scope relative to these future possibilities. Various swing-contract examples are used to illustrate, in concrete terms, the manner in which swing contracts could be formulated to meet the needs of wholesale power systems as currently structured and managed in the United States. As will be shown, swing contracts can be used by generation units to offer dispatchable power injections and by distributed-resource aggregators and storage facilities to offer dispatchable down/up power-paths as ancillary service.

A New Swing-Contract Design for Wholesale Power Markets, First Edition, Leigh Tesfatsion.

Each illustrated swing contract is assumed to be submitted by a dispatchable resource m into an independent system operator (ISO)-managed swing-contract market M(T) for a future operating period $\mathrm{T} = [t^{\mathrm{s}}, t^{\mathrm{e}})$. Each submitted swing contract offers reserve (dispatchable power-paths) for delivery at m's grid location b during period T. The start-time t^{s} and end-time t^{e} for T, as well as all swing-contract exercise times t^{ex}, are specific calendar times expressed at the granularity of minutes.[1]

A key issue addressed for each illustrated swing contract is cost coverage. A dispatchable resource m planning to submit a swing contract $\mathrm{SC}_m = (\alpha_m, \mathbb{T}_m^{\mathrm{ex}}, \mathbb{PP}_m, \phi_m)$ into M(T) must first distinguish between cost incurred *regardless* of submission and cost incurred *because* of submission. The former cost is sunk cost that should not affect the submission decision. The latter cost is avoidable cost that needs to be covered by the terms of SC_m if losses are to be avoided.

To ensure coverage of its avoidable cost, resource m should first divide this avoidable cost into two parts, as follows:

$$\text{Avoidable cost} = \text{Avoidable fixed cost} + \text{Variable cost} \tag{5.1}$$

Resource m's avoidable fixed cost is the cost that m would incur if SC_m is cleared, regardless of subsequent ISO exercise or dispatch decisions for operating period T. Resource m can ensure coverage of this avoidable fixed cost by including it within its offer price α_m. In contrast, resource m's variable cost is uncertain cost whose exact extent depends on subsequent ISO exercise and/or dispatch decisions for T. Resource m can ensure the ex post coverage of this variable cost by appropriate specification of its performance payment method ϕ_m.

For concreteness and simplicity of exposition, the market M(T) is assumed to close at time 0, and the start-time and end-time for the operating period T are assumed to satisfy $0 < t^{\mathrm{s}} < t^{\mathrm{e}}$. The exercise times t^{ex} for each illustrated swing contract are assumed to satisfy $0 \leq t^{\mathrm{ex}} < t^{\mathrm{s}}$. The exercise set $\mathbb{T}^{\mathrm{ex}}$ is omitted from all illustrated swing contracts in firm form; that is, from all illustrated swing contracts whose exercise sets consist of the single exercise time $t^{\mathrm{ex}} = 0$.[2] Also, resource-issuer subscripts are omitted from all illustrated swing-contract components.

Finally, a careful consideration of linkage issues for the illustrated swing contracts is delayed until Chapter 10. In this section, it is assumed that the dispatchable resource m issuing each illustrated swing contract SC_m is in an off-line state at the close of M(T) and needs to return to an off-line state as soon as possible following the close of the subsequent operating period T. As will be seen, this assumption affects the determination of avoidable fixed cost for SC_m.

---

[1]More precisely, the complete expression of these times would indicate the year, month, day, hour, and minute during which these times take place.

[2]Recall from Section 4.3 that a firm swing contract submitted into M(T) is defined to be a swing contract whose clearing at the close of M(T) immediately triggers its exercise.

5.2 A SIMPLE ENERGY-BLOCK SWING CONTRACT IN FIRM FORM

Consider the following simple energy-block swing contract SC_m in firm form, submitted by a dispatchable generator m into an ISO-managed swing-contract market M(T) to facilitate routine net load balancing for a transmission system during a future operating period $T = [t^s, t^e)$.

$$SC_m = [\alpha,\ \mathbb{PP},\ \phi] \tag{5.2}$$

where

$$\begin{aligned}
\alpha &= \text{offer price} \\
\mathbb{PP} &= (b,\ t^s,\ p^{disp},\ t^e) \\
b &= \text{delivery location} \\
t^s &= \text{start-time for energy block E} \\
p^{disp} &= \text{maintained power injection for energy block E} \\
t^e &= \text{end-time for energy block E} \\
\phi &= \text{pre-specified price } \pi \text{ for delivered energy.}
\end{aligned}$$

The offer price α ($), assumed to be nonnegative, denotes the amount that must be paid to m if swing contract (5.2) is cleared. The power-path set $\mathbb{PP}$ consists of a single power-path **p** in energy-block form, characterized by four attributes: (b, t^s, p^{disp}, t^e). Attribute b is generator m's grid location, hence the delivery location for any power that m injects into the transmission grid. The start-time t^s and end-time t^e characterize generator m's offered delivery period, which here coincides with the entire operating period T. Attribute p^{disp} (MW) denotes the power injection that generator m is offering to maintain at b during this delivery period, where p^{disp} lies within m's feasible sustainable power range $[0, P^{max}]$.

Finally, the performance payment method ϕ requires generator m to be compensated ex post for total delivered energy at a pre-specified energy price π ($/MWh). If swing contract (5.2) is cleared, and m subsequently fulfills its delivery obligations, m's required ex post compensation ($) for total delivered energy is given by $(\pi \cdot p^{disp} \cdot [t^e - t^s] \cdot 1\ \text{hour}/60\ \text{minutes})$.

The energy requirements that must be met by generator m, should its offered swing contract (5.2) be cleared, will now be illustrated in more concrete terms. However, an accurate characterization of these energy requirements necessitates that a careful distinction be made between m's power injection and m's total power generation at any given point in time. *Power injection* refers to the amount of power that a generator is injecting into a power grid at a point in time. In contrast, *total power generation* refers to the total power that a generator is producing at a point in time.

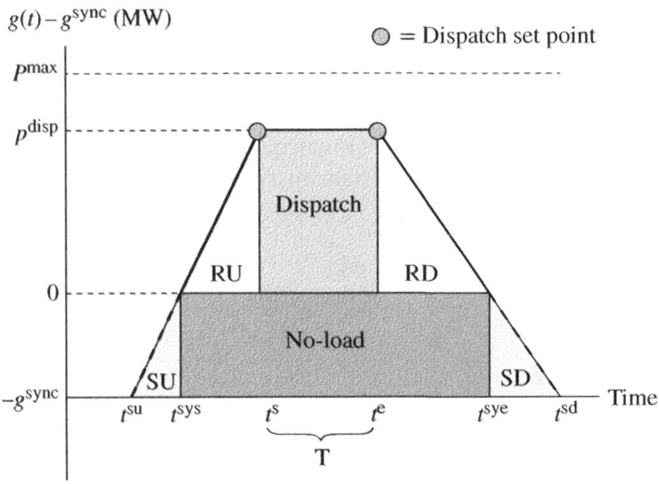

Figure 5.1 Illustration of the energy requirements for a firm swing contract (5.2) issued by a dispatchable generator m for a future operating period T. Ensuring the availability of the offered energy block ("Dispatch") during T requires m to generate start-up (SU), ramp-up (RU), no-load, ramp-down (RD), and shut-down (SD) energy before, during, and after period T.

This total power can include behind-the-meter power that a generator is expending to ramp up its generation units to a synchronized state, or to maintain its generation units in a synchronized state, prior to any actual injection of power into the grid.

For simplicity of exposition, the particular total power generation level at which generator m attains a synchronized state is assumed to be independent of time; this total power generation level is denoted by $g^{\rm sync}$. The power injection level of m at any given time t can then be expressed as $p(t) = \max\{g(t) - g^{\rm sync}, 0\}$, where $g(t) \geq 0$ denotes m's total power generation at time t.

This notational convention is used in Figure 5.1 to provide an illustrative depiction of the energy requirements that result from the clearing of m's offered swing contract (5.2) in market M(T). First, this clearing requires m to deliver an energy block ("Dispatch") during the future operating period T. However, by assumption, generator m is in an off-line state at time 0, the closing time for M(T); and m must return to this off-line state as soon as possible following the close of T. Thus, Figure 5.1 also depicts the energy that m needs to generate before, during, and after T to ensure the availability of this energy block, conditional on m's initial off-line state and required terminal off-line state.

More precisely, generator m is off-line at time 0, i.e. m's total power generation level $g(0)$ is 0. Suppose the ISO clears m's offered swing contract (5.2) at time 0, which obligates m to inject power into the grid at location b at a maintained dispatch level $p^{\rm disp}$ during the subsequent operating period $T = [t^s, t^e)$. In preparation for this delivery, m starts up at time $t^{\rm su}$; and m continues to ramp up its total power generation until time t^{sys}, the time when *m* first attains its synchronized total generation power level g^{sync}. At t^{sys}, generator *m* is spinning at synchronous speed, ready to inject power into the grid; but *m* is not yet injecting any power into the grid.

Generator *m* then continues to ramp up until its power *injection* level reaches its cleared dispatch level p^{disp} at the start-time t^{s} for operating period T. At t^{s}, *m*'s total power generation level is $g(t^{s}) = p^{disp} + g^{sync}$; however, only the portion p^{disp} injected into the transmission grid is visible to the ISO.

Generator *m* maintains the power injection level p^{disp} during period T. At the end-time t^{e} for period T, generator *m* initiates a ramp-down process. During the initial phase of this ramp-down process, from time t^{e} to time t^{sye}, *m* is still injecting power into the transmission grid. However, once *m*'s total power generation level falls to its synchronized total power generation level g^{sync}, at time t^{sye}, *m* is no longer injecting power into the grid. Generator *m* then continues to ramp down until *m* reaches a shut-down state $g(t^{sd}) = 0$ at time t^{sd}.

The power *injected* by *m* into the transmission grid during operating period T to meet its dispatch obligations to the ISO results in delivered energy taking an energy-block form, labeled "Dispatch" in Figure 5.1. The cost of this delivered energy will be referred to as *m*'s *dispatch cost*.

The power generated by *m* during the start-up interval $[t^{su}, t^{sys})$ in which *m* is transiting from an off-line state to a synchronized state is labeled SU in Figure 5.1; the cost of SU will be called *m*'s *start-up cost*. The behind-the-meter power generated by *m* to maintain itself in a synchronized state during the time interval $[t^{sys}, t^{sye})$ is labeled "No-load" in Figure 5.1; the cost of No-load will be called *m*'s *no-load cost*. The power generated by *m* during the time interval $[t^{sye}, t^{sd})$ to transit to a shut-down state at time t_{sd}, starting from a synchronized state at time t_{sye}, is labeled "SD" in Figure 5.1; the cost of SD will be called *m*'s *shut-down cost*.

Finally, note that Figure 5.1 also depicts two additional power-injection regions ("RU" and "RD") that are not included in the power-generation regions associated with *m*'s start-up, no-load, dispatch, and shut-down costs. The ramp-up region RU reflects the need for *m* to ramp up its power injection level to p^{disp}, starting from zero power injection at t^{sys}, in order to meet its dispatch obligations. The ramp-down region RD reflects the need for *m* to ramp down its power injection level to 0 after these dispatch obligations are met at time t^{e}, as part of *m*'s required progression toward a terminal off-line state. The cost of the ramp-up power injection region RU will be called *m*'s *ramp-up cost*, and the cost of the ramp-down power injection region RD will be called *m*'s *ramp-down cost*.

As a general economic rule of thumb, generator *m* should offer swing contract (5.2) into market M(T) only if *m* can ensure coverage of all of the avoidable cost it would incur should this offer be cleared. The two-part pricing form of this swing contract permits *m* to ensure this coverage.

For example, for the case depicted in Figure 5.1, *m*'s *availability* cost is the summation of its start-up cost, no-load cost, shut-down cost, ramp-up cost, and ramp-down cost; and *m* can ensure coverage of this availability cost by setting its offer price α equal to the summation of these costs. Moreover, *m*'s *performance* cost is its dispatch cost; and m can ensure coverage ex post of this dispatch cost by specifying an appropriately high price π for period-T delivered energy.

Nevertheless, the firm energy-block swing contract (5.2) is problematic on two grounds. First, it does not account for generator m's ramp requirements. Second, it offers zero swing (flexibility) to the ISO for real-time reserve deployment.

With regard to the first issue, real-world generators do not have infinite ramp-rates. Thus, as illustrated in Figure 5.1, the delivery of a period-T energy block at a maintained power injection level $p^{\rm disp} > 0$ could require m to ramp to this power level prior to the start of T and to ramp away from this power level after the close of T, depending on m's contractual obligations and/or physical requirements outside of T. This ramping could entail power injections into the transmission grid outside of T that are not visible to the ISO from the contractual terms of (5.2), thus complicating the ISO's need to maintain continual grid balance. The linked swing-contract market design proposed in Chapter 10 is meant to handle these issues.

Second, the swing contract (5.2) offers zero swing (flexibility) to the ISO for real-time reserve deployment. Specifically, the clearing of this contract obligates the ISO to dispatch a particular energy-block power-path during period T, and m is obligated to deliver this power-path in response to the ISO's dispatch signals during period T. This rigid form of contract is most suitable as an instrument for the balancing of net load that will occur at a sustained level during a designated operating period with a high degree of certainty.

5.3 AN ENERGY-BLOCK SWING CONTRACT IN OPTION FORM

Consider the following energy-block swing contract SC_m in option form, submitted by a dispatchable generator m into a swing-contract market M(T) as contingent reserve for a transmission system during a future operating period $\mathrm{T} = [t^{\rm s}, t^{\rm e})$.

$$\mathrm{SC}_m = [\alpha,\ t^{\rm ex},\ \mathbb{PP},\ \phi] \tag{5.3}$$

where

$$\alpha = \text{offer price}$$

$$t^{\rm ex} = \text{exercise time}$$

$$\mathbb{PP} = (b,\ t^{\rm s},\ p(\mathrm{E1}),\ t^{\rm R},\ r,\ t^{\rm E2},\ p(\mathrm{E2}),\ t^{\rm e})$$

$$b = \text{delivery location}$$

$$t^{\rm s} = \text{start-time for energy block E1}$$

$$p(\mathrm{E1}) = \text{maintained power injection for energy block E1}$$

$$t^{\rm R} = \text{start-time for ramp interval R}$$

r = ramp-rate during ramp interval R

t^{E2} = start-time for energy block E2

$p(E2)$ = maintained power injection for energy block E2

t^{e} = end-time for energy block E2

ϕ = pre-specified price π for delivered energy.

The exercise time t^{ex} occurs later than time 0, the M(T) closing time, but strictly prior to the start-time t^{s} for operating period T. The power-path set $\mathbb{PP}$ consists of a single power-path **p** to be delivered by m at its grid location b during period T. The power-path **p** is composed of two energy-block intervals separated by a ramp interval. The start-times and end-times for these intervals are assumed to satisfy $t^{s} < t^{R} < t^{E2} < t^{e}$. The maintained power injection levels $p(E1)$ and $p(E2)$ for energy blocks E1 and E2 lie within m's feasible sustainable power range $[0, P^{max}]$.

By assumption, generator m is in an off-line state at time 0, the closing time for M(T); and m must return to an off-line state as soon as possible following the close of the subsequent operating period T. Figure 5.2 provides an illustrative depiction of a power-path **p** that m could offer for period T by means of swing contract (5.3).

Also depicted are the additional SU, RU, no-load, RD, and SD energy amounts that m would need to generate before, during, and after T to ensure the availability of this power-path **p** during period T, conditional on m's initial off-line state and required terminal off-line state. However, in contrast to the previous firm swing contract (5.2), the clearing of the option swing contract (5.3) does not guarantee that the contract will in fact be exercised. If the contract is not exercised, all availability cost is avoided. Consequently, the probability of exercise could affect the extent to which m decides to include the cost of this additional generation in its swing-contract offer price.

The exercise time t^{ex} in the option swing contract (5.3) provides some degree of flexibility to the ISO for reserve deployment during period T even though only one power-path **p** is offered for dispatch during period T. Moreover, the inclusion of a ramp interval between the two energy blocks offered by **p** helps to ensure that these energy blocks can be feasibly delivered by m during period T.

5.4 SWING-CONTRACT IMPLEMENTATION OF STANDARD SUPPLY OFFERS

The supply offers used in the seven current US regional transmission organization/independent system operator (RTO/ISO)-managed day-ahead markets (DAMs) are similar in form. For concreteness, this section focuses on the three-part supply offer formulation used by the Electric Reliability Council of Texas (ERCOT). As will be shown below, these three-part supply offers can be implemented as firm step-function swing contracts providing swing (flexibility) in offered power-steps.

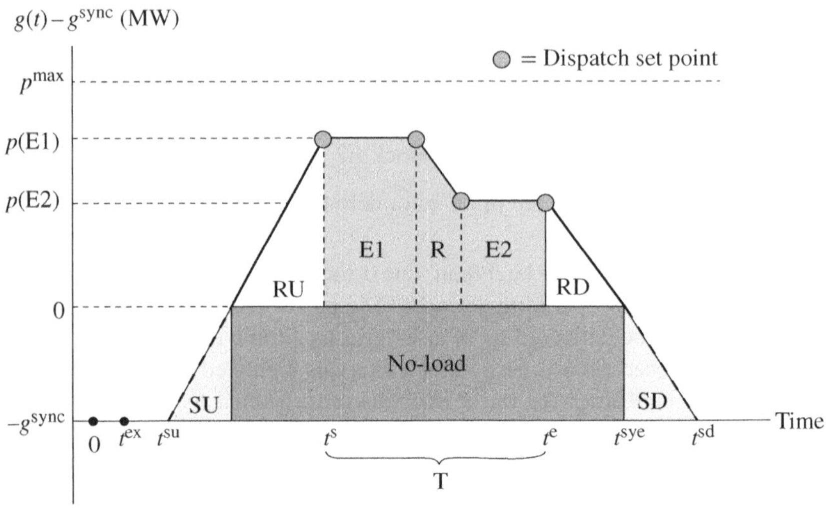

Figure 5.2 Illustration of the energy requirements for an option swing contract (5.3) issued by a dispatchable generator m for a future operating period T. Ensuring the availability of energy blocks E1 and E2, with supporting interim ramp R, for dispatch during operating period T requires m to generate start-up (SU), ramp-up (RU), no-load, ramp-down (RD), and shut-down (SD) energy before, during, and after T. The probability of contract exercise at time t^{ex} could affect the manner in which m ensures cost coverage of this additional generation through its offer price.

As detailed in ERCOT documentation [47–49], *qualified scheduling entities (QSEs)* can submit hourly supply offers into the ERCOT DAM for all twenty-four hours of the following day. For a currently off-line QSE, an hourly supply offer typically consists of three parts: a *Startup Offer* ($/start); a *Minimum-Energy Offer* consisting of a price ($/MWh) together with a power level (MW) referred to as a *low sustained limit (LSL)*; and a non-decreasing piecewise linear *Energy Offer Curve* in the MW-$/MWh plane consisting of a finite collection of linearly-connected dispatch points dp(H) whose power levels commence at LSL.[3]

Suppose an off-line QSE at a grid location b submits a three-part supply offer into the ERCOT DAM on day D for hour H of day D+1. The Startup Offer permits the QSE to report its total start-up cost SUC(b,H) ($) for hour H (if any). The Minimum-Energy Offer permits the QSE to report its minimum-energy cost MEC(b,H) ($) for hour H (if any).[4] The Energy Offer Curve permits the QSE to

[3]A QSE can submit an Energy Offer Curve without Startup and/or Minimum-Energy Offers. However, a QSE cannot submit a Startup Offer and/or a Minimum-Energy Offer without also submitting an Energy Offer Curve.

[4]MEC(b,H) can include all energy cost incurred by m to maintain a minimum feasible sustained power injection level during H, including energy cost for behind-the-meter generation.

report, in approximate form, its marginal production cost curve for injected power at grid location *b* during hour H of day D+1.

Suppose the QSE is scheduled for a particular dispatch point dp*(H) at grid location *b* for hour H of day D+1 as a result of the Security-Constrained Unit Commitment/Security-Constrained Economic Dispatch (SCUC/SCED) optimization conducted for the ERCOT DAM on day D. The day-D DAM settlement received by the QSE for hour H of day D+1 is then calculated as follows.

First, the QSE is paid the DAM LMP ($/MWh) determined at grid location *b* for hour H of day D+1 for each MWh of its scheduled delivered energy at *b* during hour H of day D+1; let this DAM LMP be denoted by LMP(*b*,H). The resulting energy payment ($) is the QSE's *Energy Revenue* at grid location *b* for hour H, denoted below as ER(*b*,H).[5] For example, suppose the QSE's dispatch set point at grid location *b* for hour H is dp*(H) = (80 MW, $40/MWh), and LMP(*b*,H) = $50/MWh. Then the QSE's scheduled dispatched energy would be 80 MW × 1 hour and its energy payment would be 80 MWh × $50/MWh = $4000.

Next, the QSE's Energy Offer Curve is used to calculate the QSE's *average incremental energy cost (AIEC)*, an approximation for the QSE's true production cost for its scheduled energy delivery at its grid location *b* during hour H; see [49, Sec. 4.6.5] for AIEC calculation details. Let this approximate production cost be denoted by AIEC(*b*,H) ($). The total cost ($) that the QSE would have to incur to fulfill its scheduled energy delivery obligations at grid location *b* during hour H of day D+1 is then approximated as TC(*b*,H) = SUC(*b*,H) + MEC(*b*,H) + AIEC(*b*,H).

Finally, Let ASR(*b*,H) ($) denote the *ancillary service revenue* that is awarded to the QSE for hour H.[6] If TC(*b*,H) > [ER(*b*,H) + ASR(*b*,H)], the QSE receives a *Make-Whole-Payment* ($) equal to [TC(*b*,H) − ER(*b*,H) − ASR(*b*,H)] in addition to ER(*b*,H) + ASR(*b*,H). Otherwise, the QSE receives ER(*b*,H) + ASR(*b*,H).

It will next be shown how an ERCOT DAM three-part supply offer can be implemented as a special case of a step-function swing contract in firm form that provides swing (flexibility) in offered ramp-rates and power-steps. For increased clarity of exposition, the general form of this swing contract will first be illustrated for an operating day that is subdivided into an initial ramp interval and two energy-block intervals separated by a second ramp interval. The extension of this illustrative step-function swing contract to include 24 energy blocks is straightforward. It will then be shown how an ERCOT DAM three-part supply offer can be represented as a special case of this step-function swing contract with 24 energy blocks by omitting the ramp intervals and by appropriately modifying the performance payment method.

---

[5]Note that ER(*b*,H) depends only on the QSE's total scheduled energy delivery at grid location *b* during hour H, not on the particular power-path used within H by the QSE to achieve this energy delivery. As emphasized in [47, Module 3, p. 12], the ERCOT DAM does not consider ramp-rates; it is the responsibility of the individual participant QSEs to manage their ramp constraints in order to fulfill their scheduled energy delivery obligations as closely as possible.

[6]As detailed in ERCOT [49, Secs. 4.51 & 4.64], ancillary service (MW) for hour H of day D+1 is settled in ERCOT on the basis of its *market clearing price for capacity (MCPC)* ($/MW) for hour H, calculated in the day-D DAM as the dual variable for its hour-H reserve requirement constraint.

Suppose a dispatchable generator m is a participant in an ISO-managed day-ahead market M(D+1) held during the morning of day D. The purpose of M(D+1) is to procure reserve (dispatchable power-paths) for the future operating day D+1. Let the start-time for day D+1 be denoted by t^s, and let the end-time for day D+1 be denoted by t^e. Suppose generator m submits the following firm step-function swing contract SC_m into M(D+1).

$$SC_m = [\alpha,\ \mathbb{PP},\ \phi] \tag{5.4}$$

where

$$\begin{aligned}
\alpha &= \text{offer price}\\
\mathbb{PP} &= (b,\ t^s,\ p^s,\ \mathbb{RR}(\text{R1}),\ t^{\text{E1}},\ \mathbb{P}(\text{E1}),\ t^{\text{R2}},\ \mathbb{RR}(\text{R2}),\ t^{\text{E2}},\ \mathbb{P}(\text{E2}),\ t^e)\\
b &= \text{delivery location}\\
t^s &= \text{start-time for ramp interval R1}\\
p^s &= \text{power injection level at start-time } t^s\\
\mathbb{RR}(\text{R1}) &= \text{set of feasible ramp-rates } r(p^s, p_i(\text{E1})) \text{ for R1}\\
t^{\text{E1}} &= \text{start-time for energy block E1}\\
\mathbb{P}(\text{E1}) &= \text{set of feasible maintained power-steps } p_i(\text{E1}) \text{ for E1}\\
t^{\text{R2}} &= \text{start-time for ramp interval R2}\\
\mathbb{RR}(\text{R2}) &= \text{set of feasible ramp-rates } r(p_i(\text{E1}), p_j(\text{E2})) \text{ for R2}\\
t^{\text{E2}} &= \text{start-time for energy block E2}\\
\mathbb{P}(\text{E2}) &= \text{set of feasible maintained power-steps } p_j(\text{E2}) \text{ for E2}\\
t^e &= \text{end-time for E2}\\
\phi &= \text{payment for ramp and delivered energy calculated}\\
&\quad\ \text{by means of power-path mileage and a pre-specified price } \pi(p)\\
&\quad\ \text{for each } p\ \in \mathbb{P}(\text{E1})\ \cup \mathbb{P}(\text{E2}).
\end{aligned}$$

The swing contract (5.4) offers an initial ramp interval R1 plus two energy-block intervals E1 and E2 that are separated by a second ramp interval R2. The start-times and end-times for these intervals span the entire operating day D+1 $= [t^s, t^e)$.

The possible maintained power-steps p (MW) for energy blocks E1 and E2 lie within generator m's feasible sustainable power range $[P^{\min}, P^{\max}]$, where $0 \leq P^{\min}$. The energy-block prices $\pi(p)$ (\$/MWh) that are pre-specified in ϕ for each energy block satisfy $\pi(p) \leq \pi(p')$ if $p \leq p'$.

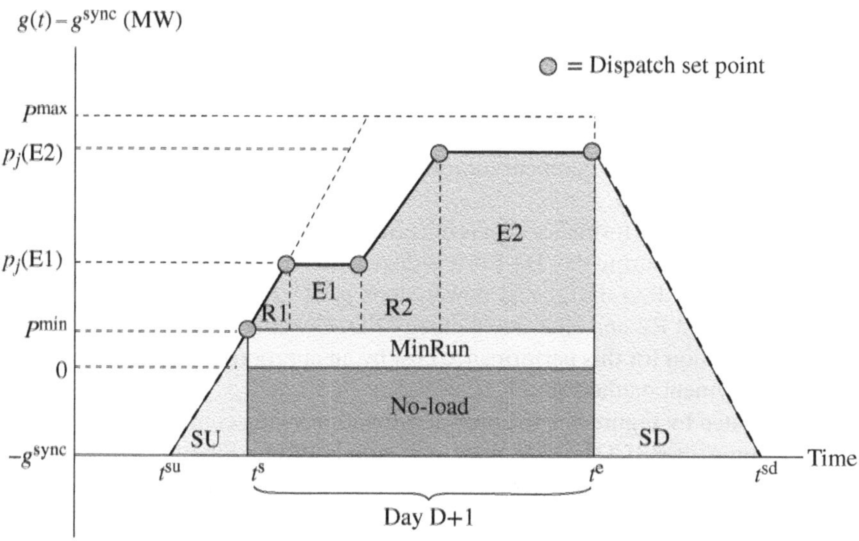

Figure 5.3 Illustration of one among multiple possible dispatched power-paths on day D+1 for generator m if m's firm step-function swing contract (5.4) with swing (flexibility) in its offered energy blocks is cleared in the day-ahead market M(D+1). Ensuring the availability of energy blocks E1 and E2, with supporting ramp R1 and R2, for dispatch during day D+1 requires m to generate start-up (SU), no-load, MinRun, and shut-down (SD) energy before, during, and after day D+1.

Finally, the ramp-rates r (MW/min) permitting feasible transitions between p^s and E1, and between E1 and E2, are determined as functions of generator m's maintained power-step levels for E1 and E2. These ramp-rates all lie within m's feasible down/up ramp-rate range $[-R^D, R^U]$, where $-R^D < 0 < R^U$.

Suppose generator m's swing contract (5.4) is cleared by the ISO. One possible power-path in $\mathbb{PP}$ that the ISO could then signal m to deliver during day D+1 is illustrated in Figure 5.3.[7]

In this illustration, m's required initial power level p^s at time t^s is m's minimum sustainable power injection level P^{min}. Taking this starting point as given, the ISO dispatches m at the particular ramp-rate in $\mathbb{RR}$(R1) that permits m to attain at time t^{E1} the power-step $p(E1) \in \mathbb{P}(E1)$ that the ISO has selected for the first energy block E1. Similarly, at time t^{R2} the ISO dispatches m at the particular ramp-rate in $\mathbb{RR}$(R2) that permits m to attain at time t^{E2} the power-step $p(E2) \in \mathbb{P}(E2)$ that the ISO has selected for the second energy block E2.

This illustration reveals the need for generator m to consider with care the specification of its offer price α and its performance payment method ϕ in its

[7]The full coverage of the piecewise linear power-paths in $\mathbb{PP}$ is indicated in Figure 5.3 as a trapezoid with power injections ranging from P^{min} to P^{max} during day D+1.

swing-contract reserve offer (5.4) in order to guarantee the full recovery of its availability and performance costs.

Generator m's availability cost is the cost that m knows prior to dispatch that it must incur to ensure the availability of its offered power-paths during day D+1. This cost includes m's cost for start-up (SU), no-load, and MinRun energy provision. Generator m can guarantee full compensation for this availability cost by appropriately specifying its offer price α.

In contrast, m's performance cost is uncertain prior to dispatch; it is the variable cost that m will incur during day D+1 if it is dispatched by the ISO to deliver energy blocks E1 and E2 at ISO-designated power-steps p(E1) and p(E2) with supportive ramp energy R1 and R2 and shut-down energy SD. Generator m can guarantee full ex post compensation for this performance cost by an appropriate specification of its performance payment method ϕ.

As indicated by Figure 5.3, the firm step-function swing contract (5.4) captures three key features of ERCOT DAM three-part supply offers. First, it permits coverage of start-up and minimum energy cost. Second, it permits multiple consecutive energy blocks to be offered with some degree of swing (flexibility) in the power-step for each block. Third, it permits a pre-specified price to be assigned to each possible power-step for each offered energy block, with higher prices assigned to higher possible power-steps for each block.

The extension of the step-function swing contract (5.4) to include 24 energy blocks with interspaced ramp intervals is straightforward. The ERCOT DAM three-part supply offer can then be obtained as a modified version of this extended swing contract as follows:

(a) Set the offer price α equal to start-up cost plus minimum-energy cost.
(b) Omit all ramp intervals and ramp-rate restrictions.
(c) Omit ramp compensation from the performance payment method.
(d) Specify a performance payment method that accords with ERCOT's method for determining compensation by means of DAM LMPs, ancillary service revenues, and make-whole payments, and that "takes back" all offer price payments.

With regard to modification (d), the process for determining locational marginal prices (LMPs) in the day-D DAM for day D+1 depends in part on the Energy Offer Curve submitted by each participant QSE, as does the calculation of make-whole payments. Consequently, each participant QSE's Energy Offer Curve must still be included as part of its performance payment method specification.

As indicated by modifications (a)–(d), the step-function swing-contract formulation (5.4) enables five important departures from standard supply offers in current RTO/ISO-managed DAMs.

First, the offer prices requested by dispatchable resources are not restricted to coverage of start-up and minimum-energy cost; all avoidable fixed cost can be included. All included avoidable fixed cost is then taken into account in the determination of optimal contract clearing. Second, dispatchable resources are paid their offer prices (if positive); hence, dispatchable resources are fully compensated for all of their included avoidable fixed cost. Third, swing contracts permit the incorporation of ramp requirements in offered power-path sets $\mathbb{PP}$, and the incorporation of compensation ex post for ramp services in performance payment methods ϕ. Any such ramp aspects are thus also taken into account in the determination of optimal contract clearing. Fourth, each dispatchable resource m can submit a performance payment method ϕ_m that is tailored to guarantee compensation for its own idiosyncratically determined variable costs; performance payment methods are not dictated by the ISO. Fifth, no performance payments are made in advance of actual verified performance.

Comparisons between the swing-contract DAM design and standard US DAM designs will be addressed at greater length in Chapter 14.

5.5 A SWING CONTRACT OFFERING CONTINUOUS SWING (FLEXIBILITY) IN POWER AND RAMP

Suppose a dispatchable resource m is a participant in a day-ahead market M(D+1) held during day D to facilitate routine net load balancing on day D+1 = $[t^s, t^e)$. Suppose m submits the following firm swing contract SC_m into M(D+1) that offers continuous swing (flexibility) in power and ramp.

$$\mathrm{SC}_m = [\alpha,\ \mathbb{PP},\ \phi] \tag{5.5}$$

where

$$\begin{aligned}
\alpha &= \text{offer price} \\
\mathbb{PP} &= (b,\ t^s,\ p^s,\ \mathbb{P},\ \mathbb{RR},\ t^e) \\
b &= \text{delivery location} \\
t^s &= \text{start-time for power delivery} \\
p^s &= \text{initial power level at time } t^s \\
\mathbb{P} &= [P^{\min},\ P^{\max}] = \text{range of feasible down/up power levels } p \\
\mathbb{RR} &= [-R^{\mathrm{D}},\ R^{\mathrm{U}}] = \text{range of feasible down/up ramp-rates } r \\
t^e &= \text{end-time for power delivery} \\
\phi &= \text{performance payment method for power-path delivery.}
\end{aligned}$$

The delivery location b denotes the grid location at which m can inject or withdraw power. The initial power level p^s for m is an element of m's feasible sustainable

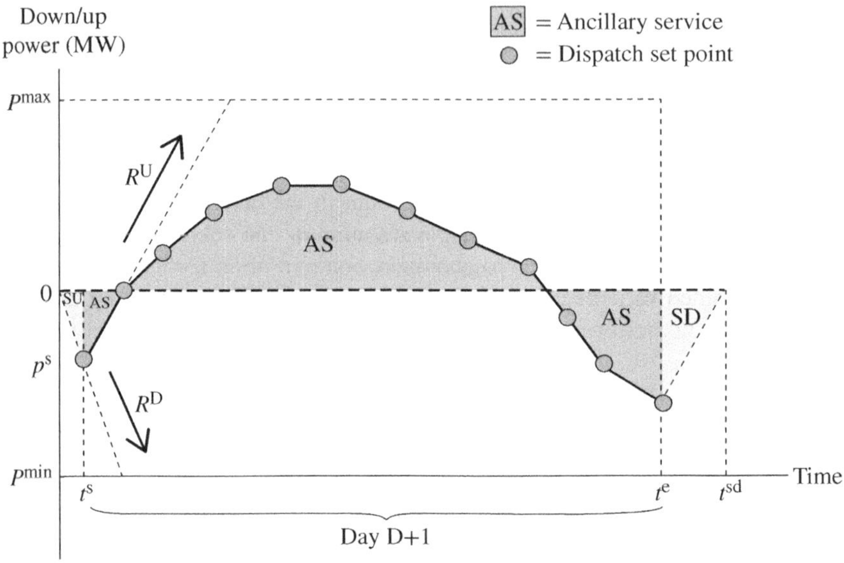

Figure 5.4 Illustration of one of many possible dispatched power-paths on day D+1 for ancillary service provider *m*, given *m*'s swing contract (5.5) is cleared in the day-ahead market M(D+1). The fulfillment of this dispatch obligation requires *m* to make start-up (SU) and shut-down (SD) power withdrawals before and after day D+1, given its initial shut-down state at time 0 and its need to return to a shut-down state after day D+1.

power range $\mathbb{P} = [P^{min}, P^{max}]$. This power range consists of pure power injections if $P^{min} \geq 0$, in which case *m* is participating in M(D+1) as a dispatchable generator. Alternatively, $\mathbb{P}$ consists of pure power withdrawals (absorptions) if $P^{max} \leq 0$, in which case *m* is participating in M(D+1) as a dispatchable demand-response resource. Finally, $\mathbb{P}$ provides bi-directional power capabilities if $P^{min} < 0 < P^{max}$, in which case *m* is participating in M(D+1) as a dispatchable provider of ancillary service (down/up power-paths). The down/up limits $-R^D$ and R^U for *m*'s feasible ramp-rates *r* (MW/min) satisfy $-R^D \leq 0 \leq R^U$.

The manner in which *m* specifies its offer price α and performance payment method ϕ will depend on whether *m* is participating in M(D+1) as a pure generator, a demand-response resource, or a down/up power provider. If *m* is participating as a pure generator, *m* could specify α and ϕ as in previously presented swing-contract examples. However, in the other two cases, *m* will typically not incur no-load or minimum-run cost; and *m* should request compensation ex-post for dispatched down-power delivery as well as for dispatched up-power delivery.

For example, suppose *m* is an ancillary service provider that submits into M(D+1) a swing contract (5.5) with a feasible down/up power range $P^{min} < 0 < P^{max}$ and an initial power withdrawal (absorption) level $p^s < 0$ at the start-time t^s for day D+1. In addition, suppose *m* functions as a grid-edge resource aggregator for a distribution grid connected to the transmission system at grid location b; and that m fulfills its ancillary service role in the transmission system by harnessing down/up power from the equipment and appliances of businesses and households directly connected to this distribution grid.

Figure 5.4 depicts one of many possible power-paths in $\mathbb{PP}$ that the ISO could then signal m to deliver during day D+1, in accordance with the contractual terms of m's swing contract (5.5), assuming this swing contract is cleared. Note that m has to engage in start-up and shut-down power withdrawals SU and SD in order to fulfill its dispatch obligations, given its initial off-line state at the close of M(D+1) and its need to return to an off-line state as soon as possible following day D+1. However, given the source of its down/up power delivery, m does not have any no-load energy requirements.

Since the power withdrawals SU and SD are not dispatched, they could result in start-up and shut-down costs charged to m. The start-up cost can be calculated by m prior to dispatch; hence, m can include this start-up cost in its offer price α. However, the precise extent of m's shut-down cost will depend on m's dispatched power-path during day D+1. To ensure compensation ex post for any incurred shut-down cost, m can include a shut-down cost recovery proviso in its performance payment method ϕ.

Finally, Figure 5.4 depicts m's dispatched delivery of ancillary service (down/up power) as regions marked "AS." The ex post compensation that m requires for this delivered ancillary service in its performance payment method ϕ can take many different forms. For example, m could simply require that each megawatt hour of its delivered AS energy injections and withdrawals, measured in absolute value terms, be compensated ex post at a pre-specified energy price π. In addition, m could require compensation ex post for delivered AS ramp, as measured by a mileage metric. Moreover, m could require compensation ex post for the duration of its delivered down/up power-path.

5.6 A SWING CONTRACT OFFERING BATTERY SERVICES

Suppose a dispatchable battery resource m is a participant in an ISO-managed swing-contract market M(H) held in advance of an operating hour $\text{H} = [t^s, t^e)$ to facilitate routine net load balancing during H. Suppose m submits the following firm swing contractSC_m into M(H) that offers continuous swing (flexibility) in power and ramp.

$$\text{SC}_m = [\alpha,\ \mathbb{PP},\ \phi] \tag{5.6}$$

where

$$\alpha = \text{offer price}$$

$$\mathbb{PP} = (b,\ ECap^{\max},\ \eta,\ t^s,\ \mathbb{SOC}^s,\ \mathbb{RR},\ \mathbb{P},\ t^e,\ \mathbb{SOC}^e)$$

$$b = \text{delivery location}$$

$ECap^{\max}$ = maximum energy storage capacity

η = round-trip efficiency

t^s = start-time for power discharge/charge

$\mathbb{SOC}^s$ = set of feasible state-of-charge percentages at t^s

$\mathbb{P} = [P^{\min}, P^{\max}]$ = range of feasible discharge/charge levels p

$\mathbb{RR} = [-R^D, R^U]$ = range of feasible ramp-rates r

t^e = end-time for power discharge/charge

$\mathbb{SOC}^e$ = set of feasible state-of-charge percentages at t^e

ϕ = performance payment method for down/up power-path delivery.

The delivery location b denotes the grid location where m is able to discharge (inject power) or charge (absorb power). The maximum energy storage capacity (MWh) of m is denoted by $ECap^{\max}$. The round-trip efficiency (decimal percent) of m, i.e. the fraction of energy put into storage that can be retrieved, is denoted by η. The range $\mathbb{P} = [P^{\min}, P^{\max}]$ of m's feasible discharge/charge levels p (MW) satisfies $P^{\min} < 0 < P^{\max}$. The range $\mathbb{RR} = [-R^D, R^U]$ of m's feasible down/up ramp-rates r (MW/min) satisfies $-R^D < 0 < R^U$.

Resource m's stored energy at any time t during hour H = $[t^s, t^e)$, i.e. m's *state-of-charge (SoC)* (MWh), must lie in $[0, ECap^{\max}]$. The elements of $\mathbb{SOC}^s$ and $\mathbb{SOC}^e$ are decimal percentages in [0,1] to be applied to $ECap^{\max}$ (MWh). Resource m's SoC at the start-time t^s for hour H must take the form $ss \cdot ECap^{\max}$ for some element $ss \in \mathbb{SOC}^s$; and resource m's SoC at the end-time t^e for hour H must take the form $se \cdot ECap^{\max}$ for some element $se \in \mathbb{SOC}^e$.

The swing contract (5.6) thus permits m to offer into M(H) a collection $\mathbb{PP}$ of down/up power-paths for hour H, each of which m could feasibly deliver in response to dispatch signals. However, in contrast to the ancillary service swing contract depicted in Figure 5.4, the amount of swing (flexibility) in these offered power-paths is constrained by *global* hour-H conditions imposed on m's SoC during hour H.

The manner in which the battery resource m specifies its offer price α and performance payment method ϕ can take a variety of forms, depending on the exact nature of m's availability cost and expected ex-post performance cost. However, as a provider of ancillary services, m would presumably want to request compensation ex post for power absorption (charging) as well as for power injection (discharging).

For example, suppose a battery resource m^* offers a swing contract (5.6) into M(H) for which $\mathbb{SOC}^s = \mathbb{SOC}^e = \{1.0\}$. Thus, m^* is offering to be in its maximum feasible SoC at the start-time t^s for hour H; and m^* is requiring that any dispatch signals it receives during hour H be such that it is returned to this maximum feasible

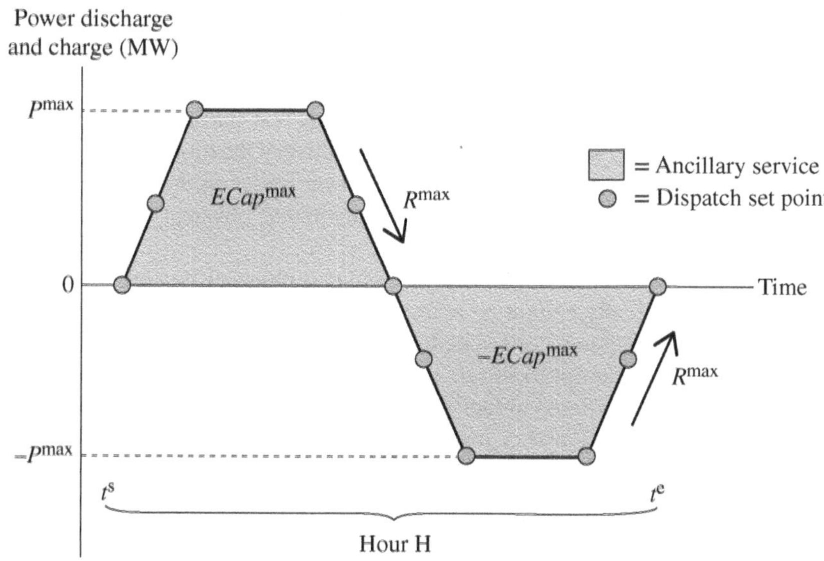

Figure 5.5 One of many possible dispatchable power-paths the ISO could signal the battery resource m^* to deliver during hour H, if the ISO clears m^*'s swing-contract offer for market M(H).

SoC by the end-time t^e for hour H. For simplicity of exposition, suppose also that $\eta = 1.0$, $R^D = R^L \equiv R^{max} > 0$ and $P^{min} = -P^{max} < 0$.

Figure 5.5 depicts one of many possible power-paths in $\mathbb{PP}$ that the ISO could then signal the battery resource m^* to deliver during hour H, if the ISO clears m^*'s swing-contract offer for M(H). For the first 30 minutes of hour H, resource m^* is signaled to inject power in a manner that reduces its SoC from $ECap^{max}$ to 0. For the final 30 minutes of hour H, resource m^* is signaled to withdraw (absorb) power in a manner that increases its SoC from 0 to $ECap^{max}$.

More precisely, resource m^* begins hour H at a discharge/charge level $p = 0$, with its initial SoC (stored energy) equal to $ECap^{max}$. For the first 10 minutes of hour H, resource m^* is ramped up at rate R^{max} from $p = 0$ to $p = P^{max}$. During minutes 11 through 20 of hour H, m^* is maintained at the discharge (power injection) level $p = P^{max}$. During minutes 21 through 30 of hour H, resource m^* is ramped down at rate $-R^{max}$ from $p = P^{max}$ to $p = 0$. By the end of minute 30, m^*'s initial SoC has been reduced to 0. During minutes 31 through 40, resource m^* continues to be ramped down at rate $-R^{max}$ until m^* reaches its maximum charge (power absorption) level $p = -P^{max}$. During minutes 41 through 50, m^* is maintained at the charge level $p = -P^{max}$. Finally, during minutes 51 through 60, resource m^* is ramped up at rate R^{max} from $p = -P^{max}$ to $p = 0$. By the end of minute 60, m^*'s SoC (stored energy) has been restored to $ECap^{max}$.

5.7 SWING-CONTRACT FACILITATION OF PRIVATE BILATERAL CONTRACTING

A physical bilateral contract is an agreement between two parties to exchange a product under contractually specified terms such as delivery location, delivery time, amount, and price. A large portion of wholesale power trades in the United States are currently undertaken by means of privately negotiated physical bilateral contracts [74].

The outcomes resulting from privately negotiated physical bilateral contracts for wholesale power typically cannot be realized without transmission grid support. Rather, these outcomes need to be self-scheduled with transmission service providers such as RTO/ISOs in an attempt to secure this support. This section considers how swing contracts can be formulated to facilitate this self-scheduling.

Suppose a generator G and a load-serving entity L are participants in an ISO-managed swing-contract market M(D+1) held on day D to procure reserve (dispatchable power-paths) for the normal balancing of net load on day D+1. Market M(D+1) operates over a grid in which neither G nor L has any ownership stake. Generator G is able to inject power into the grid at location A, and the customers serviced by L are able to withdraw power from the grid at a different location B.[8]

Let $\mathbf{p}^*$ denote L's day-D forecast for the aggregate 24-hour load profile of its customers on day D+1. Load-serving entity (LSE) L would like to negotiate a physical bilateral contract with generator G on day D to ensure the delivery of power-path $\mathbf{p}^*$ at grid location B during day D+1 in a manner that does not entail any price risk for itself. However, since neither G nor L has transmission facilities, this power-path delivery must be self-scheduled through M(D+1).

How might this self-scheduling be done? One possibility is that G and L could be permitted to submit *coupled* (offer,bid) contracts into M(D+1).

To illustrate, suppose G and L enter into a bilateral contract at the start of day D that commits G to deliver power-path $\mathbf{p}^*$ at grid location B during day D+1 in return for appropriate compensation, as follows. First, L agrees to pay G a contractually-stipulated assurance fee α^* (\$) on day D. Second, L agrees to compensate G ex post for the verified withdrawal of power-path $\mathbf{p}^*$ at grid location B during day D+1 by its customers, where this compensation is in accordance with a contractually stipulated performance payment method ϕ^*. Third, G agrees to compensate L ex post for any transmission service cost charged to L by the ISO on day D+1.

Under the terms of this bilateral contract, LSE L does not face any price risk or transmission service cost risk. In contrast, generator G faces risk from three sources: production, fuel cost, and transmission service cost. Generator G can attempt to hedge these risks through the contractual terms it negotiates for the assurance fee α^* and

[8]Since grid location A differs from grid location B, any power injection at A can affect the flow of power on all lines of the transmission grid, even if this power injection is matched by a power withdrawal at B. This dispersed power flow could result in costly transmission congestion and power losses.

the performance payment method ϕ^*. The extent of this hedging will depend on the extent to which L and G are risk averse.[9] For the purposes of this illustration, it will be assumed that L and G have come to an agreement regarding all terms of their bilateral contract.

As will next be explained, generator G and LSE L could be permitted to submit into M(D+1) a coupled pairing of a swing-contract reserve offer and a reserve bid that permits the implementation of their bilateral contract's terms. First, generator G could submit a *tagged* swing contract into M(D+1) taking the following form:

$$\mathrm{SC_G} = [\mathrm{Tag}^*,\ \alpha^*,\ \mathbb{PP}_\mathrm{G},\ \phi^*] \tag{5.7}$$

where

$$\begin{aligned}
\mathrm{Tag}^* &= \text{coupled contract identification number} \\
\alpha^* &= \text{offer price} \\
\mathbb{PP}_\mathrm{G} &= \{\mathrm{A}, t^\mathrm{s}, \mathbf{p}^*, t^\mathrm{e}\} \\
\mathrm{A} &= \text{power delivery location} \\
t^\mathrm{s} &= \text{start-time for day D+1} \\
\mathbf{p}^* &= \text{24-hour power-path for delivery during day D+1} \\
t^\mathrm{e} &= \text{end-time for day D+1} \\
\phi^* &= \text{performance payment method for power-path delivery.}
\end{aligned}$$

Second, LSE L could submit a *tagged* must-service (non-price-sensitive) reserve bid taking the following form.

$$\mathrm{RB_L} = [\mathrm{Tag}^*, \mathbb{PP}_\mathrm{L}] \tag{5.8}$$

where

$$\begin{aligned}
\mathrm{Tag}^* &= \text{coupled contract identification number} \\
\mathbb{PP}_\mathrm{L} &= \{\mathrm{B}, t^\mathrm{s}, \mathbf{p}^*, t^\mathrm{e}\} \\
\mathrm{B} &= \text{power withdrawal location} \\
t^\mathrm{s} &= \text{start-time for day D+1} \\
\mathbf{p}^* &= \text{24-hour load profile for withdrawal during day D+1} \\
t^\mathrm{e} &= \text{end-time for day D+1.}
\end{aligned}$$

[9]For the case at hand, the risk aversion of L can be measured by the maximum amount that L is willing to pay to avoid price risk and transmission service cost risk; and the risk aversion of G can be measured by the minimum amount that G must be compensated to take on production, fuel cost, and transmission service cost risk.

The ISO then treats the commonly tagged contracts (5.7) and (5.8) as a single coupled contract for clearing purposes; either both tagged contracts are cleared or neither tagged contract is cleared. If both tagged contracts are cleared, the following settlements take place.

First, the offer price α^* to be paid to generator G is charged to LSE L. Second, the compensation to be paid ex post to G in accordance with performance payment method ϕ^* is charge to L; this is the only reserve procurement cost charged to L. Third, any transmission service cost arising on day D+1 as a result of the contracts cleared in market M(D+1) on day D is allocated across market participants in accordance with the power imbalance at each transmission grid bus; see Section 6.7. Finally, to complete the implementation of their bilateral contract's terms, generator G and LSE L must augment their bilateral and market contracts with one additional contract: namely, a side-payment contract that requires G to reimburse L for any transmission service cost allocated to L by the ISO for day D+1 operations.

CHAPTER 6

SWING-CONTRACT MARKET DESIGN

6.1 CHAPTER PREVIEW

The basic contention of this study is that current US RTO/ISO-managed wholesale power markets are not correctly focused on the actual source of value for market participants and for the RTO/ISOs tasked with maintaining the efficiency and reliability of transmission system operations. This actual source of value is dispatchable power *paths* available for delivery at specific transmission grid locations during designated future operating periods.

More precisely, it is the attributes of these power-paths that provide value, such as start-time, duration, ramp-rate, and down/up power range. Each individual power-path provides a correlated combination of these attributes; changing any one of its attributes can require changes in all others. In standard economic terms, these attributes are joint products. However, *flexibility* in these attributes can be obtained by having dispatchable resources offer *diverse collections* of power-paths for possible dispatch during future operating periods.

As discussed in Section 3.2, the market-clearing optimization routines currently used in US RTO/ISO-managed wholesale power markets do not permit proper cost and benefit valuation of this flexibility because they do not take into account its joint-product genesis and its joint-product use. This study contends that an alternative swing-contract market design would permit a more appropriate valuation for the flexible provision of reserve (dispatchable power-paths) in centrally managed wholesale power markets. The purpose of the present chapter is to explain the general formulation of this proposed new design.

6.2 GENERAL SWING-CONTRACT MARKET FORMULATION

A swing-contract market M(T) is an ISO-managed forward reserve market for a future operating period T, where reserve takes the form of dispatchable power-paths for T.

A New Swing-Contract Design for Wholesale Power Markets, First Edition, Leigh Tesfatsion.

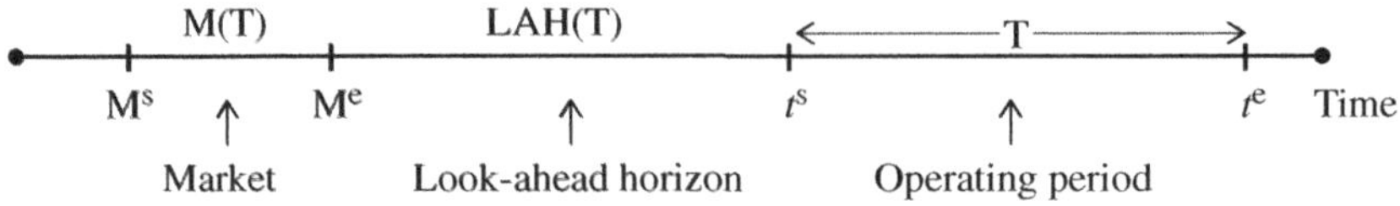

Figure 6.1 Time-line for a swing-contract market M(T) in relation to a future operating period T.

As shown in Figure 6.1, the *look-ahead horizon* for M(T) is the length of time between the close of M(T) and the start of T. As will be carefully motivated and explained in Chapter 10, this look-ahead horizon can range from very long term (multiple years) to very short term (minutes). This chapter focuses on properties of M(T) that hold generically for any look-ahead horizon.

The participants in M(T) consist of load-serving entities (LSEs), dispatchable resources, and variable energy resources (VERs). LSEs are intermediaries that service the power-usage requirements of retail customers. Dispatchable resources include dispatchable generators, dispatchable distributed-resource aggregators, and dispatchable storage devices. VERs include wind farms and solar power facilities that are not fully firmed by storage.[1]

Reserve offers submitted into M(T) by dispatchable resources take the form of portfolios of swing contracts for operating period T. As explained and illustrated in Chapters 4 and 5, a swing contract SC_m offered by a dispatchable resource m consists of four components. The first component is an offer price α_m that must be paid to m (if positive) or by m (if negative) if and only if SC_m is cleared. The second component is an exercise set $\mathbb{T}_m^{ex}$ whose specification determines whether SC_m is a firm or option contract. The third component is a physically characterized set $\mathbb{PP}_m$ of power *paths* for period T, each of which m could feasibly deliver at a designated grid location during T in response to dispatch signals. The fourth component is a performance payment method ϕ_m stipulating how m must be compensated ex post for the dispatched delivery of any power-path in $\mathbb{PP}_m$.

Reserve bids submitted into M(T) by LSEs are demands for power-path delivery during period T. As explained more carefully in Section 6.4, these reserve bids can take either a fixed or price-sensitive form. Fixed reserve bids express the LSEs' forecasts for must-service (non-price-sensitive) customer power usage during period T. Price-sensitive reserve bids express the LSEs' maximum willingness to pay for different amounts of customer power usage during T.

In addition, market participants can submit reserve offers and reserve bids into M(T) in coupled form in an attempt to secure transmission support for the

[1] As in existing US RTO/ISO-managed wholesale power markets, financial traders with no physical generation or customer service obligations could be "virtual" swing-contract market participants whose quantity positions have to be liquidated (reduced to zero) prior to scheduled delivery times. However, the efficacy of virtual offers/bids in these existing markets remains highly controversial; see [14]. The present study focuses attention on market participants able to make physically covered offers and bids.

outcomes of privately negotiated physical bilateral contracts.[2] A simple way to accomplish this coupling through the use of common tags was illustrated in Section 5.7.

Finally, the ISO inputs into M(T) forecasts for non-dispatched power injection during operating period T. The *forecasted net must-service load* for any region of the transmission grid during T is defined to be forecasted must-service load minus forecasted non-dispatched power injection for this region during T.

As will be carefully covered in Chapters 7–9, the ISO tasked with the management of M(T) then solves a contract-clearing optimization problem to determine which reserve offers and price-sensitive reserve bids to clear for operating period T. The ISO's objective is to maximize the expected total net benefit of the market participants, conditional on initial state conditions and subject to system constraints.

Total net benefit is given by total benefit net of total avoidable cost. The system constraints include power balance, transmission line, and reserve constraints. These constraints incorporate, as exogenous inputs: (i) all fixed reserve bids; (ii) all ISO forecasts for non-dispatched power injection; (iii) all of the power-path attributes included by dispatchable resources in their reserve offers; and (iv) system-wide and zonal reserve requirements. The reserve requirements consist of net load uncertainty-set coverage requirements, set by the ISO, whose purpose is to provide a robust means of protection against net load forecast errors.

At the close of M(T), the ISO announces which reserve offers and price-sensitive reserve bids have been cleared for the future operating period T. The clearing of reserve offers triggers immediate offer-price payment obligations.

At the close of period T, the ISO must use the performance payment methods included in cleared reserve offers to determine ex post compensation for any dispatched power-path delivery during T. In addition, LSEs with cleared price-sensitive reserve demands are obligated to pay ex post for the delivery of power-paths during T that are in accordance with these demands, at prices determined by these demands.

In summary, the swing-contract market M(T) functions purely as a period-T planning device for the ISO and market participants. In accordance with the pay-for-actual-performance principle underlying FERC Order 755 [54], no performance payments are made in advance of actual period-T performance.

---

[2]Market participants can also engage in privately negotiated *financial* contracting in order to hedge their financial risks [199, 200]. For example, in return for a negotiated flat fee, a dispatchable generator g might receive a contract from an insurance firm that obligates this firm to provide a make-whole payment to g should the performance payment that results from g's dispatch during T fall below some contractually-stipulated level. Since purely financial contracts do not entail physical power delivery, they do not require transmission support; hence, they do not have to be reported to the ISO responsible for the management of M(T).

6.3 FINANCIAL AND PHYSICAL FEASIBILITY OF SWING-CONTRACT OFFERS

A dispatchable resource m that offers a swing-contract $SC_m = (\alpha_m, \mathbb{T}_m^{ex}, \mathbb{PP}_m, \phi_m)$ into a swing-contract market M(T) for a future operating period T is responsible for ensuring that SC_m is financially and physically feasible for m to implement.

Ensuring the financial feasibility of SC_m entails two aspects. First, the offer price α_m should be sufficient to cover all avoidable fixed cost that m would have to incur to guarantee the availability of the power-paths in $\mathbb{PP}_m$ for possible dispatch during period T. This availability cost could include: amortized (pro-rated) acquisition cost for needed but not-yet-acquired generation facilities; transaction cost (e.g. licensing and legal fees); unit commitment cost (e.g. start-up cost, no-load cost); and opportunity cost. Second, the performance payment method ϕ_m should ensure coverage, ex post, for any variable cost that m incurs for the dispatched delivery of a power-path in $\mathbb{PP}_m$ during period T. This performance cost could include fuel cost, labor cost, equipment depreciation, and short-term equipment rental charges.

Ensuring the physical feasibility of SC_m also entails two aspects. First, each power-path in $\mathbb{PP}_m$ should be physically feasible for m to deliver during period T. Thus, the physical characterizations of the power-paths in $\mathbb{PP}_m$ must embody m-specific attributes such as m's delivery location, ramp-rate limits, power amplitude limits, and minimum down/up-times. Second, m should be capable of following a possibly time-varying dispatch signal during period T within operationally acceptable limits as long as the signaled power-path is an element of $\mathbb{PP}_m$.

6.4 RESERVE BIDS

A swing-contract market M(T) is an ISO-managed reserve market for a future operating period T, with reserve taking the form of dispatchable power-paths. As explained and illustrated in previous sections, reserve supply is expressed by means of reserve offers, where each reserve offer consists of a portfolio of swing contracts. However, the precise manner in which reserve demand is expressed through reserve bids has not yet been addressed.

A crisp demarcation between reserve offers and reserve bids submitted into M(T) cannot be made on the basis of whether their clearing would result in power injection or power withdrawal. As seen in Chapter 5, reserve offers can be submitted by dispatchable resources (e.g. storage devices) capable of both power injection and power withdrawal. Rather, the fundamental distinction between reserve offers and reserve bids has to do with their purpose. This, in turn, determines the direction of their compensation as well as the control of their real-time deployment.

Reserve offers submitted into M(T) by dispatchable resources are contracts that offer power-paths for possible period-T dispatched delivery at designated grid locations. Period-T dispatch instructions are typically required to resolve swing in the offered power-paths. This period-T dispatch is entirely controlled by the ISO for the purpose of carrying out its fiduciary responsibilities to the transmission system as a whole. Reserve offers stipulate how a resource must be separately compensated for availability and delivery.

In contrast, reserve bids submitted into M(T) by LSEs are contracts that demand the period-T delivery of power-paths at designated grid locations for use by customers. As in current US RTO/ISO-managed markets, these reserve bids can take a fixed or price-sensitive form. A *fixed reserve bid* demands the period-T delivery of a designated power-path at a designated grid location, as must-service load, regardless of required payment. A *price-sensitive reserve bid* demands the period-T delivery of a power-path at a designated grid location, where this power-path must be selected from a designated power-path set. Each of the power-paths in this set is accompanied by information indicating willingness to pay for its delivery. This willingness to pay can reflect benefit valuations determined by the issuing LSE or benefit valuations communicated to the issuing LSE by its customers.

Apart from emergency conditions, if an LSE submits a fixed reserve bid into M(T), the power-path (load profile) designated by this fixed reserve bid for period T is automatically incorporated into the system constraints for M(T) as must-service load for period T. Hence, there is no market-clearing process for fixed reserve bids. Rather, LSEs submitting fixed reserve bids into M(T) are obligated to pay ex post for the delivery of power to meet the load of their customers during period T as expressed in these bids. This payment is determined ex post as a share of realized period-T net reserve procurement cost.[3]

As will be carefully explained in Chapter 9, price-sensitive reserve bids can vary in form from simple demand functions with time-of-use pricing to reserve bids directly expressed in terms of benefit functions. An LSE that submits a price-sensitive reserve bid into M(T) must be able to use real-time telemetry to communicate period-T price or power-usage set points to the customers whose power demand is represented by this bid.

Let RBid_j denote a price-sensitive reserve bid submitted into M(T) by LSE j. The reserve bid RBid_j is *cleared* for M(T) if the ISO incorporates the power-path set and willingness-to-pay information designated in this bid into the optimization formulation for M(T). If RBid_j is cleared for M(T), then LSE j must pay ex post compensation for any power delivered in period T to service the power-usage demands of the customers represented by this bid. This ex post compensation is determined in accordance with the willingness-to-pay information included in RBid_j.

As will be seen in Chapter 9, if a price-sensitive reserve bid RBid_j is incorporated into an M(T) optimization, the possible price-sensitive power usage levels designated in this bid appear as ISO decision variables in this optimization. However, M(T) is for planning purposes only. If RBid_j includes a zero-power-path in its designated power-path set, the ISO can always effectively exclude RBid_j from real-time implementation by dispatching this zero-power-path for T. Thus, price-sensitive reserve bids submitted into M(T) whose designated power-path sets include a zero-power-path can be automatically cleared. The same is not true for reserve offers since the clearing of a reserve offer triggers an offer price payment obligation even if no power-path delivery is subsequently required.

---

[3]An ISO must pass through to market participants all net costs (i.e. costs minus revenues) incurred through market operations in order to preserve its independent status. An ISO cost allocation method for swing-contract markets is proposed in Section 6.7.

6.5 HANDLING OF FIXED RESERVE BIDS AND NON-DISPATCHED POWER

As explained in Section 6.4, reserve bids submitted into a swing-contract market M(T) for a future operating period T can take a fixed or price-sensitive form. A fixed reserve bid is a demand for a specific power-path (load profile) to be delivered at a specific grid location during period T, with no accompanying price information signaling willingness to pay for this delivered power.

In addition, VERs are permitted to inject non-dispatched power into the transmission grid at specific grid locations during period T. This power injection is not accompanied by any price information signaling required compensation.

Fixed reserve bids submitted into M(T) and forecasts for non-dispatched power injection during operating period T are directly entered into the power balance constraints for the ISO's optimization problem for M(T) as must-service and must-run obligations, respectively; the benefit and cost of these obligations are not accounted for in the ISO's objective function. However, these must-service and must-run power obligations can be curtailed by the ISO during real-time operations if this curtailment becomes necessary to ensure real-time net load balancing.[4]

6.6 PERFORMANCE PENALTIES AND INCENTIVES

In current US ISO/RO-managed wholesale power markets, penalties for failure to follow dispatch instructions are administratively determined. For example, during 2015 the California Independent System Operator (CAISO) used a comparison of automatic generation control (AGC) set points to actual telemetry in order to judge the accuracy with which dispatch instructions were followed [78]. Resources that failed to provide the power movements called for by dispatch instructions within designated accuracy limits were required to make penalty payments.

As discussed in Section 6.3, this study assumes that each dispatchable resource is responsible for ensuring the physical feasibility of any reserve offer it submits into an ISO-managed swing-contract market M(T) for a future operating period T. In particular, each dispatchable resource must be capable of responding with acceptable accuracy to any dispatch signals it receives from the ISO during period T that are *admissible*, i.e. that are in accordance with the terms of its reserve offer.

Realistically, however, non-performance penalties and eligibility requirements might need to be instituted to ensure that issuers of cleared reserve offers follow

[4] As will be seen in Chapter 7, the ISO's optimization formulation for M(T) includes a slack variable in the power balance constraint for each grid bus and each period-T time-step, and a penalty term in the objective function that penalizes any non-zero solution value for this slack variable. A non-zero slack-variable solution is a warning to the ISO that currently cleared reserve offers are not sufficient to cover fully the ISO's forecasted net load uncertainty sets for operating period T; hence, unless further reserve offers are cleared in a subsequent market prior to T, non-market curtailment actions (e.g. load shedding, wind spillage) might need to be taken during T.

admissible real-time dispatch signals with sufficient accuracy. For example, standardized penalties could be imposed on dispatchable resources for uninstructed deviations from dispatch signals; and standardized incentive payments could be offered as rewards for close following of dispatch signals. In addition, basic performance validation tests could be conducted in advance of market participation to check a resource's claimed ability to deliver particular types of power-paths in response to dispatch signals.

Standardized failure-to-perform penalties and incentive payments could also be required aspects of the performance payment methods included within reserve offers. Alternatively, dispatchable resources might find it advantageous to voluntarily include failure-to-perform penalties as part of the performance payment methods included within their reserve offers in order to signal to the ISO the quality of their offered services relative to those of their competitors.

In addition, LSEs submitting fixed-demand reserve bids into M(T) for a future operating period T could be penalized for systematic deviations[5] between the actual period-T must-service load of their customers and the forecasted amount of period-T must-service load implied by these bids. Similarly, LSEs submitting price-sensitive reserve bids into M(T) on behalf of a designated collection of retail customers could be penalized for significant deviations between the ISO-dispatched price-sensitive power usage of these customers during period T and the actual power usage of these customers during period T.

Finally, as explained more carefully in Section 6.7, penalty and incentive payments can also take the form of appropriate adjustments in the amount of reserve procurement cost allocated by the ISO to each M(T) participant.

6.7 ISO COST ALLOCATION

As in current US RTO/ISO-managed wholesale power markets, the ISO managing a swing-contract market M(T) for a future operating period T is constituted as an *independent* system operator. Roughly stated, independence means that the ISO is prohibited from having any material or financial stake in M(T) operations that could hinder its ability to carry out its management duties in an impartial manner.

As part of its fiduciary responsibilities, the ISO must function as a neutral clearing house for the reserve offers and price-sensitive reserve bids submitted into M(T). In addition to ensuring that M(T) is cleared in an efficient and reliable manner, the ISO must also ensure that all ensuing payment obligations are fulfilled.

First, dispatchable resources with cleared reserve offers receive from the ISO immediate assurance of availability (offer price) payments if their offer prices are

[5] A *forecast error* is a deviation between a realized value and a forecasted value for some outcome. Roughly, a forecast error is *systematic* if its average value over an observation period, conditional on available information, differs significantly from zero. Such a non-zero value indicates the forecast error is correlated with available information; hence, this information has not been fully exploited to produce the best possible forecast.

positive. If the offer price for a cleared reserve offer is negative, the resource is instead obligated to pay this offer price.

In addition, dispatchable resources with cleared reserve offers receive from the ISO immediate assurance that the performance payment methods they have included in these offers will be used ex post to compensate them for any dispatched power-path delivery they are called upon to make during period T. Deviations between ISO-dispatched power-path delivery and actual power-path delivery by these resources during period T are treated as increments or decrements to *non-dispatched* power by these resources.

Second, all LSEs with cleared price-sensitive reserve bids for designated collections of retail customers are obligated to make ex post price payments for the ISO-dispatched power usage of these customers during period T, in accordance with the willingness to pay expressed in these bids. Deviations between ISO-dispatched power usage and actual power usage by these customers during period T are treated as increments or decrements to *must-service* load for these customers.

The *net reserve cost* arising from M(T) operations is the sum of the net availability cost and net performance cost arising from M(T) operations. The *net availability cost* is the sum of all offer price payments to submitters of cleared reserve offers with positive offer prices net of any offer price revenues received from submitters of cleared reserve offers with negative offer prices. The *net performance cost* is the sum of all performance compensation paid ex post to submitters of cleared reserve offers who are dispatched by the ISO during period T, net of all incoming revenues from period-T operations. These incoming revenues include price payments by LSEs with cleared price-sensitive reserve bids for period T as well as any standardized penalties paid by participants in M(T) for deviations from ISO dispatch instructions during period T.

In general, there is no guarantee that the ISO's net reserve cost will be zero either for individual markets M(T) or averaged over successive markets for successive operating periods. Thus, without further regulation, the ISO's market clearing operations could result in persistent earnings or losses over time.

Consequently, in order to maintain its independent status, the ISO managing M(T) is required to pass through to market participants any net reserve cost arising from M(T) operations. Moreover, for similar reasons, the ISO is required to pass through to market participants any *transmission service cost* arising from M(T) operations, i.e. any cost arising from the need to maintain the reliable physical operation of the transmission grid during period T.

The following question thus arises: How should the ISO allocate net reserve cost and transmission service cost across the participants in M(T) in a manner that appropriately takes into account must-service load and non-dispatched power injection? As outlined in summary form in Table 6.1, this study proposes three principles for the determination of this allocation.[6]

[6]The first two ISO cost allocation principles were formulated in collaboration with Swathi Battula.

TABLE 6.1 ISO cost allocation principles

M(T) operational costs	ISO cost allocation principle
Net availability cost	Allocate in accordance with each participant's historically recorded or anticipated net must-service load volatility for similar operating periods
Net performance cost	Allocate ex post in accordance with each participant's realized net must-service load during operating period T
Transmission service cost	Allocate ex post in accordance with the power imbalance recorded at each participant's grid location during operating period T

The *first ISO cost allocation principle* asserts that the net availability cost arising from M(T) operations should be allocated across M(T) participants in accordance with the historically observed volatility of their "net must-service load" during similar operating periods. A resource's *net must-service load* at any given point in time is defined to be the resource's non-dispatched power withdrawal (if any) minus the participant's non-dispatched power injection (if any). Thus, a resource's net must-service load is the non-dispatched portion (if any) of the resource's net power withdrawal or injection.

This first ISO cost allocation principle explicitly recognizes that net must-service load during an operating period T poses a risk for transmission system operations during period T to the extent that it is uncertain at the start of M(T). An unexpected *positive* net must-service load during period T must be balanced by additional power injection, and an unexpected *negative* net must-service load during period T must be balanced by additional power withdrawal. This power imbalance risk must be hedged by appropriate M(T) reserve procurement to ensure period-T coverage of possible net must-service load realizations above and below expected net must-service load.

The first ISO cost allocation principle would transfer this power imbalance risk to the M(T) participants who give rise to this risk.[7] This, in turn, would encourage these participants to take actions to reduce or eliminate this risk.

For example, LSEs would have an incentive to express their customer loads as price-sensitive reserve demands rather than as must-service loads, to the largest extent possible. LSEs would also have an incentive to improve the reliability of their price-sensitive reserve demands, resulting in smaller deviations between ISO-dispatched price-sensitive power usage and actual power usage. VERs would have an incentive to firm their power output into dispatchable power injection by

[7] In contrast to this focus on risk, FERC Order 844 [60, Sec. III.29] recommends that the *consequences* (harmful vs. helpful) of a market participant's real-time actions be used to determine the participant's allocated share of out-of-market (OOM) make-whole payment ("uplift") costs. The gaming opportunities created by this ex post focus on consequences will be addressed in Section 15.8.

means of coupled storage capabilities. Finally, all dispatchable resources with cleared reserve offers would have an incentive to closely follow ISO dispatch instructions.

The *second ISO cost allocation principle* asserts that net performance cost arising from M(T) operations should be allocated across M(T) participants in proportion to their realized net must-service load during period T. The rationale for this second principle is that M(T) participants whose net must-service load realizations during period T require the dispatched delivery of down/up power during period T in order to maintain power balance should be charged ex post for the cost of this dispatch.[8]

Finally, the *third ISO cost allocation principle* asserts that transmission service cost arising from M(T) operations should be allocated ex post across M(T) participants in proportion to the power imbalance at their particular grid locations during period T. *Power imbalance* is said to occur at a particular grid location b at a particular time t if there is a non-zero net power withdrawal or injection at b at time t that requires the transmission of power from or to other grid locations in order to ensure power balance across the transmission grid as a whole at time t.

The rationale for this third ISO cost allocation principle is that power withdrawals and/or injections at a particular grid location contribute to transmission service cost if and only if they result in power imbalance at this location. The allocation of transmission service cost is a well-studied problem within the power systems literature; see [118]. The proposed use of power imbalance to allocate transmission service cost across grid locations is in general accordance with the Z_{bus} method developed by Conejo et al. [36]. The further allocation of transmission service cost across the market participants at each particular grid location is a classic Shapley Value problem [93, pp. 385-410], [164].

One special case warrants separate consideration. Consider an ISO-managed swing-contract market M(T) for which, as usual, the objective of the ISO is the maximization of total net benefit (i.e. total benefit minus total avoidable cost) for operating period T. *Suppose net must-service load is zero for period T.* That is, suppose no fixed-demand reserve bids are submitted into M(T), no non-dispatched power injection is forecasted or realized for period T, and no penalty or incentive payments are paid for performance during period T. In this case:

1. The ISO can ensure total net benefit is zero for period T without violation of any system constraints simply by choosing not to clear any reserve offers or price-sensitive reserve bids for period T.
2. The ISO will clear and dispatch a price-sensitive reserve bid for period T only if, as a result, total net benefit for period T either increases or remains the same.

[8] Note that all period-T LSE price payments for ISO-dispatched power usage resulting from LSE price-sensitive reserve bids cleared in M(T) are subtracted from performance cost to obtain *net* performance cost. Under the second ISO cost allocation principle, this net performance cost is then allocated across M(T) participants in accordance with their net *non-dispatched* power withdrawal or injection during period T. This assures that an LSE with a cleared price-sensitive reserve bid is not double-charged for any ISO-dispatched power usage that results under this bid.

3. The prices paid for power usage during period T under the terms of any price-sensitive reserve bids the ISO chooses to clear and dispatch for period T reflect the benefit (willingness to pay) of the bid submitters. Hence, by (2), these prices must be sufficient to cover the total avoidable cost (availability plus performance) incurred by the submitters of reserve offers that the ISO clears and dispatches to service this power usage.
4. It follows from (1) to (3) that all avoidable cost incurred for period T will be fully covered by price-payment revenue for period T, implying total net benefit for period T will be nonnegative.

Consequently, for the special case in which net must-service load is zero for operating period T, the resulting nonnegative total net benefit can be used to partially or fully offset transmission service cost prior to the allocation of this transmission service cost back to the market participants.

Although improbable for wholesale power markets, absence of must-service load is in fact the normal case for markets around the world. As a rule, customers do not submit demand bids into markets in an open "willing to pay anything" form. A major exception is the market for health services with its notoriously inefficient pricing structure, hardly a market to be emulated.

CHAPTER 7

SWING-CONTRACT MARKET OPTIMIZATION: BASE-CASE MILP FORMULATION

7.1 CHAPTER PREVIEW

As demonstrated in Chapter 5, swing-contract reserve offers can take a wide variety of forms. Consequently, formulations for the optimal clearing of a swing-contract market can also take a wide variety of forms.

This chapter develops a *mixed integer linear programming (MILP)* optimization formulation for a base-case ISO-managed swing-contract market M(T). The reserve offers submitted into M(T) by dispatchable resources provide swing (flexibility) in both power amplitude and ramp and request time-of-use performance payments for delivered down/up power and ramping services. All reserve bids take a fixed (nonprice sensitive) form.

Given the assumption of fixed reserve bids, the benefit of power usage (as measured by maximum willingness to pay) is effectively infinite. The ISO's maximization of expected total net benefit thus reduces to the minimization of expected total avoidable cost, conditional on initial state conditions and subject to system constraints.

This chapter provides a complete analytical MILP representation for this base-case optimization problem. This MILP optimization formulation generalizes the MILP optimization formulations developed for swing-contract markets in previous work [72, 113, 115] in five key ways, as follows:

- The length of the look-ahead horizon for M(T) and the duration of the future operating period T are not restricted;
- Ramping constraints are reformulated to accommodate down/up power offers and the imposition of start-up and shut-down ramping restrictions;
- Performance cost for ramping services is included in the objective function;
- Power balance constraints are augmented with slack variables;
- Imbalance cost is included in the objective function.

A New Swing-Contract Design for Wholesale Power Markets, First Edition, Leigh Tesfatsion.

Imbalance cost is measured by a penalty imposed for nonzero slack variables in the power balance constraints. A positive imbalance cost warns the ISO that current reserve offers are not sufficient to meet period-T net load balancing needs as currently forecasted.

A nomenclature table providing a consolidated list of symbols and symbol definitions for the base-case analytical MILP optimization formulation developed in this chapter can be found in Table A.4 in Appendix A.

7.2 GENERAL ASSUMPTIONS AND NOTATION

The base-case swing-contract market M(T) considered in this chapter satisfies the following 11 assumptions:

(A1) The market M(T) operates over a transmission grid with bus set $\mathbb{B} = \{1, \ldots, NB\}$ for which at most one transmission line connects any two distinct buses.[1] Let $\mathbb{L}$ denote the set of all distinct transmission lines ℓ that directly connect pairs of distinct buses in $\mathbb{B}$. For each $\ell \in \mathbb{L}$, the originating bus for ℓ is denoted by O (ℓ) and the end bus for ℓ is denoted by E (ℓ).

(A2) The bus set $\mathbb{B}$ is divided into NZ reserve zones[2] $\mathbb{B}(z) \subseteq \mathbb{B}$ indexed by $z \in \mathbb{Z} = \{1, \ldots, NZ\}$. These reserve zones form a partition of $\mathbb{B}$, i.e. $\cup_{z\in\mathbb{Z}} \mathbb{B}(z) = \mathbb{B}$, and $\mathbb{B}(z_i) \cap \mathbb{B}(z_j) = \emptyset$ for any z_i and z_j in $\mathbb{Z}$ with $z_i \neq z_j$.

(A3) The participants in M(T) consist of a set $\mathbb{M}$ of dispatchable resources m, a set $\mathbb{NG}$ of nondispatchable generators n, and a set $\mathbb{LS}$ of LSEs j that service the power usage requirements of retail customers.

(A4) Each dispatchable resource $m \in \mathbb{M}$ injects or withdraws power at a particular bus b_m called *m's location*. Each nondispatchable generator $n \in \mathbb{NG}$ injects power at a particular bus b_n called *n's location*. Each LSE $j \in \mathbb{LS}$ services the power usage requirements of retail customers that withdraw power at a particular bus b_j called *j's location*.

(A5) Each dispatchable resource $m \in \mathbb{M}$ submits into M(T) a reserve offer consisting of a firm swing contract SC_m with an offer price α_m and with m-specific offered amounts of swing (flexibility) in power amplitude and ramp[3] for delivery at bus b_m during a designated *delivery period* $T_m \subseteq T$. Time-of-use

[1]Multiple transmission grid branches directly connecting two distinct buses can equivalently be represented by a single appropriately configured transmission line.

[2]For the purposes of this study, a *reserve zone* is defined to be an ISO-designated collection of buses whose connecting transmission lines have experienced relatively infrequent congestion (binding line-limit constraints) during past operating periods.

[3]Swing contracts offering flexibility in power amplitude and ramp are illustrated in Chapter 5; see, for example, Section 5.5 and Figure 5.4.

compensation is requested for delivered down/up power and ramping services.

(A6) Each dispatchable resource m is in a zero-power shut-down state at time 0, the closing time for M(T). If m's reserve offer SC_m is cleared for the future operating period T, m ramps to a synchronized state to ensure the availability of its offered services during its offered delivery period T_m and returns to a zero-power shut-down state as rapidly as possible after the close of T_m.

(A7) All of the forecasted power usage requirements to be serviced by the LSEs $j \in \mathbb{LS}$ during the future operating period T are expressed as fixed reserve bids for T, i.e. as forecasted must-service load profiles for T with no accompanying price information signaling willingness to pay.

(A8) Nondispatched power injection is treated as negative load. The *forecasted net must-service load* for a region is defined to be the forecasted must-service load for this region minus the forecasted nondispatched power injection for this region.

(A9) The ISO's *expected total avoidable cost* for the future operating period T is the summation of availability cost, expected performance cost, and expected imbalance cost. The objective of the ISO is to clear reserve offers for M(T) that minimize expected total avoidable cost subject to system constraints.

(A10) The system constraints for M(T) include power balance constraints, transmission line limits, and system-wide and zonal down/up reserve requirement constraints whose purpose is to provide a robust means of protection against net load forecast errors. In addition, the system constraints for M(T) include constraints reflecting the physical attributes of market participants (e.g. ramping restrictions, power capacity limits, and power-usage limits) as reported in their submitted reserve offers.

(A11) The MILP optimization formulation for M(T) uses a standard *DC optimal power flow (DC-OPF)* approximation to represent power flow constraints [175, 176]. Thus, the following three technical assumptions are made: (i) The resistance for each transmission line is negligible compared to the reactance; hence, this resistance is set to 0; (ii) the voltage magnitude at each bus is equal to a common base voltage magnitude; and (iii) the voltage angle difference $\Delta\theta(\ell)$ across any line ℓ is sufficiently small that the following approximations hold: $\cos(\Delta\theta(\ell)) \approx 1$ in size and $\sin(\Delta\theta(\ell)) \approx \Delta\theta(\ell)$ in size.

7.3 DISCRETIZATION OF THE ISO's OPTIMIZATION PROBLEM

As noted in [165, Slide 10], a market operator tasked with determining an optimal market-clearing solution for a wholesale power market would ideally like to determine an accurate solution for an accurately posed problem. In practice, however, this is not a feasible option. Consequently, market operators typically seek to determine accurate solutions for approximated problems.

This section shows how the swing-contract market M(T) described in Section 7.2 can be analytically expressed in an approximate discretized MILP form to which any standard MILP solver can be applied. This discretization consists of two successive steps. First, the operating period T is partitioned into a finite number of time-steps. Second, the power-path set $\mathbb{PP}_m$ and performance payment method ϕ_m included in the swing-contract reserve offer SC_m submitted into M(T) by each dispatchable resource m are discretized in a manner that comports with this partitioning of T. These two successive steps will now be carefully explained.

First, let the future operating period $T = [t^s, t^e)$ be partitioned into NK consecutive time-steps k with equal length Δt measured in hours (e.g. 2 hours, 1 hour/12, ...). Let $\mathbb{K} = \{1, \ldots, NK\}$ denote the index set for these time-steps. Time-step k is then given by the half-open interval $[k^s, k^e)$, where $k^s \equiv t^s + [k-1]\Delta t$ and $k^e \equiv t^s + k\Delta t$ for $k = 1, \ldots, NK$. The look-ahead horizon $[0, t^s)$ between the closing time 0 for M(T) and the start-time t^s for T will be referred to as *time-step 0*. The augmented set $\{0\} \cup \mathbb{K} = \{0, 1, \ldots, NK\}$ will be denoted by $\mathbb{K}^0$.

Second, extending the assumptions (A1) through (A6) in Section 7.2, suppose that the firm swing-contract SC_m with swing in power amplitude and ramp that is submitted into M(T) by each dispatchable resource $m \in \mathbb{M}$ has the form

$$SC_m = [\alpha_m, \mathbb{PP}_m, \phi_m] \tag{7.1}$$

where

α_m = offer price

$\mathbb{PP}_m = (b_m, t^s_m, \mathbb{P}_m, \mathbb{RR}_m, t^e_m)$

b_m = delivery location

t^s_m = start-time for power delivery

$\mathbb{P}_m = [P^{\min}_m, P^{\max}_m]$ = range of feasible down/up power levels p

$\mathbb{RR}_m = [-R^D_m, R^U_m]$ = range of feasible down/up ramp-rates r

t^e_m = end-time for power delivery

ϕ_m = performance payment method for power-path delivery.

The delivery location b_m denotes the bus at which m can inject or withdraw power. The range $\mathbb{P}_m = [P^{\min}_m, P^{\max}_m]$ for m's feasible sustainable power levels p (MW) represents pure power injections if $0 \le P^{\min}_m$, pure power withdrawals (absorptions) if $P^{\max}_m \le 0$, and bidirectional power capabilities if $P^{\min}_m < 0 < P^{\max}_m$. The down/up limits for m's feasible ramp-rates r (MW/Δt) satisfy $-R^D < 0 < R^U$.

Let $T_m \equiv [t^s_m, t^e_m)$ denote the offered *delivery period* for dispatchable resource m. The delivery period T_m consists of a sequence $F_m, \ldots, L_m$ of consecutive[4] time-steps in $\mathbb{K}$. The start-time t^s_m for T_m coincides with the start-time F^s_m for the first time-step F_m, and the end-time t^e_m for T_m coincides with the end-time L^e_m for the last time-step L_m. Let this collection of consecutive time-steps be denoted by $\mathbb{K}_m \subseteq \mathbb{K}$. The augmented set $\{0\} \cup \mathbb{K}_m = \{0, F_m, \dots, L_m\}$ will be denoted by $\mathbb{K}^0_m$.

Each power-path $\mathbf{p}_m \in \mathbb{PP}_m$ has the following general form:

$$\mathbf{p}_m = \{p_m(t) \mid p_m(t) \in \mathbb{P}_m \ ; \ t \in T_m\} \tag{7.2}$$

Let $p_m(k)$ denote the power level $p_m(k^s)$ at the start of time-step $k = [k^s, k^e)$, for each $k \in \mathbb{K}_m$. Each power-path (7.2) is then approximated by the discretized power sequence

$$\{p_m(k) \mid p_m(k) \in \mathbb{P}_m \ ; \ k \in K_m\} \tag{7.3}$$

that provides a step-function approximation for $\mathbf{p}_m$.[5] For brevity, $p_m(k)$ will hereafter be called the power level for m at time-step k.

Finally, the performance payment method ϕ_m included in SC_m consists of two parts: (i) a collection of prices $\pi^E_m(k)$ (\$/MWh) for the dispatched delivery of down/up energy during T_m, one price for each time-step $k \in \mathbb{K}_m$; and (ii) a collection of prices $\pi^R_m(k)$ (\$/MW) for ramping as measured by means of a power-path mileage metric, one price for each time-step $k \in \mathbb{K}^0_m$.

More precisely, the performance-cost compensation (\$) that m requests for each dispatchable power-path $\mathbf{p}_m \in \mathbb{PP}_m$ is a weighted average

$$\phi_m(\mathbf{p}_m) = C^E_m(\mathbf{p}_m) + \mu_m C^R_m(\mathbf{p}_m) \tag{7.4}$$

of down/up energy cost $C^E_m(\mathbf{p}_m)$ (\$) and down/up ramp cost $C^R_m(\mathbf{p}_m)$ (\$), with scalar unit-free weight factor $\mu_m \geq 0$, where

$$C^E_m(\mathbf{p}_m) = \sum_{k \in \mathbb{K}_m} [\pi^E_m(k)|p_m(k)|\Delta t] \tag{7.5}$$

$$C^R_m(\mathbf{p}_m) = \sum_{k \in \mathbb{K}^0_m} [\pi^R_m(k) \cdot |p_m(k+1) - p_m(k)|] \tag{7.6}$$

---

[4]The assumption that m's offered delivery period T_m consists of *consecutive* time-steps is made for ease of exposition. The case of successive but not necessarily consecutive time-steps is handled by Li and Tesfatsion [113, 115]. More generally, Heo and Tesfatsion [76, 77] permit reserve offers to take the form of portfolios of swing contracts, where each swing contract can have its own distinct offered delivery period.

[5]Step functions are universal approximators for the class of all continuous real-valued functions $f:[a,b] \to R$ defined over compact intervals $[a,b]$ of the real line. For example, given any such function f, and any $\epsilon > 0$, it is straightforward to establish the existence of a step function $f_\epsilon:[a,b] \to R$ with finitely many time-steps for which the maximum absolute approximation error $|f(x) - f_\epsilon(x)|$ over $x \in [a,b]$ is less than ϵ. This assertion follows immediately from the Heine–Cantor theorem, which establishes that any continuous function $f:X \to Y$ between metric spaces X and Y, with X compact, is uniformly continuous. An interesting corollary of this assertion is that different types of dispatchable resources might find it advantageous to be able to use partitions $\mathbb{K}$ of T with different time-step lengths Δt tailored to permit the best expression of their specific attributes. For further discussion of this issue, see [84, Sec. 3.1.1] and [129, p. 10].

Note that the ramp cost (7.6) includes a charge for any start-up ramping that m might have to undertake to ramp from a zero-power shut-down state prior to its delivery period T_m to its dispatched power level at the start of T_m, or from its dispatched power level at the end of T_m to a zero-power shut-down state subsequent to T_m.[6]

Expanding upon assumptions (A7) and (A8), forecasted net must-service loads for buses, for reserve zones, and for the transmission system as a whole are constructed as follows for each time-step $k \in \mathbb{K}$:

$\mathbb{LS}(b) \subseteq \mathbb{LS}$	Subset of LSEs located at bus b;
$\mathbb{LS}(z) \subseteq \mathbb{LS}$	Subset of LSEs located in reserve zone z;
$\hat{p}_j^f(k)$	Forecast (MW) for the must-service load of LSE j's retail customers at time-step k;
$\hat{L}_b^f(k)$	Forecast (MW) for the must-service load at bus b at time-step k, given by $\sum_{j\in\mathbb{LS}(b)} \hat{p}_j^f(k)$;
$\hat{L}_z^f(k)$	Forecast (MW) for the must-service load in reserve zone z at time-step k,given by $\sum_{j\in\mathbb{LS}(z)} \hat{p}_j^f(k)$;
$\hat{L}^f(k)$	Forecast (MW) for the system-wide must-service load at time-step k, given by $\sum_{j\in\mathbb{LS}} \hat{p}_j^f(k)$;
$\mathbb{NG}(b) \subseteq \mathbb{NG}$	Subset of non-dispatchable generators located at bus b;
$\mathbb{NG}(z) \subseteq \mathbb{NG}$	Subset of non-dispatchable generators located in zone z;
$\hat{p}_n(k)$	Forecast (MW) for the nondispatched power injection of generator n at time-step k;
$\widehat{NG}_b(k)$	Forecast (MW) for the nondispatched power injection at bus b at time-step k, given by $\sum_{n\in\mathbb{NG}(b)} \hat{p}_n(k)$;
$\widehat{NG}_z(k)$	Forecast (MW) for the nondispatched power injection in reserve zone z at time-step k, given by $\sum_{n\in\mathbb{NG}(z)} \hat{p}_n(k)$;
$\widehat{NG}(k)$	Forecast (MW) for the system-wide nondispatched power injection at time-step k, given by $\sum_{n\in\mathbb{NG}} \hat{p}_n(k)$;
$\widehat{NL}_b^f(k)$	Forecast (MW) for the net must-service load at bus b at time-step k, given by $[\hat{L}_b^f(k) - \widehat{NG}_b(k)]$;
$\widehat{NL}_z^f(k)$	Forecast (MW) for the net must-service load in reserve zone z at time-step k, given by $[\hat{L}_z^f(k) - \widehat{NG}_z(k)]$;
$\widehat{NL}^f(k)$	Forecast (MW) for the system-wide net must-service load at time-step k, given by $[\hat{L}^f(k) - \widehat{NG}(k)]$.

[6]By construction, m's delivery period $T_m \subseteq T$ consists of a set $\mathbb{K}_m$ of time-steps $\{F_m, \ldots, L_m\}$ that are elements of $\mathbb{K} = \{1, \ldots, NK\}$, the set of time-steps that comprise T. As will be seen in Section 7.5, the constraints for the ISO's optimization problem for the base-caseswing-contract market M(T) require m to be in a zero-power shut-down state at the start of any time-step in $\mathbf{K}^0 = \{0, 1, \ldots, NK\}$ that is *not* an element of the time-step set $\mathbb{K}_m$ that comprises T_m.

Finally, recall from previous discussion that the operating period T consists of the union of the time-steps $k \in \mathbb{K}$, and the delivery period T_m designated by m in its reserve offer SC_m in (7.1) consists of the union of the time-steps in $\mathbb{K}_m \subseteq \mathbb{K}$. For each $k \in \mathbb{K}$, let $A_m(k)$ denote the binary 1/0 *availability offer indicator* for SC_m indicating whether (1) or not (0) time-step k lies within the delivery period $T_m \subseteq T$ designated in SC_m.

Note that $A_m(k)$ can be calculated by the ISO from the information provided in SC_m. Suppose, for example, that M(D+1) is a day-ahead market held on day D for all 24 hours k of the following day D+1. Suppose SC_m offers the availability of power-path delivery only between the hours of 8:00 and 16:00 on day D+1; that is, suppose $\mathbb{K}_m = \{8, \dots, 15\}$. Then

$$A_m(k) = \begin{cases} 1 & \text{if } k = 8, \dots, 15 \\ 0 & \text{if } k = 1, \dots, 7 \text{ or } k = 16, \dots, 24 \end{cases}$$

In the remainder of this study, it is assumed that the ISO managing a swing-contract market M(T) calculates the availability offer indicators $\{A_m(k) \mid k \in \mathbb{K}\}$ for each reserve offer SC_m as soon as it is received. Hence, these indicators are treated as exogenous inputs for the ISO's optimization problem.

7.4 ISO OBJECTIVE FUNCTION

The objective of the ISO tasked with managing the base-case swing-contract market M(T) for the future operating period T is minimization of expected total avoidable cost ($), subject to system constraints. In this section, certainty equivalence is used to approximate expectation; that is, terms with uncertain values are replaced with their forecasted values. Thus, *expected* total avoidable cost ($) is replaced by *forecasted* total avoidable cost ($), expressed as follows[7]:

$$\hat{C}(T) = \sum_{m \in \mathbb{M}} [c_m \alpha_m + \phi_m(\mathbf{p}_m)] + \sum_{b \in \mathbb{B}} \sum_{k \in \mathbb{K}} [\Lambda^- \beta_b^-(k) + \Lambda^+ \beta_b^+(k)] \Delta t \qquad (7.7)$$

The term c_m in (7.7) is a binary 1/0 *contract clearing indicator* that indicates whether (1) or not (0) the ISO has cleared the reserve offer (7.1) submitted by dispatchable resource m into M(T) for operating period T. If the ISO clears this reserve offer, i.e. if $c_m = 1$, then the ISO is obligated to ensure the offer price α_m ($) is either paid to m (if $\alpha_m \geq 0$) or received from m (if $\alpha_m < 0$). The expression $\phi_m(\mathbf{p}_m)$ in (7.7) denotes m's requested performance payment ($) for the period-T delivery of the power-path $\mathbf{p}_m \in \mathbb{PP}_m$.

[7]Although power levels for all dispatchable resources nominally appear in the ISO's objective function (7.7), it will be seen in Section 7.5 that the system constraints for the ISO's optimization problem force a dispatchable resource m's power levels $p_m(k)$ to be zero for all time-steps $k \in \mathbb{K}^0$ if $c_m = 0$, i.e. if the swing contract SC_m submitted by m is not cleared.

The double-summed term in (7.7) is the *imbalance cost* for period T. This imbalance cost expresses the cost (if any) that the ISO anticipates will have to be incurred after the close of M(T) to avoid down/up power imbalance during period T, given each possible choice of its decision variables.[8]

As will be seen in Section 7.5, down/up power imbalance at each bus b at each time-step k is measured by a slack variable $\beta_b(k)$ (MW) in the power balance constraint for bus b at time-step k. The expressions $\beta_b^-(k)$ (MW) and $\beta_b^+(k)$ (MW) in the imbalance cost are down/up power imbalance slack variables derived as the negative and positive parts of $\beta_b(k)$. Finally, the terms Λ^- ($/MWh) and Λ^+ ($/MWh) in the imbalance cost are imbalance penalty weights.

7.5 COMPLETE ANALYTICAL MILP FORMULATION

ISO Objective:

Select decision variables to minimize forecasted total avoidable cost, subject to system constraints, where forecasted total avoidable cost is given by:

$$\hat{C}(T) = \sum_{m \in \mathbb{M}} [c_m \alpha_m + \phi_m(\mathbf{p}_m)] + \sum_{b \in \mathbb{B}} \sum_{k \in \mathbb{K}} [\Lambda^- \beta_b^-(k) + \Lambda^+ \beta_b^+(k)] \Delta t$$

ISO Binary Decision Variables:

- *Contract clearing indicators*: $c_m \in \{0, 1\}, \ \forall m \in \mathbb{M}$

ISO Continuously Valued Decision Variables:

- *Power dispatch levels*: $p_m(k), \ \forall m \in \mathbb{M}, \ k \in \mathbb{K}$
- *Bus voltage angles*: $\theta_b(k), \ \forall b \in \mathbb{B}/\{1\}, \ k \in \mathbb{K}$

Variables Determined by ISO Decisions and System Constraints:

- *Run-time minimum power levels*: $\underline{p}_m(k), \ \forall m \in \mathbb{M}, \ k \in \mathbb{K}$
- *Run-time maximum power levels*: $\overline{p}_m(k), \ \forall m \in \mathbb{M}, \ k \in \mathbb{K}$
- *Unit availability indicators*: $v_m(k) \in \{0, 1\}, \ \forall m \in \mathbb{M}, \ k \in \mathbb{K}$
- *Transmission line power flows*: $w_\ell(k), \ \forall \ell \in \mathbb{L}, \ k \in \mathbb{K}$
- *Power balance slack variables*: $\beta_b(k), \beta_b^-(k), \beta_b^+(k), \ \forall b \in \mathbb{B}, \ k \in \mathbb{K}$
- *Bus voltage angle for reference bus* $1 : \theta_1(k), \ \forall k \in \mathbb{K}$

[8]For example, these costs could be payments for additional reserve, or estimated losses associated with forced load shedding or wind spillage.

System Constraints

Unit availability indicator constraint for each $m \in \mathbb{M}$ and $k \in \mathbb{K}$:

$$v_m(k) = c_m \cdot A_m(k) \tag{7.8}$$

$$v_m(0) = 0 \tag{7.9}$$

Bus voltage angle bounds for each $b \in \mathbb{B}/\{1\}$ and $k \in \mathbb{K}$:

$$-\pi \leq \theta_b(k) \leq \pi \tag{7.10}$$

Bus voltage angle at reference bus 1 for each $k \in \mathbb{K}$:

$$\theta_1(k) = 0 \tag{7.11}$$

Transmission line power flow constraints[9] for each $\ell \in \mathbb{L}$ and $k \in \mathbb{K}$:

$$w_\ell(k) = S_o B(\ell)\ [\theta_{O(\ell)}(k) - \theta_{E(\ell)}(k)] \tag{7.12}$$

$$-F_\ell^{\max} \leq w_\ell(k) \leq F_\ell^{\max} \tag{7.13}$$

Power balance constraint (with slack variable) for each $b \in \mathbb{B}$ and $k \in \mathbb{K}$:

$$\sum_{m \in \mathbb{M}(b)} p_m(k) + \left[\sum_{\ell \in \mathbb{L}_{E(b)}} w_\ell(k) - \sum_{\ell \in \mathbb{L}_{O(b)}} w_\ell(k) \right] = \widehat{NL}_b^f(k) + \beta_b(k) \tag{7.14}$$

Slack variable constraints for each $b \in \mathbb{B}$ and $k \in \mathbb{K}$:

$$\beta_b^-(k) = \max\{0, -\beta_b(k)\} \tag{7.15}$$

$$\beta_b^+(k) = \max\{0,\ \beta_b(k)\} \tag{7.16}$$

Power capacity constraints for each $m \in \mathbb{M}$ and $k \in \mathbb{K}^0$:

$$\underline{p}_m(k) \leq p_m(k) \leq \overline{p}_m(k) \tag{7.17}$$

$$\overline{p}_m(k) \leq P_m^{\max} v_m(k) \tag{7.18}$$

$$\underline{p}_m(k) \geq P_m^{\min} v_m(k) \tag{7.19}$$

[9] As usual in DC-OPF formulations, the bracketed term appearing on the right-hand side of the line power flow constraint (7.12) denotes the *unit-free size* of the indicated voltage angle difference $\Delta\theta$, not the difference in radians. In particular, this bracketed term reflects the DC-OPF formulation assumption (A11) in Section 7.2 that the voltage angle difference $\Delta\theta$ is small enough to permit the approximation $\sin(\Delta\theta) \approx \Delta\theta$ *in size*.

Down/up ramp constraints for each $m \in \mathbb{M}$ and $k \in \mathbb{K}$:

$$p_m(k-1) - \underline{p}_m(k) \leq R^{\mathrm{D}}_m \Delta t v_m(k-1) - P^{\mathrm{min}}_m [1 - v_m(k-1)] v_m(k) \tag{7.20}$$

$$\overline{p}_m(k) - p_m(k-1) \leq R^{\mathrm{U}}_m \Delta t v_m(k) - P^{\mathrm{min}}_m [1 - v_m(k)] v_m(k-1) \tag{7.21}$$

System-wide down/up reserve requirement constraints for each $k \in \mathbb{K}$:

$$\sum_{m \in \mathbb{M}} \underline{p}_m(k) \leq [1 - \mathrm{RRD}(k)] \cdot \widehat{NL}^f(k) \tag{7.22}$$

$$\sum_{m \in \mathbb{M}} \overline{p}_m(k) \geq [1 + \mathrm{RRU}(k)] \cdot \widehat{NL}^f(k) \tag{7.23}$$

Zonal down/up reserve requirement constraints for each $z \in \mathbb{Z}$ and $k \in \mathbb{K}$:

$$\sum_{m \in \mathbb{M}(z)} \underline{p}_m(k) \leq [1 - \mathrm{RRD}(z,k)] \cdot \widehat{NL}^f(z,k) \tag{7.24}$$

$$\sum_{m \in \mathbb{M}(z)} \overline{p}_m(k) \geq [1 + \mathrm{RRU}(z,k)] \cdot \widehat{NL}^f(z,k) \tag{7.25}$$

7.6 ADDITIONAL DISCUSSION OF OPTIMIZATION ASPECTS

The absolute value terms $|p_m(k)|$ appear in the objective function (7.7) because a dispatchable resource m might be called upon to provide either power withdrawals $p_m(k) < 0$ or power injections $p_m(k) > 0$ in support of net load balancing requirements for a time-step k. For the illustrative case at hand, it is assumed that each dispatchable resource m with bidirectional power capabilities asks to be compensated at the same price $\pi^{\mathrm{E}}_m(k)$ (\$/MWh) for the dispatched delivery of a power level $p_m(k)$ whether this power level is a withdrawal or an injection.

The absolute value terms $|p_m(k)|$ and $|p_m(k+1) - p_m(k)|$ that appear in the objective function (7.7) do not pose any computational difficulty. Because the goal is to minimize this objective function, these absolute value terms can equivalently be represented by linear inequality constraints.

For example, the terms $|p_m(k)|$ can be handled as follows. First, introduce new decision variables for the ISO: $p^a_m(k), \forall m \in \mathbb{M}$, $k \in \mathbb{K}$. Second, in the objective function (7.7), replace $|p_m(k)|$ by $p^a_m(k)$, $\forall m \in \mathbb{M}$, $k \in \mathbb{K}$. Third, include the following additional linear inequality constraints in the set of system constraints: $p^a_m(k) \geq p_m(k)$ and $p^a_m(k) \geq -p_m(k)$, $\forall m \in \mathbb{M}, k \in \mathbb{K}$. Any solution for the resulting constrained minimization problem will then require $p^a_m(k) = |p_m(k)|, \forall m \in \mathbb{M}, k \in \mathbb{K}$.

The unit availability constraints (7.8) take the form

$$v_m(k) = c_m \cdot A_m(k), \ \forall \ m \in \mathbb{M}, \ k \in \mathbb{K} \tag{7.26}$$

The term $v_m(k)$ in (7.26) is a *unit availability indicator* whose binary 1/0 value is determined by two factors:

(a) Has m's reserve offer SC_m been cleared by the ISO or not?

(b) Does SC_m offer power-path delivery for time-step k or not?

The contract clearing indicator $c_m \in \{0, 1\}$ represents condition (a). As explained and illustrated in Section 7.3, the availability offer indicator $A_m(k) \in \{0, 1\}$, directly determined from the delivery period T_m designated in SC_m, represents condition (b). If conditions (a) and (b) are both met, m is available to provide power-path delivery during time-step k. Otherwise, if at most one of these conditions is met, m is not available to provide power-path delivery during time-step k.

The variables $\{\underline{p}_m(k), \overline{p}_m(k) \mid k \in \mathbb{K}\}$ appearing in the power capacity constraints (7.17)–(7.19) and the ramping constraints (7.20) and (7.21) denote *run-time* lower and upper bounds on power availability for dispatchable resource m at each time-step $k \in \mathbb{K}$. These run-time bounds can differ from the absolute minimum and maximum power capacity limits $P_m^{\min}$ and $P_m^{\max}$ included by m in its submitted swing-contract SC_m depicted in (7.1).

The precise restrictions imposed on the run-time variables $\{\underline{p}_m(k), \overline{p}_m(k) \mid k \in \mathbb{K}\}$ by run-time operating conditions depend on the unit availability indicators $\{v_m(k) \mid k \in \mathbb{K}\}$ as determined in (7.26). To see this, first note that the power capacity constraints (7.17)–(7.19) have the following implication:

$$\text{If } v_m(k) = 0, \text{ then } \underline{p}_m(k) = p_m(k) = \overline{p}_m(k) = 0 \tag{7.27}$$

That is, recalling the definition of the discretized power level $p_m(k)$, dispatchable resource m must be at a zero power injection level at the start-time for any time-step k for which $v_m(k) = 0$. Conversely,

$$\text{If } v_m(k) = 1, \text{ then } P_m^{\min} \le \underline{p}_m(k) \le p_m(k) \le \overline{p}_m(k) \le P_m^{\max} \tag{7.28}$$

Second, note that the vector $(v_m(k-1), v_m(k))$ of unit availability indicators for two consecutive time-steps $k-1$ and k can take on only one of four possible configurations: namely, (0,0), (1,0), (0,1), or (1,1). Configurations (0,0) and (1,1) correspond to cases in which the availability status of m remains unchanged from $k-1$ to k. Configuration (1,0) requires m to be at a zero power injection level at the start-time for time-step k following possible dispatch at time-step $k-1$. Conversely, configuration (0,1) requires m to be at a zero power injection level at the start-time for time-step $k-1$ yet be available for dispatch at time-step k.

Taking as given the power capacity restrictions (7.27) and (7.28), the ramping constraints (7.20) and (7.21) impose ramping restrictions on $\underline{p}_m$ and $\overline{p}_m(k)$ during time-steps $k-1$ and k for each nonzero unit-availability vector $(v_m(k-1), v_m(k))$ with $k \in \mathbb{K}$. These ramping restrictions are shown in Table 7.1.

TABLE 7.1 Restrictions on minimum and maximum run-time power availability for dispatchable resource m during time-steps $k-1$ and k under possible unit availability $(v_m(k-1), v_m(k))$

$v_m(k-1)$	0	1	0	1
$v_m(k)$	0	0	1	1
$\underline{p}_m(k-1)$	$\underline{p}_m(k-1) = 0$	$\underline{p}_m(k-1) \leq R_m^{\mathrm{D}} \Delta t$	$\underline{p}_m(k-1) = 0$	
$\overline{p}_m(k-1)$	$\overline{p}_m(k-1) = 0$	$P_m^{\min} \leq \overline{p}_m(k-1)$	$\overline{p}_m(k-1) = 0$	
$\underline{p}_m(k)$	$\underline{p}_m(k) = 0$	$\underline{p}_m(k) = 0$	$P_m^{\min} \leq \underline{p}_m(k)$	$p_m(k-1) - R_m^D \Delta t \leq \underline{p}_m(k)$
$\overline{p}_m(k)$	$\overline{p}_m(k) = 0$	$\overline{p}_m(k) = 0$	$\overline{p}_m(k) \leq R_m^{\mathrm{U}} \Delta t$	$\overline{p}_m(k) \leq p_m(k-1) + R_m^U \Delta t$

Finally, for each reserve zone $z \in \mathbb{Z}$ and time-step $k \in \mathbb{K}$, define:

$$RR^{\min}(z,k) = \sum_{m \in \mathbb{M}(z)} \underline{p}_m(k); \quad RR^{\max}(z,k) = \sum_{m \in \mathbb{M}(z)} \overline{p}_m(k) \tag{7.29}$$

The power levels $RR^{\min}(z,k)$ and $RR^{\max}(z,k)$ are the run-time minimum and maximum power amounts that are available within reserve zone z at the start-time for time-step k along the solution path for the ISO's MILP optimization problem set out in Section 7.5. For later purposes, let this *Inherent Reserve Range* be denoted by

$$\mathrm{IRR}(z,k) = [\ RR^{\min}(z,k), RR^{\max}(z,k)\ ] \tag{7.30}$$

7.7 FIVE-BUS TEST CASE

A 5-Bus SC DAM Test Case is reported in [122] for an ISO-managed swing-contract day-ahead market (SC DAM). A base-case MILP optimization formulation is used to determine the SCUC/SCED solutions for this test case under a variety of assumptions regarding reserve zones and adaptive reserve-zone updating. This section briefly summarizes the key features and results for this test case for treatments in which a single fixed reserve zone is considered.

The 5-Bus SC DAM Test Case assumes that an SC DAM takes place on three successive days: D_0, D_1, and D_2. For each day D, the SC DAM is conditioned on hourly net must-service load forecasts for day D+1, where net must-service load is must-service load minus nondispatched wind generation. ERCOT hourly load and wind generation data [51] are used to construct a set of 90 three-day net must-service load scenarios for days D_1, D_2, and D_3, each assigned equal probability of occurrence.

As depicted in Figure 7.1, the Lally grid [108] used for this test case consists of five buses B1–B5 and six transmission lines L1–L6. The participants include five dispatchable thermal generators G1–G5 located at buses B1, B3, B4, and B5; one nondispatchable wind farm located at bus B3; and three LSEs servicing load at buses B2, B3, and B4. The designated reference bus is B4. The single fixed reserve zone

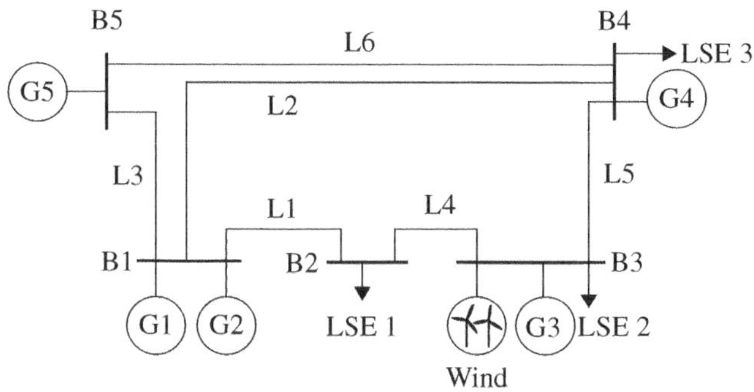

Figure 7.1 Transmission grid for the 5-Bus SC DAM Test Case.

TABLE 7.2 Transmission line data for the 5-Bus SC DAM Test Case

Line	From bus	To bus	X (p.u.)	Limit (MW)
L1	B1	B2	0.0281	450
L2	B1	B4	0.0304	300
L3	B1	B5	0.0064	250
L4	B2	B3	0.0108	200
L5	B3	B4	0.0297	150
L6	B4	B5	0.0297	240

consists of the collection of all five buses B1–B5. The specified physical attributes of the six transmission lines are given in Table 7.2.

The swing contracts submitted into the SC DAM by generators G1–G5 all take the form of (7.1); hence, swing (flexibility) is offered in power and ramp-rate levels. Each generator m requests a single energy price $\phi_{m,j}$ be paid in compensation for delivered energy on each successive day D_j; the ramp-cost weight μ_m is set to 0. The specific attribute values for each offered swing contract are shown in Table 7.3.

The number of time-steps k for the SC DAM optimization is set at $NK = 24$, and the length of each time-step k is set at $\Delta t = 1$ hour. The ISO's estimate for the maximum down/up deviations RRD(k) and RRU(k) of system-wide net load from its forecasted value are commonly set at the decimal percentage 0.05. The imbalance penalty weights Λ^- and Λ^+ are both set to \$1000/MWh, and the base power S_o is set equal to 100 MW.

The ISO's objective function consists of three summed cost components: (i) offer cost; (ii) expected ex post performance cost; and (iii) expected ex post imbalance cost. The costs (ii) and (iii) are "expected" costs because the SC DAM optimization is conditioned on net must-service load forecasts for the following day.

TABLE 7.3 Swing contracts submitted by the five generators over three successive days for the 5-Bus SC DAM Test Case

Day	Gen	Delivery period	Power range	Ramp-rate range	Offer price	Performance price
D_j	m	T_m	$[P_m^{min}, P_m^{max}]$ (MW)	$[-R_m^D, R_m^U]$ (MW/h)	α_m ($)	ϕ_m ($/MWh)
	G1	{1, … ,24}	[10,380]	[−150,150]	1400	10
	G2	{4, … ,22}	[10,325]	[−150,150]	1200	12
D_0	G3	{1, … ,24}	[50,385]	[−240,240]	1150	11
	G4	{4, … ,22}	[20,100]	[−50, 50]	1100	18
	G5	{1, … ,24}	[40,600]	[−320,320]	1800	30
	G1	{3, … ,24}	[10,340]	[−150,150]	1200	10
	G2	{1, … ,24}	[10,160]	[−80, 80]	1328	12
D_1	G3	{1, … ,24}	[50,320]	[−240,240]	1550	20
	G4	{1, … ,22}	[20,400]	[−200,200]	1600	18
	G5	{1, … ,24}	[40,600]	[−320,320]	1780	25
	G1	{1, … ,24}	[10,380]	[−150,150]	1200	10
	G2	{3, … ,22}	[10,325]	[−150,150]	1350	13
D_2	G3	{1, … ,24}	[50,385]	[−240,240]	1000	10
	G4	{3, … ,24}	[20,100]	[−50, 50]	1100	15
	G5	{1, … ,24}	[40,600]	[−320,320]	1800	28

As detailed in [122, Appendix], the following three metrics are used to summarize cost outcomes for the SC DAM MILP optimization on each day D_j, where $j = 0, 1, 2$:

- **OC(D_j).** *Offer cost*: Payment of offer prices to sellers of cleared SCs on day D_j in return for ensuring the availability of services on day D_{j+1};
- **E^{xp}[PC(D_j)].** *Expected ex-post performance cost*: Expected compensation to be paid ex post to sellers of cleared SCs for actual performance of services on day D_{j+1}, where the expectation is taken with respect to all possible net must-service load scenarios for day D_{j+1};
- **E^{xp}[IC(D_j)].** *Expected ex-post imbalance cost*: Expected cost for day D_{j+1} from imbalance adjustments (e.g. load curtailments), where the expectation is taken with respect to all possible net must-service load scenarios for day D_{j+1}.

Cost outcomes for the 5-Bus SC DAM Test Case for the single reserve zone treatment are reported in Table 7.4 for the three successive simulated days D_0, D_1, and D_2. Note that expected imbalance cost $E^{xp}[IC(D_j)]$ for each successive day D_j is fairly substantial. This finding reflects the lack of attention paid to the possibility

TABLE 7.4 Cost outcomes over three successive days for the 5-Bus SC DAM Test Case, assuming a single reserve zone consisting of all five transmission grid buses

Day D_j	Gen contract clearing $[c_{G1}, c_{G2}, c_{G3}, c_{G4}, c_{G5}]$	$OC(D_j)$	$E^{xp}[PC(D_j)]$	$E^{xp}[IC(D_j)]$
D_0	[1, 1, 1, 1, 0]	$4,580.00	$221,434.26	$93,221.92
D_1	[1, 1, 1, 1, 0]	$5,730.00	$288,859.21	$71,695.40
D_2	[1, 1, 1, 0, 1]	$5,350.00	$238,877.33	$2,466.25

TABLE 7.5 Computational requirements for the SC DAM MILP optimization conducted on each successive day for the 5-Bus SC DAM Test Case

Case	Number of binary decision variables	Number of continuous decision variables	Number of constraints
5-Bus	5	216	1602

of reserve deliverability problems arising in real-time operations due to transmission congestion. As shown in [122], and briefly reviewed in Section 7.8, imbalance cost sharply declines when SC DAM operations are extended to include a method for the adaptive updating of reserve zones based on past observed line congestion.

Although all simulations for the 5-Bus SC DAM Test Case were conducted on a high-performance computer (HPC), the computational requirements reported in Table 7.5 indicate that a personal computer could have been employed. Pyomo 5.3.0 was used to implement the ISO's SC DAM optimization problem on each day D, and CPLEX Python API 12.6 was employed as the MILP solver.

7.8 THIRTY BUS TEST CASE WITH ADAPTIVE RESERVE ZONES

The basic goal of the swing-contract market design proposed in this study is to promote flexible reserve provision to handle increased penetration of VERs. The base-case swing-contract market presented and implemented in previous sections of this chapter demonstrates how flexible reserve provision can feasibly be facilitated by permitting reserve offers to take a swing-contract form.

However, the base-case swing-contract market assumes a fixed configuration of reserve zones. As many previous researchers have noted, the increased penetration of VERs is leading to increased volatility in grid congestion over time, which in turn hinders the ability of system operators to ensure delivery of reserve to zones where power is needed to balance real-time net load.

In response to this reserve deliverability concern, Ma et al. [122] propose an extension of the base-case swing-contract market design to include a method permitting the adaptive updating of reserve zones based on changing system conditions and net load uncertainties. This section briefly reviews this proposed extension.

Ma et al. [122] focus on a base-case swing-contract day-ahead market M(D+1) held on the morning of each day D to plan for net load balancing during the following day D+1. Their adaptive reserve zone method consists of four steps to be carried out immediately prior to the opening of M(D+1) on each day D, as follows:

Step 1. *Dual variable solutions*: Given a set of possible net load scenarios for day D+1, together with estimated probabilities, use off-line preliminary optimizations to determine dual variable solutions for each transmission line.

Step 2. *Line congestion risk indices*: Given these dual variable solutions, construct a congestion risk index for each transmission line.

Step 3. *Shift factor index*: Given these congestion risk indices, construct a matrix SFWA whose elements are weighted averages of shift-factor differences.

Step 4. *Reserve zone specification*: Given SFWA, use a hierarchical clustering algorithm to partition the set of buses into reserve zones for use in M(D+1).

Ma et al. [122] use 5-bus and 30-bus test cases to evaluate the performance of their adaptive reserve zone method. This section summarizes the outcomes of the 30-bus test case, referred to below as the *30-Bus SC DAM Test Case*.

As depicted in Figure 7.2, the transmission grid for the 30-Bus SC DAM Test Case consists of 30 buses B1–B30 and 41 transmission lines L1–L41. The market participants include six dispatchable thermal generators G1–G6 located at buses B1, B2, B23, B27, B22, B13; four nondispatchable wind farms located at buses B7, B8, B21, B24; and 20 LSEs servicing load at buses B2, B3, B4, B6, B7, B10, B12, B14, B15, B16, B17, B18, B19, B20, B21, B23, B24, B26, B29 and B30. The designated reference bus is B1.

As for the 5-Bus SC DAM Test Case reviewed in Section 7.7, the length of the time-step k for the 24-hour MILP SCUC/SCED optimization for each swing-contract DAM held on each successive day D_0, D_1, and D_2 is set at $\Delta t = 1$ hour. ERCOT hourly load and wind generation data [51] are used to construct a set of 150 three-day net must-service load scenarios for days D_1, D_2, and D_3, each assigned equal probability of occurrence.

The swing-contract reserve offers submitted by generators G1–G6 into M(D+1) on each successive day D = D_0, D_1, and D_2 all take the general form (7.1) offering swing (flexibility) in power and ramp-rate levels with ramp-rate cost set to zero. The ISO's estimates RRD(k), RRU(k), RRD(z, k), and RRD(z, k) for the maximum down/up deviations of net load from its forecasted value, both system-wide and by reserve zone, are commonly set at the decimal percentage 0.05. Finally, the imbalance penalty weights Λ^- and Λ^+ for down/up power

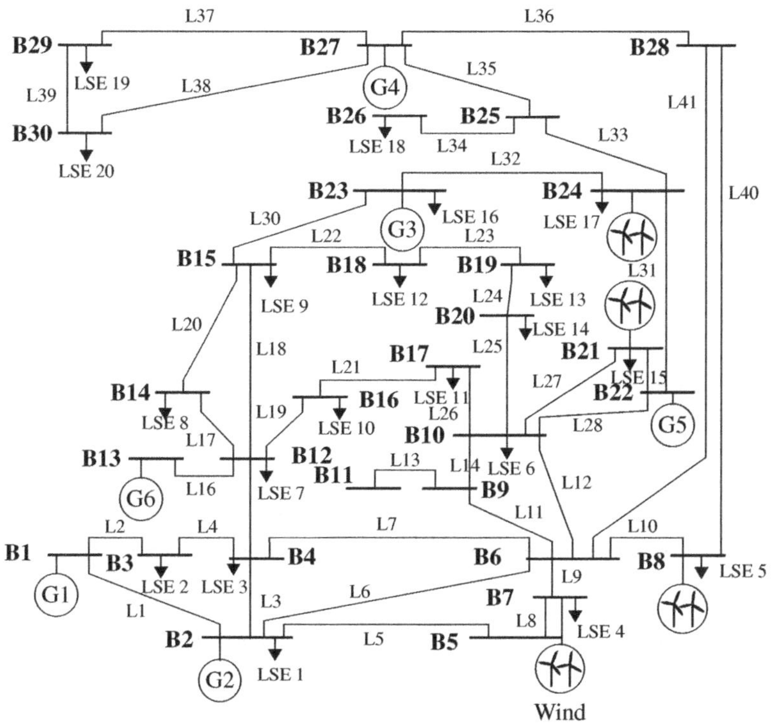

Figure 7.2 Transmission grid for the 30-Bus SC DAM Test Case.

imbalance are commonly set at \$1000/MWh, and the base power S_o is set equal to 100 MW.

Using the same performance metrics as reported for the 5-Bus SC DAM Test Case in Section 7.7, performance outcomes are reported in Table 7.6 for the 30-Bus SC DAM Test Case for three successive simulated days D_0, D_1, D_2 under two different reserve-zone treatments labeled "Single Zone" and "Proposed." The "Single Zone" treatment groups all buses into a single reserve zone on each successive day. The "Proposed" treatment adaptively updates the configuration of reserve zones on each successive day in accordance with the method proposed by Ma et al. [122].

As seen, each reserve zone treatment results in a different set of cleared swing contracts as well as different resulting values for the three performance metrics on each successive day. Note, in particular, that a substantial amount of reserve zone updating takes place from one day to the next under the "Proposed" reserve zone treatment. The "Single Zone" treatment is observed to result in lower offer cost OC(Z,D_j) on each successive day D_j; this unsurprising result reflects the absence of attention paid to potential real-time reserve deliverability problems in the determination of optimal (least-cost) cleared contracts. However, the "Single Zone" treatment is also seen to result in substantially higher expected ex post performance

TABLE 7.6 30-Bus SC DAM performance for three successive days for two reserve zone treatments

Day D_j	Treatment	Reserve zones zj	Gen contract clearing $c_{G1}, c_{G2}, c_{G3}, c_{G4}, c_{G5}, c_{G6}$	OC (Z,D_j)	E^{xp}[PC (Z,D_j)]	E^{xp}[IC (Z,D_j)]
D_0	Proposed	$z1$: Bus 23 $z2$: Bus 27 29 30 $z3$: Bus 1–22 24 25 26 28	[1, 1, 1, 1, 1, 1]	$10,750	$100,555.65	$194.22
	Single zone	$z1$: Bus 1–30	[1, 1, 1, 1, 0, 1]	$8,750	$106,420.12	$5,371.73
D_1	Proposed	$z1$: Bus 23 $z2$: Bus 21 22 24–27 29 30 $z3$: Bus 1–20 28	[1, 1, 1, 1, 1, 1]	$10,700	$98,012.73	$10,359.74
	Single zone	$z1$: Bus 1–30	[1, 1, 1, 1, 0, 1]	$9,100	$99,996.96	$13,990.73
D_2	Proposed	$z1$: Bus 23 24 25 26 $z2$: Bus 27 29 30 $z3$: Bus 1–22 28	[1, 1, 1, 1, 1, 1]	$9,410	$104,494.04	$10,597.97
	Single zone	$z1$: Bus 1–30	[1, 1, 1, 1, 0, 1]	$7,810	$105,077.11	$13,282.30

TABLE 7.7 Computational requirements for the SC DAM MILP optimization conducted on each successive day for the 30-Bus SC DAM Test Case

Case	Number of binary decision variables	Number of continuous decision variables	Number of constraints
30-Bus	6	840	4888

cost $E^{xp}[PC(Z,D_j)]$ and imbalance cost $E^{xp}[IC(Z,D_j)]$ than the "Proposed" zone treatment on each successive day D_j.

As for the 5-Bus SC DAM Test Case reported in Section 7.7, the computational requirements reported in Table 7.7 for the 30-Bus SC DAM Test Case indicate that a personal computer would have sufficed to conduct the simulations. Pyomo 5.3.0 was used to implement the ISO's SC DAM optimization problem on each day D, and CPLEX Python API 12.6 was employed as the MILP solver.

CHAPTER 8

INCLUSION OF RESERVE OFFERS WITH PRICE SWING

8.1 CHAPTER PREVIEW

No restrictions are placed *a priori* on the performance payment methods ϕ_m included by dispatchable resources m in the swing-contract reserve offers $SC_m = (\alpha_m, \mathbb{PP}_m, \phi_m)$ they submit into a swing-contract market M(T). Ideally, ϕ_m should be a functional $\phi_m:\mathbb{PP}_m \rightarrow R_+$ such that each valuation $\phi_m(\mathbf{p})$ reflects the true avoidable cost that would be incurred by m for the dispatched delivery of power-path $\mathbf{p}$ during T, whatever form this avoidable cost takes. However, as a practical matter, the performance payment methods ϕ_m have to be constrained by the current computational limitations of ISO optimization tools.

The performance payment method ϕ_m included within each reserve offer SC_m submitted into the illustrative base-case swing-contract market M(T) presented in Chapter 7 takes a relatively simple time-of-use form that is easily handled by standard mixed integer linear programming (MILP) solvers. For each time-step $k \in \mathbb{K}_m$,[1] ϕ_m stipulates that resource m's compensation for down/up power-path delivery and ramp service is to be calculated by means of a single specified energy price $\pi_m^{\mathsf{E}}(k)$ (\$/MWh) and a single specified ramp price $\pi_m^{\mathsf{R}}(k)$ (\$/MW). These prices do not depend on the specific dispatch set point conveyed to m for each time-step k.

The purpose of this chapter is to demonstrate that the optimization of a swing-contract market M(T) can still be formulated as a MILP optimization, solvable by standard MILP solvers, when performance payment methods included in reserve offers provide "price swing." Here *price swing* means that the compensation requested by a dispatchable resource m cleared for power-path delivery during a time-step k depends on m's dispatch set point for k.

For concreteness, attention will be focused on reserve offers submitted into M(T) by dispatchable resources m operating as pure generators.

[1]Recall from Section 7.3 that $\mathbb{K}_m$ is a partitioning of a dispatchable resource m's offered delivery period $T_m \subseteq T$ into time-steps k, each of length Δt.

A New Swing-Contract Design for Wholesale Power Markets, First Edition, Leigh Tesfatsion.

8.2 COST FUNCTION PRELIMINARIES

Consider a swing-contract market M(T) for a future operating period T that consists of successive time-steps $k = 1, \ldots, NK$. Each time-step k has the same length Δt, measured in hours. Let m denote a dispatchable resource that is participating in M(T) through the submission of a reserve offer for period T. The following assumptions are maintained for m:

- The feasible sustainable power range $\mathbb{P}_m(k)$ for m during each time-step k is an interval $[P_m^{\min}(k), P_m^{\max}(k)]$ with $0 \leq P_m^{\min}(k)$. Thus, m is a pure generator.
- If the ISO managing M(T) commits generator m for a time-step k, the ISO's dispatch instructions conveyed to m at the start-time k^s for k consist of a single dispatch set point $p_m(k) \in \mathbb{P}_m(k)$. This dispatch set point indicates the power level that m is to provide at k^s.
- The *avoidable cost function* (\$/h) of generator m for each time-step $k \in \mathbb{K}_m$, formulated at the start of M(T), is a nondecreasing convex function

$$\mathrm{C}_{m,k} : \mathbb{P}_m(k) \rightarrow R_+ \tag{8.1}$$

For example, the avoidable cost function (8.1) could take the quadratic form:

$$\mathrm{C}_{m,k}(p) = a_m(k) + b_m(k)p + c_m(k)p^2, \ \forall p \in \mathbb{P}_m(k) \tag{8.2}$$

where the cost coefficient $a_m(k) \geq 0$ has units \$/h, the cost coefficient $b_m(k) \geq 0$ has units \$/MWh, and the cost coefficient $c_m(k) \geq 0$ has units \$/(MW)^2 h.

Suppose, for simplicity of exposition, that m's avoidable cost function (8.1) is a differentiable function of p. Then m's *supply function* for time-step k can be represented as follows[2]:

$$\pi = \mathrm{S}_{m,k}(p) \equiv \frac{\partial \mathrm{C}_{m,k}(p)}{\partial p} \geq 0, \ \forall p \in \mathbb{P}_m(k) \tag{8.3}$$

Each power-price combination (p, π) satisfying (8.3) has the following interpretation: π (\$/MWh) is the minimum price that generator m is willing to accept in payment for an incremental increase in its dispatch set point at the start of k, given that its current dispatch set point at the start of k is p. Power economists typically presume this minimum price is equal to the additional cost that m would have to incur in order to produce power at this incrementally higher dispatch set point during time-step k. Thus, power economists refer to the supply function (8.3) as m's *marginal cost function* for time-step k.

[2]Note that the supply function (8.3) is a nondecreasing function of p; i.e. $\partial S_{m,k}(p)/\partial p \geq 0$. This follows from the assumption that m's avoidable cost function (8.1) is a convex function of p.

The avoidable cost function (8.1) can then be expressed as the sum of two terms, *avoidable fixed cost* ($/h) and *variable cost* ($/h), as follows[3]:

$$C_{m,k}(p) = \left[C_{m,k}(P_m^{\min}(k)) + \int_{P_m^{\min}(k)}^{p} S_{m,k}(q)\, dq\right], \ \forall p \in \mathbb{P}_m(k) \qquad (8.4)$$

The first term on the right-side of (8.4),

$$\text{AvoidFC}_{m,k} \equiv C_{m,k}(P_m^{\min}(k)) \qquad (8.5)$$

is the cost ($/h) that m must incur in order to inject power at its minimum feasible sustainable level $P_m^{\min}(k)$ during time-step k; this is a fixed cost for m because it is independent of m's dispatch set point p. The second term on the right-side of (8.4),

$$\text{VC}_{m,k}(p) \equiv \left[\int_{P_m^{\min}(k)}^{p} S_{m,k}(q)\, dq\right], \ \forall p \in \mathbb{P}_m(k) \qquad (8.6)$$

is m's nondecreasing *variable cost function* ($/h) for time-step k. It is a *variable* cost function because it expresses cost as a function of m's dispatch set point p for time-step k. Note both (8.5) and (8.6) are avoidable costs for m at the start of M(T) because m can avoid these costs by deciding not to submit a reserve offer into M(T).

For example, suppose m's avoidable cost function (8.1) takes the quadratic form (8.2). Generator m's avoidable fixed cost (8.5) is then given by:

$$C_{m,k}(P_m^{\min}(k)) = a_m(k) + b_m(k)P_m^{\min} + c_m(k)[P_m^{\min}(k)]^2 \qquad (8.7)$$

Generator m's supply function (8.3) takes the simple linear form:

$$\pi = [b_m(k) + 2c_m(k)p\ ], \ \forall p \in \mathbb{P}_m(k) \qquad (8.8)$$

Finally, the integration of the supply function (8.8) over intervals $[P_m^{\min}(k), p]$ gives generator m's variable cost function:

$$\text{VC}_{m,k}(p) \equiv \left[\int_{P_m^{\min}(k)}^{p} [b_m(k) + 2c_m(k)q]\, dq\right], \ \forall p \in \mathbb{P}_m(k) \qquad (8.9)$$

8.3 MILP TRACTABLE FORM OF RESERVE OFFERS WITH PRICE SWING

It will next be shown how the ISO's optimization problem for the base-case swing-contract market M(T) presented in Chapter 7 can be extended to permit the participation of a dispatchable generator m with a general nondecreasing convex avoidable cost function (8.1) for each time-step k. The resulting modified optimization problem can be solved using a standard MILP solver.

[3]See Section 3.2.7 for an extended discussion of these cost concepts.

Specifically, suppose generator m submits a reserve offer $\mathrm{SC}_m = (\alpha_m, \mathbb{PP}_m, \phi_m)$ into M(T). It will be shown below how an approximate expression for m's total variable cost ($) can be derived using a piecewise-linear approximation[4] for m's nondecreasing convex variable cost function (8.6) for each time-step $k \in \mathbb{K}_m$. This approximate total variable cost is then used to determine m's required compensation for power-path delivery, as stipulated in m's performance payment method ϕ_m. In addition, for each $k \in \mathbb{K}_m$, the avoidable fixed cost portion $\mathrm{C}_{m,k}(P_m^{\mathrm{min}}(k))$ of m's avoidable cost function (8.1) is used to determine the appropriate total avoidable fixed cost ($) to be compensated through m's offer price α_m.

As illustrated in Figure 8.1, a piecewise linear approximation for m's variable cost function (8.6) for each time-step $k \in \mathbb{K}_m$ is obtained by connecting finitely many power-cost points $\{(P_i, C_i) \mid i = 0, \ldots, NS_m(k)\}$ in the power-cost plane satisfying

$$P_m^{\mathrm{min}}(k) = P_0 < P_1 < P_2 \cdots < P_{NS_m(k)} = P_m^{\mathrm{max}}(k) \tag{8.10}$$

and $C_i = \mathrm{VC}_{m,k}(P_i)$ for each i. This results in a nondecreasing convex piecewise-linear approximation for m's variable cost function (8.6) whose plot in the power-cost plane consists of connected line segments $i = 1, \ldots, NS_m(k)$.

Given any dispatch set point $p_m(k) \in \mathbb{P}_m(k)$ for generator m at the start of time-step $k \in \mathbb{K}_m$, generator m's resulting total variable cost ($) is approximated by $\phi_m(k)$, where $(p_m(k), \phi_m(k)/\Delta t)$ lies on the plot of m's approximate variable cost

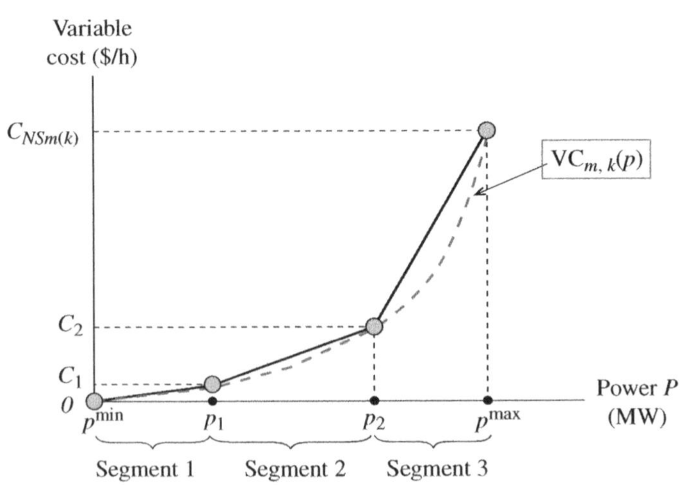

Figure 8.1 Illustration of the method for determining a piecewise-linear approximation for the variable cost function (8.6) of a dispatchable generator m for any time-step $k \in \mathbb{K}_m$. The depicted number of line segments is $NS_m(k) = 3$.

[4]The method for deriving this piece-wise linear approximation is adapted from Ref. [30, Sec. II.A].

function.[5] The minimum price that m is willing to accept in payment for an incremental increase in its dispatch set point $p_m(k)$ at the start of time-step k is then approximated by the slope of the particular line segment i that includes $(p_m(k), \phi_m(k)/\Delta t)$.

The modifications needed to the base-case M(T) optimization formulation in order to determine an optimal dispatch set point $p_m(k)$ for generator m for each time-step $k \in \mathbb{K}_m \subseteq \mathbb{K}$, given these piecewise-linear approximations for m's variable cost functions, will next be explained in detail.

The ISO's optimal selection for m's dispatch set point $p_m(k)$ at time-step $k \in \mathbb{K}_m$, with corresponding approximate total variable cost $\phi_m(k)$, is determined by means of the linear constraints (8.12) through (8.16); these constraints must be incorporated into the system constraints for M(T). The physical attributes of the power-paths $\mathbb{PP}_m$ offered by m in its reserve offer SC_m, e.g. power capacity limits and ramping restrictions, must also be incorporated into the system constraints for M(T). Finally, the dispatch set point $p_m(k)$ for m must be incorporated into the power balance constraint for m's bus location b_m for each $k \in \mathbb{K}_m$.

Also, for each $k \in \mathbb{K}_m$, the MW levels $\{\delta_{i,m}(k) \mid i = 1, \ldots, NS_m(k)\}$ that appear in constraints (8.12) through (8.16), must be incorporated into the M(T) optimization formulation as continuously valued ISO decision variables; and the dispatch set point $p_m(k)$ for m must be included among the derived variables determined by the ISO's decision variables and the system constraints. The ISO's objective function must then be extended to include generator m's approximate total variable cost $\phi_m(\mathbf{p}_m)$ as a function of m's discretized power-path $\mathbf{p}_m = \{p_m(k) \mid k \in \mathbb{K}_m\}$. This approximate total variable cost ($) is given by

$$\phi_m(\mathbf{p}_m) = \sum_{k \in \mathbb{K}_m} \phi_m(k) \tag{8.11}$$

Finally, the ISO's objective function must be extended to include $c_m \alpha_m$, where c_m is the contract clearing indicator for m's reserve offer SC_m and α_m is m's offer price for this reserve offer that ensures coverage for m's total avoidable fixed cost $\mathrm{C}_{m,k}(P_m^{\mathrm{min}}(k))\Delta t$ for each $k \in \mathbb{K}_m$.

A method for automatically generating power-cost points (P_i, C_i) satisfying restriction (8.10) with $C_i = \mathrm{VC}_{m,k}(P_i)$ for each i will be given below. Suppose, for the moment, that these power-cost points have already been specified.

Total variable cost ($) approximation for generator m at time-step k:

$$\phi_m(k) = \sum_{i=1}^{NS_m(k)} (MC_i \cdot \delta_{i,m}(k))\Delta t \tag{8.12}$$

[5] Generator m's variable cost function $\mathrm{VC}_{m,k}(p)$ for time-step k measures cost in $/h units. Recall that Δt is the length of a time-step k measured in hours, e.g. 1 hour/4. Consequently, if m maintains a constant power level $p_m(k)$ during time-step k, the total variable cost $\phi_m(k)$ ($) incurred by m during k is $\mathrm{VC}_{m,k}(p_m(k)) \times \Delta t$.

$$p_m(k) = P_0 v_m(k) + \sum_{i=1}^{NS_m(k)} \delta_{i,m}(k) \tag{8.13}$$

$$\delta_{i,m}(k) \leq P_i - P_{i-1},\ \forall i = 1, \cdots, NS_m(k) \tag{8.14}$$

$$\delta_{i,m}(k) \geq 0,\ \forall i = 1,\ \ldots\ , NS_m(k) \tag{8.15}$$

where

$$MC_i = \frac{C_i - C_{i-1}}{P_i - P_{i-1}},\ \forall i = 1,\ \ldots\ , NS_m(k) \tag{8.16}$$

The term Δt in (8.12) is the length of time-step k, measured in hours. The term $\phi_m(k)$ ($) in (8.12) is an approximated value for the total variable cost ($) that generator m incurs during time-step k.

The unit availability indicator $v_m(k)$ in (8.13) indicates whether (1) or not (0) generator m is available for service in time-step k. The power level $p_m(k)$ in (8.13) denotes the ISO's choice of a dispatch set point for m for time-step k. If $v_m(k) = 0$, then $p_m(k) = \phi_m(k) = 0$.[6] The marginal cost ($/MWh) of m at any dispatch set point p lying between P_{i-1} and P_i is approximated by MC_i in (8.16).

For example, suppose $v_m(k) = 1$, and suppose there exists a line segment whose index n lies strictly between 1 and $NS_m(k)$ such that $\delta_{i,m}(k)$ takes on its maximum possible value for $i = 1, \ldots, n$ and $\delta_{i,m}(k) = 0$ for $i > n$. Then $p_m(k) = P_n$ and $\phi_m(k) = C_n \Delta t = \mathrm{VC}_{m,k}(P_n)\Delta t$. On the other hand, suppose $\delta_{i,m}(k) = 0$ for all $i = 1, \ldots, NS_m(k)$. Then, $p_m(k) = P_0 = P_m^{\min}(k)$ and $\phi_m(k) = 0$.[7]

Finally, an automatic method for constructing power-cost points $\{P_i, C_i\}$ satisfying restriction (8.10) with $C_i = \mathrm{VC}_{m,k}(P_i)$ for each i is given below for any $k \in \mathbb{K}_m$. This method partitions generator m's power domain $\mathbb{P}_m(k) = [P_m^{\min}(k), P_m^{\max}(k)]$ into power segments having equal lengths. The resulting power-cost points should then be treated as exogenous inputs for the M(T) optimization.

Power-cost point construction method for generator m at a time-step k. This automated method requires the following inputs for each $k \in \mathbb{K}_m$: (i) the analytical form of generator m's variable cost function (8.6); (ii) generator m's minimum and maximum power capacity limits $P_m^{\min}(k)$ and $P_m^{\max}(k)$; and (iii) a positive integer value $NS_m(k)$ determining the total number of line segments i to be used in the approximation of generator m's variable cost function (8.6). The inputs (i)–(iii) can be used to compute power-cost points $\{P_i, C_i\}$ satisfying restriction (8.10) with $C_i = \mathrm{VC}_{m,k}(P_i)$ for each i, as follows:

(a) The initial power-cost points are set to $P_0 = P_m^{\min}(k)$ and $C_0 = \mathrm{VC}_{m,k}(P_0)$;

---

[6]The system constraints for the M(T) optimization presented in Chapter 7 imply that $p_m(k) = 0$ for any time-step $k \in \mathbb{K}$ for which m's availability indicator $v_m(k)$ is zero. Hence, constraints (8.12), (8.13), and (8.15) together with the assumption $P_m^{\min}(k) \geq 0$ imply that $\phi_m(k) = 0$ as well.

[7]Note that the following fuller set of relationships actually holds in this case: $\phi_m(k) = 0 = C_0 \Delta t = \mathrm{VC}_{m,k}(P_0)\Delta t = \mathrm{VC}_{m,k}(P_m^{\min}(k))\Delta t$.

(b) The width of each power segment is set equal to

$$PSW_m(k) = \frac{[P_m^{\max}(k) - P_m^{\min}(k)]}{NS_m(k)} \tag{8.17}$$

(c) For each $i = 1, \ldots, NS_m(k)$, the power point P_i is set equal to $P_0 + i \cdot PSW_m(k)$, and the cost point C_i is set equal to $\mathrm{VC}_{m,k}(P_i)$.

Figure 8.2 illustrates the energy requirements for a dispatchable generator m that maintains a constant power output $p_m(k) = P_2$ during a time-step $k \in \mathbb{K}_m$ in accordance with dispatch instructions. In this figure, g^{sync} denotes the nonnegative power generation level at which m attains a synchronized state, ready to inject power into the grid but not yet injecting any power into the grid. The total power generation sustained by m during time-step k is thus given by $g(t) = P_2 + g^{\mathrm{sync}}$ for all $t \in k$.

The amount of energy that generator m expends locally (behind-the-meter) during time-step k in order to maintain itself in a synchronized state is denoted by NoLoad in Figure 8.2. The amount of energy that m injects into the grid during time-step k to maintain its minimum sustainable power output $P^{\min}$ during time-step k is denoted by MinRun. The remaining amount of energy that m injects into the grid during time-step k is denoted by NetDispatch.

In keeping with the cost definitions provided in Chapter 4 (Footnote 3), generator m's total *no-load cost* for time-step k is all of the avoidable fixed cost that m incurs behind-the-meter in order to maintain its minimum feasible sustainable power-injection level $P^{\min}$ during k. Generator m's *minimum-run cost* for time-step k is all of the avoidable fixed cost that m incurs front-of-the-meter in order to maintain $P^{\min}$ during k. Generator m's *opportunity cost* for time-step k is all of the avoidable fixed cost (lost net earnings) that m incurs by not deploying its generation capacity in

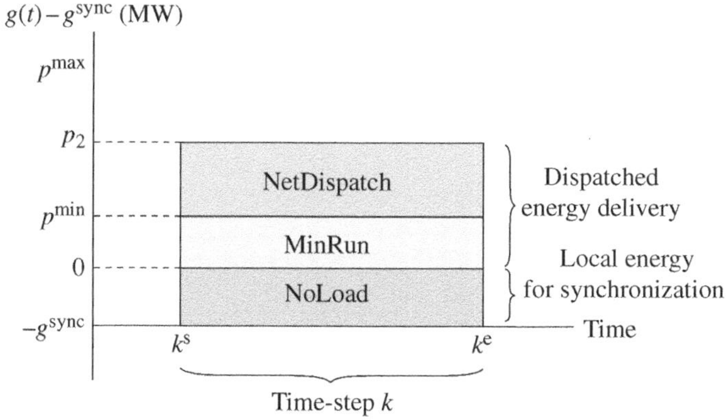

Figure 8.2 Illustrative depiction of the energy requirements for a dispatchable generator m that is dispatched at a power level P_2 at the start of a time-step $k \in \mathbb{K}_m$.

a next-best alternative use during k. Finally, generator m's *variable cost* for time-step k is all of the avoidable cost that m incurs for the dispatched delivery of power during k within its feasible sustainable power range $[P^{\min}, P^{\max}]$.

For the case depicted in Figure 8.2, generator m's no-load cost for time-step k includes the energy cost incurred for NoLoad, and its minimum-run cost includes the energy cost incurred for MinRun. The avoidable cost function (8.1) can account for these costs, together with any opportunity cost for time-step k, by appropriate specification of the avoidable fixed cost portion of (8.4). Generator m can ensure coverage of this avoidable fixed cost by including it within the offer price for its reserve offer.

Generator m's variable cost for time-step k includes the energy cost incurred for NetDispatch in Figure 8.2. This variable cost can be accounted for by appropriate specification of the variable cost portion of (8.4). Generator m can ensure compensation for this variable cost by appropriate specification of the performance payment method included within its reserve offer.

CHAPTER 9

INCLUSION OF PRICE-SENSITIVE RESERVE BIDS

9.1 CHAPTER PREVIEW

Demand bids submitted by load-serving entities (LSEs) into current US RTO/ISO-managed day-ahead markets (DAMs) are demands for the delivery of power on behalf of retail customers, with or without accompanying price information indicating willingness to pay. If a demand bid submitted by an LSE into a SCUC/SCED optimization for a DAM is cleared, the LSE is obligated to pay for the resulting scheduled delivery of power to its customers. These LSE payments are determined in part through LMP assessments (which take into account any LSE submitted price information) and in part through subsequent RTO/ISO allocations of net operational costs across LSEs on the basis of their load shares.[1]

In current US DAMs, most LSE demand bids take a fixed-demand form.[2] A *fixed demand* submitted into a DAM held on day D specifies for each hour H of day D+1 a non-price-sensitive power usage level $\widehat{p}^{\mathrm{f}}(\mathrm{H})$ (MW) to be maintained during H, with no accompanying price information indicating willingness to pay.

In economic terms, for each hour H, the power usage level $\widehat{p}^{\mathrm{f}}(\mathrm{H})$ effectively represents a vertical demand curve in a power-price plane, as if customers had an infinite willingness to pay for this power usage. Although the effective[3] maximum willingness to pay for this power usage is necessarily finite, it cannot be determined. Thus, the presence of fixed-demand bids hinders an RTO/ISO's ability to ensure that a SCUC/SCED optimization will result in an efficient allocation of resources.

[1]To preserve its independent status, an RTO/ISO cannot have a financial stake in the market operations it manages. Thus, an RTO/ISO must pass through to market participants all net costs incurred as a result of market operations.

[2]For example, the percentage of cleared price-sensitive demand in ISO-New England's DAM remained nearly constant at around 27% from 2012 to 2016; see [89, Fig. 3-20].

[3]*Effective* willingness to pay is willingness to pay backstopped by actual purchasing power.

A New Swing-Contract Design for Wholesale Power Markets, First Edition, Leigh Tesfatsion.

These concerns also arise for the base-case swing-contract market formulated in Chapter 7. All LSE reserve bids are assumed to take a fixed form. Consequently, all LSE reserve bids are entered into power balance constraints as must-service load obligations. Since benefits are then not measurable, the usual market-clearing goal – maximization of expected total net benefit – is replaced by the minimization of expected total avoidable cost.

This chapter shows how the base-case swing-contract market formulated in Chapter 7 can be generalized to allow LSEs to submit reserve bids with accompanying price (valuation) information permitting the measurement of customer benefits.[4] Three cases are considered: reserve bids with *time-of-use (TOU)* pricing; reserve bids in the form of quantity-price demand schedules; and reserve bids directly expressed in terms of benefit functions.[5]

9.2 INCORPORATION OF BENEFITS

The base-case swing-contract market M(T) formulated in Chapter 7 is managed by an ISO that clears reserve offers to secure the availability of net load balancing services for a future operating period T. This operating period T is assumed to be partitioned into successive time-steps $k \in \mathbb{K} = \{1, \ldots, NK\}$, where each time-step k has equal length Δt measured in hours.

Each LSE $j \in \mathbb{LS}$ services the power usage of retail customers at its bus location b_j during each time-step k. Each LSE j submits into M(T) a fixed reserve bid expressed as a forecasted power-usage amount $\hat{p}_j^{\text{f}}(k)$ (MW) at bus b_j for each time-step k, with no accompanying price information.

Now suppose that each LSE j also submits into M(T) a price-sensitive reserve bid for each time-step k on behalf of a collection $\mathbb{C}_j$ of retail customers. This bid expresses possible aggregate power-usage levels $p_j^{\text{s}}(k)$ (MW) that the customers in $\mathbb{C}_j$ could maintain at bus b_j during time-step k together with some type of metric for measuring the benefit to these customers of each of these power-usage levels.

Let the benefit ($) assigned by LSE j to each possible power-usage sequence

$$\mathbf{p}_j^{\text{s}} = \{p_j^{\text{s}}(k) \mid k \in \mathbb{K}\} \tag{9.1}$$

[4] As Spence [168, Sec. II] correctly notes, reliance on willingness-to-pay metrics (such as price-sensitive bids) as a measure of customer benefits can be criticized on the grounds that the preferences customers reveal through their purchase requests are constrained by their income and wealth. The typical economic response to this criticism is that disparities in the distribution of income or wealth that are judged to be unfair, based on principles of distributive justice, ought to be handled by direct income or wealth redistributions, not by price distortions.

[5] For the latter two cases, participant retail customers would need to have installed real-time telemetry permitting them to respond to real-time price or power-usage set points communicated to them by the LSE or some other designated scheduling entity.

be denoted by $\text{ben}_j(\mathbf{p}_j^{\text{s}})$ ($). The total benefit ($) resulting from these price-sensitive reserve bids is then denoted by

$$\text{B(T)} = \sum_{j \in \mathbb{LS}} \text{ben}_j(\mathbf{p}_j^{\text{s}}) \tag{9.2}$$

In order to incorporate these price-sensitive reserve bids and benefit assessments into the MILP optimization developed in Section 7.5 for the base-case swing-contract market M(T), the following modifications must be made.

Required ISO Modifications:

First, the ISO objective function (7.7) for M(T), expressed solely in terms of forecasted total avoidable cost $\hat{\text{C}}(\text{T})$ for operating period T, must be modified to represent forecasted total net benefit for period T. This forecasted total net benefit ($) is expressed as

$$\widehat{\text{NB}}(\text{T}) = \text{B(T)} - \hat{\text{C}}(\text{T}) \tag{9.3}$$

Second, the objective of the ISO must be changed from the minimization of period-T forecasted total avoidable cost $\hat{\text{C}}(\text{T})$ to the maximization of period-T forecasted total net benefit (9.3).

Third, the price-sensitive power usage amounts $\{p_j^{\text{s}}(k) \mid j \in \mathbb{LS}, k \in \mathbb{K}\}$ need to be included among the ISO's continuously valued decision variables, subject to domain constraints of the form

$$p_j^{\text{s}}(k) \in \mathbb{P}_j(k) \ , \ \forall j \in \mathbb{LS}, \ k \in \mathbb{K} \tag{9.4}$$

where each domain $\mathbb{P}_j(k)$ includes 0, i.e. the zero-power option.[6]

Required System Constraint Modifications:

Fourth, the price-sensitive power usage amounts $p_j^{\text{s}}(k)$ must be appropriately entered into the power balance constraints (7.14). The resulting extended constraints take the following form:

Extended power balance constraints (with slack variables): For all $b \in \mathbb{B}, k \in \mathbb{K}$,

$$\left[\sum_{m \in \mathbb{M}(b)} p_m(k)\right] + \left[\sum_{\ell \in \mathbb{L}_{\text{E}(b)}} w_\ell(k) - \sum_{\ell \in \mathbb{L}_{\text{O}(b)}} w_\ell(k)\right] = \widehat{NL}_b(k) + \beta_b(k) \tag{9.5}$$

[6]If the domains $\mathbb{P}_j(k)$ are not required to include 0, the ISO would have no way of refusing to service price-sensitive reserve bids at LSE-set prices whose servicing results in a lowering of total net benefit. To avoid this forced servicing, it would be necessary to modify further the form of the MILP optimization to include an ISO binary 0/1 contract-clearing indicator for each submitted price-sensitive reserve bid, thus enlarging the number of ISO binary decision variables.

where

$$\widehat{NL}_b(k) = \widehat{NL}_b^{\mathrm{f}}(k) + \sum_{j \in \mathbb{LS}(b)} p_j^{\mathrm{s}}(k) \tag{9.6}$$

Fifth, the price-sensitive power usage amounts $p_j^{\mathrm{s}}(k)$ must be appropriately entered into the reserve requirement constraints (7.22) through (7.25). The resulting extended constraints take the following form:

Extended system-wide down/up reserve requirement constraints: For all $k \in \mathbb{K}$,

$$\sum_{m \in \mathbb{M}} \underline{p}_m(k) \leq [1 - \mathrm{RRD}(k)] \cdot \widehat{NL}(k) \tag{9.7}$$

$$\sum_{m \in \mathbb{M}} \overline{p}_m(k) \geq [1 + \mathrm{RRU}(k)] \cdot \widehat{NL}(k) \tag{9.8}$$

where

$$\widehat{NL}(k) = \widehat{NL}^{\mathrm{f}}(k) + \sum_{j \in \mathbb{LS}} p_j^{\mathrm{s}}(k) \tag{9.9}$$

Extended zonal down/up reserve requirement constraints: For all $z \in \mathbb{Z}$ and $k \in \mathbb{K}$,

$$\sum_{m \in \mathbb{M}(z)} \underline{p}_m(k) \leq [1 - \mathrm{RRD}(z,k)] \cdot \widehat{NL}_z(k) \tag{9.10}$$

$$\sum_{m \in \mathbb{M}(z)} \overline{p}_m(k) \geq [1 + \mathrm{RRU}(z,k)] \cdot \widehat{NL}_z(k) \tag{9.11}$$

where

$$\widehat{NL}_z(k) = \widehat{NL}_z^{\mathrm{f}}(k) + \sum_{j \in \mathbb{LS}(z)} p_j^{\mathrm{s}}(k) \tag{9.12}$$

9.3 MODELING OF PRICE-SENSITIVE RESERVE BIDS

9.3.1 Standard Demand Function Formulation

A *price-sensitive reserve bid* submitted into a swing-contract market M(T) by an LSE j to service the power-usage requirements of a collection $\mathbb{C}_j$ of customers during operating period T can take the form of a collection of standard economic demand functions $\{D_{j,k} \mid k \in \mathbb{K}\}$. Each demand function $D_{j,k}$ is a nonincreasing function

$$D_{j,k} : \mathbb{P}_j(k) \rightarrow R_+ \tag{9.13}$$

where $0 \in \mathbb{P}_j(k) \subset R_+$. Each $p \in \mathbb{P}_j(k)$ denotes a possible maintained aggregate power-usage level for the customers in $\mathbb{C}_j$ during time-step k that LSE$_j$ would be willing to service at bus b_j at price level $D_{j,k}(p)$.

More precisely, given (9.13), the price-sensitive demand schedule for LSE j at the start of time-step k consists of all power-price combinations (p, π) satisfying

$$\pi = D_{j,k}(p), \ p \in \mathbb{P}_j(k) \tag{9.14}$$

Each power-price combination (p, π) has the following interpretation: π (\$/MWh) is the maximum price that LSE j is willing to pay at the start of time-step k for an incremental increase in the maintained aggregate power usage of its customers in $\mathbb{C}_j$ during time-step k, given that their current aggregate power usage is at level p (MW). For example, the demand schedule (9.14) could take the simple linear form

$$\pi = e_j(k) - 2f_j(k) \cdot p \tag{9.15}$$

where the coefficients $e_j(k)$ (\$/MWh) and $f_j(k)$ (\$/[MW]^2 h) are positively valued.

9.3.2 Reserve Bids with Time-of-Use Pricing

Consider, first, the case in which the demand schedule (9.14) for each LSE j designates a single price $\pi_j(k)$ (\$/MWh) for each time-step k, regardless of the power-usage level p (MW). In this case, the benefit (\$) that LSE j assigns to a price-sensitive power usage sequence $\mathbf{p}_j^S = \{p_j^S(k) \mid k \in \mathbb{K}\}$ takes the following simple form:

$$\text{ben}_j(\mathbf{p}_j^S) = \sum_{k \in \mathbb{K}} \pi_j(k) p_j^S(k) \Delta t \tag{9.16}$$

The total benefit B(T) (\$) to be included in the ISO's objective function (9.3) for the entire future operating period T then takes the form:

$$\text{B(T)} = \sum_{j \in \mathbb{LS}} \text{ben}_j(\mathbf{p}_j^S) \tag{9.17}$$

9.3.3 Reserve Bids with Price Swing

Consider, instead, the case in which each LSE j submits a price-sensitive reserve bid consisting of a demand schedule (9.14) for each time-step k with a price π that varies in response to changes in the power usage level p.

For example, Figure 9.1 illustrates the physical aspects of such a price-sensitive reserve bid for an operating period T partitioned into twelve time-steps $k1, \ldots, k12$. Each time-step k has a common length Δt, measured in hours. For simplicity of depiction, it is assumed that the power domain $\mathbb{P}_j(k)$ for each time-step k is identical, consisting of seven possible power levels $\{0, p_1, \ldots, p_6\}$.

Figure 9.2 illustrates the form of the power-price demand schedule (9.14) that LSE j could then designate for each of the twelve time-steps k depicted in Figure 9.1. This type of price-sensitive demand schedule has the form required in current US

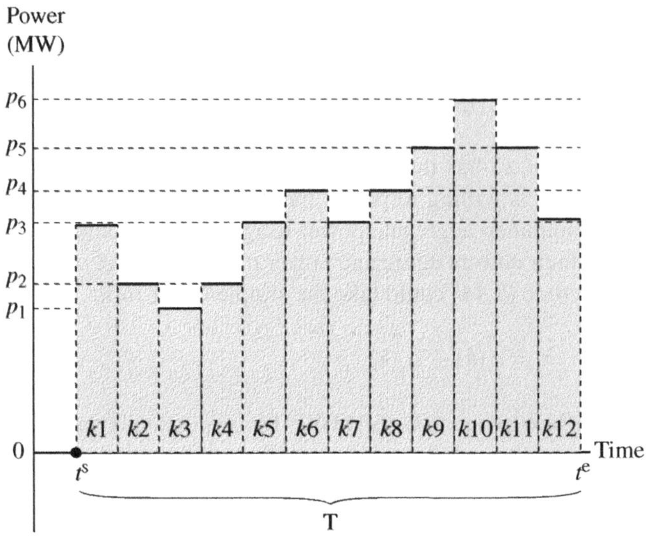

Figure 9.1 Illustration of the physical aspects of a price-sensitive reserve bid submitted by an LSE into a swing-contract market M(T) for which the operating period T is partitioned into 12 time-steps k. The shaded region denotes one possible load profile the ISO could clear for period T.

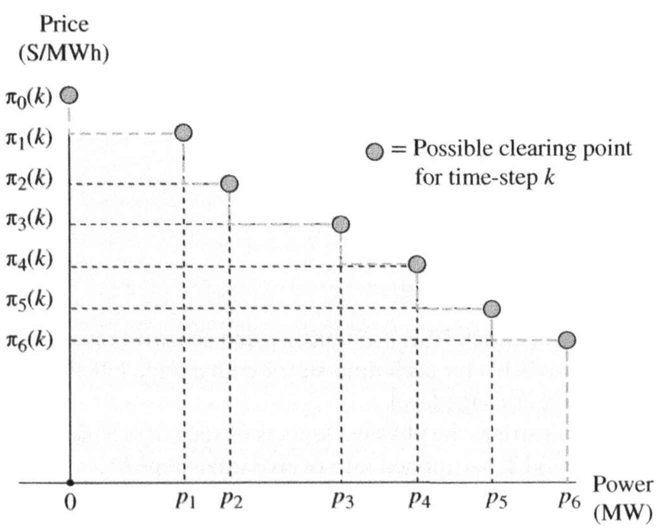

Figure 9.2 A demand schedule (9.14) with power domain $\mathbb{P}_j(k) = \{0, p_1, \ldots, p_6\}$ that could be designated for a particular time-step k by the price-sensitive reserve bid whose physical aspects are depicted in Figure 9.1.

RTO/ISO-managed DAMs. See, for example, the form required by the ERCOT DAM [47, Module 4].

Given this form of price-sensitive reserve bid, the benefit ($) that LSE j assigns to a price-sensitive power usage sequence $\mathbf{p}_j^s = \{p_{n(j,k)}^s \mid k \in \mathbb{K}\}$ takes the form:

$$\text{ben}_j(\mathbf{p}_j^s) = \sum_{k \in \mathbb{K}} \pi_{n(j,k)} p_{n(j,k)}^s \Delta t \tag{9.18}$$

where $n(j,k)$ is an index that denotes the particular power-usage level $p_{n(j,k)}^s$ (MW) selected from $\mathbb{P}_j(k)$ for time-step k, and $\pi_{n(j,k)}$ ($/MWh) denotes the corresponding price for power usage during time-step k. The total benefit B(T) ($) to be included in the ISO's objective function (9.3) for the entire future operating period T then takes the form

$$\text{B(T)} = \sum_{j \in \mathbb{LS}} \text{ben}_j(\mathbf{p}_j^s) \tag{9.19}$$

9.3.4 Reserve Bids Directly Expressed as Benefit Functions

More generally, suppose each LSE $j \in \mathbb{LS}$ assigns a benefit ($/h) to each possible power-usage level $p_j^s(k) \in \mathbb{P}_j(k)$ for each time-step $k \in \mathbb{K}$ by means of a nondecreasing concave function

$$\text{B}_{j,k} : \mathbb{P}_j(k) \to R \tag{9.20}$$

where $\mathbb{P}_j(k) = [0, P_j^{\max}(k)]$. For example, (9.20) could take the quadratic form

$$\text{B}_{j,k}(p) = d_j(k) + e_j(k) \cdot p - f_j(k) \cdot p^2 \tag{9.21}$$

with ordinate $d_j(k)$ ($/h), positive coefficients $e_j(k)$ ($/MWh) and $f_j(k)$ ($/[MW]^2 h), and

$$P_j^{\max}(k) = \frac{e_j(k)}{2f_j(k)} \tag{9.22}$$

If the benefit function (9.20) is differentiable, the maximum willingness of LSE j to pay for a maintained incremental increase in the aggregate power usage of its customers in $\mathbb{C}_j$ during time-step k, given that their current aggregate power usage is at level p, can be expressed by the *marginal benefit function*[7]

$$\pi_{j,k}(p) \equiv \frac{\partial \text{B}_{j,k}(p)}{\partial p} \geq 0 \tag{9.23}$$

Note that the price $\pi_{j,k}(p)$ ($/MWh) depends on the power usage level p.

[7] In economics, standard budget-constrained utility-maximization problems consider the maximization of utility (benefit) with respect to the choice of consumption levels for multiple possible goods subject to a budget constraint and non-negativity constraints on good purchases. For example, let $\text{U}(q,z)$ denote the utility function for a consumer measuring the benefit (utility, or "utils") gained by the consumer from the consumption of goods Q and Z in amounts q and z during some fixed time interval Δt. Consider

The benefit ($) assigned by LSE j to each possible price-sensitive power usage sequence $\mathbf{p}_j^S = \{p_j^S(k) \mid k \in \mathbb{K}\}$ then takes the form

$$\text{ben}_j(\mathbf{p}_j^S) = \sum_{k \in \mathbb{K}} \text{B}_{j,k}(p_j^S(k))\Delta t \tag{9.24}$$

Consequently, the total benefit B(T) ($) to be included in the ISO's objective function (9.3) for the entire future operating period T takes the form:

$$\text{B(T)} = \sum_{j \in \mathbb{LS}} \text{ben}_j(\mathbf{p}_j^S) \tag{9.25}$$

9.4 MILP TRACTABLE APPROXIMATION OF BENEFIT FUNCTIONS

The technique described in Chapter 8 for obtaining piecewise-linear approximations for nondecreasing *convex* variable cost functions $\text{C}_{m,k}(p)$ ($/h) can similarly be applied to obtain piecewise-linear approximations for nondecreasing *concave* benefit functions $\text{B}_{j,k}(p)$ ($/h) taking form (9.20).

Suppose the benefit function used by each LSE $j \in \mathbb{LS}$ at each time-step $k \in \mathbb{K}$ to measure the benefit ($/h) to its customers in $\mathbb{C}_j$ of different possible price-sensitive power usage levels $p_j^S(k)$ is given by a nondecreasing concave function (9.20). A piecewise-linear approximation for this benefit function can then be obtained in three simple steps.

First, select points $\{P_0, P_1, \ldots, P_{NS_j(k)}\}$ from the domain $[0, P_j^{\max}(k)]$ subject to the following restriction:

$$0 = P_0 < P_1 < P_2 < \cdots < P_{NS_j(k)} = P_j^{\max}(k) \tag{9.26}$$

Second, plot the power-benefit points $\{(P_i, B_i) \mid i = 0, \ldots, NS_j(k)\}$ in the power-benefit plane, where $B_i = \text{B}_{j,k}(P_i)$. Third, connect these power-benefit points by line segments whose slopes, by construction, are nonincreasing in i. For example, Figure 9.3 illustrates a five-segment piecewise-linear approximation for a nondecreasing concave benefit function taking the quadratic form (9.21).

---

the problem of maximizing this utility function with respect to non-negative choices of q and z subject to a budget constraint $\pi^Q q + \pi^Z z = I$, where π^Q and π^Z are given $/unit prices for goods Q and Z, and I ($) is the consumer's anticipated accumulated income for period Δt. The first-order necessary conditions for this utility maximization problem include the requirement that $\lambda \pi^Q = \partial \text{U}(q,z)/\partial q$ if $q > 0$, with a similar requirement for good Z. In this expression, λ (utils/$) denotes the consumer's *marginal utility of money*, calculated as the dual variable for the consumer's budget constraint with I taken to be the constraint constant. Thus, at optimal solution points, prices converted into utils per unit of good consumed during Δt are expressed as rates of change of benefit functions for positively consumed goods. For further discussion of these points, see, e.g. [34, 41, 102], and [101, Chapter 2]. In the present context, the benefit function $\text{B}_{j,k}(p)$ measured in $/h units corresponds to a normalized utility function $\text{U}(q,z)/\lambda$ measured in $ units with $\Delta t = 1$ hour.

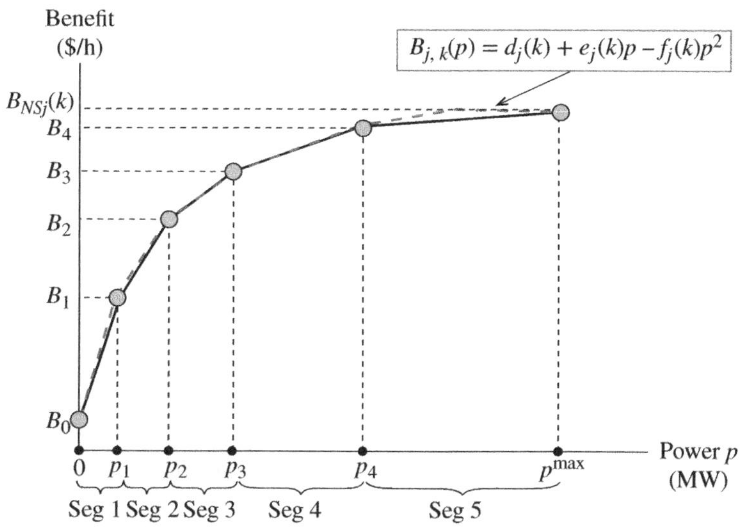

Figure 9.3 Piecewise-linear approximation of a quadratic benefit function (9.21) used by an LSE j to measure the benefit of its customers $\mathbb{C}_j$ at some time-step k. The number of line segments specified for the depicted approximation is $NS_j(k) = 5$.

Given these power-benefit points (P_i, B_i), the benefit ($) attained by the customers of each LSE j during each time-step k can be incorporated into the basic MILP optimization formulation for an ISO-managed swing-contract market M(T) in an approximate piecewise-linear form, $\text{ben}_j(k)$ ($), that preserves the MILP form of this optimization. This incorporation is similar to the incorporation of variable costs explained in Chapter 7. It proceeds as follows:

- The ISO's optimal selection of a price-sensitive power usage level $p_j^{\text{s}}(k)$ for the customers of LSE j at time-step k, with corresponding benefit approximation $\text{ben}_j(k)$ ($), is determined for each $j \in \mathbb{LS}$ and $k \in \mathbb{K}$ by means of the linear constraints (9.28) through (9.32). These constraints must be incorporated into the system constraints for M(T).
- The MW levels $\{\delta_{i,j}(k) \mid i = 1, \ldots, NS_j(k)\}$ that appear in constraints (9.28) through (9.32) must be incorporated into the ISO's optimization problem as continuously valued ISO decision variables, for each $j \in \mathbb{LS}$ and each $k \in \mathbb{K}$.
- The price-sensitive power usage levels $\{p_j^{\text{s}}(k) \mid j \in \mathbb{LS}, k \in \mathbb{K}\}$ must be included among the derived variables determined by the ISO's decision variables and the system constraints.
- As explained with care in Section 9.2, the power usage level $p_j^{\text{s}}(k)$ must be incorporated into the power balance constraint at LSE j's bus location b_j and into the forecasted net loads appearing in the system-wide and zonal reserve constraints, for each $j \in \mathbb{LS}$ and $k \in \mathbb{K}$.

- The ISO's objective for operating period T must be re-expressed as the maximization of forecasted total net benefit $\mathrm{B(T)} - \widehat{\mathrm{C}}(\mathrm{T})$, where forecasted total avoidable cost $\widehat{\mathrm{C}}(\mathrm{T})$ ($) is given by (7.7), and total benefit B(T) ($) is calculated as the summation of benefits across all LSEs $j \in \mathbb{LS}$ and time-steps $k \in \mathbb{K}$:

$$\mathrm{B(T)} = \sum_{j \in \mathbb{LS}} \sum_{k \in \mathbb{K}} \mathrm{ben}_j(k) \tag{9.27}$$

As will be shown, below, given a value for the number $NS_j(k)$ of line segments to be used for the approximation of the benefit function $\mathrm{B}_{j,k}(p)$ in (9.20) for each $j \in \mathbb{LS}$ and $k \in \mathbb{K}$, plus analytical forms for these benefit functions, a simple method permits the automatic construction of power-benefit points $\{(P_i, B_i)\}$ satisfying restriction (9.26) with $B_i = \mathrm{B}_{j,k}(P_i)$ for each i. Suppose, for the moment, that these power-benefit points have already been constructed.

The approximate total benefit $\mathrm{ben}_j(k)$ ($) attained by the customers of LSE j during time-step k is then determined by the system of equations (9.28)–(9.32), as a function of the ISO's optimal selection of the continuously valued decision variables $\{\delta_{i,j}(k) \mid i = 1, \ldots, NS_j(k)\}$:

Benefit Approximation Method for an LSE j at a Time-Step k:

$$\mathrm{ben}_j(k) = B_0 \Delta t + \sum_{i=1}^{NS_j(k)} (MB_i \cdot \delta_{i,j}(k))\Delta t \tag{9.28}$$

$$p_j^{\mathrm{S}}(k) = \sum_{i=1}^{NS_j(k)} \delta_{i,j}(k) \tag{9.29}$$

$$\delta_{i,j}(k) \le P_i - P_{i-1}\ , \ \forall i = 1, \ldots, NS_j(k) \tag{9.30}$$

$$\delta_{i,j}(k) \ge 0, \ \forall i = 1, \ldots, NS_j(k) \tag{9.31}$$

where

$$MB_i = \frac{B_i - B_{i-1}}{P_i - P_{i-1}}, \ \forall i = 1, \ldots, NS_j(k) \tag{9.32}$$

The marginal benefit ($/MWh) of LSE j's customers, evaluated at any power-usage level in the interval from P_{i-1} to P_i, is approximated by MB_i in (9.32). The benefit ($) attained by LSE j's customers at the ISO-cleared price-sensitive power usage level $p_j^{\mathrm{S}}(k)$ in (9.29) is approximated by $\mathrm{ben}_j(k)$ in (9.28). For example, suppose there exists a segment $n \in \{1, \ldots, NS_j(k)\}$ such that each $\delta_{i,j}(k)$ takes on its maximum possible value for $i = 1, \ldots, n$ and $\delta_{i,j}(k) = 0$ for $i = n+1, \ldots, NS_j(k)$. Then $p_j^{\mathrm{S}}(k) = P_n$ and $\mathrm{ben}_j(k) = B_n \Delta t = \mathrm{B}_{j,k}(P_n)\Delta t$. On the other hand, suppose $\delta_{i,j}(k) = 0$ for all $i = 1, \ldots, NS_j(k)$. Then $p_j^{\mathrm{S}}(k) = 0$ and $\mathrm{ben}_j(k) = B_0 \Delta t = \mathrm{B}_{j,k}(0)\Delta t$.

Finally, the following method permits the automatic generation of power-benefit points $\{(P_i, B_i) \mid i = 0, \ldots, NS_j(k)\}$ satisfying restriction (9.26) with $B_i = \mathrm{B}_{j,k}(P_i)$ for each i. This method partitions LSE j's benefit function domain $[0, P_j^{\mathrm{max}}(k)]$ into power segments having equal lengths. The resulting power-benefit points should then be treated as exogenous inputs for the M(T) optimization.

Power-Benefit Point Construction Method for an LSE j at a Time-Step k:

This automated method requires the following inputs: (i) the analytical form of the benefit function (9.20); (ii) the maximum power-usage level $P_j^{\max}(k)$; and (iii) a positive integer value $NS_j(k)$ determining the total number of line segments to be used for the approximation of (9.20). Given inputs (i)–(iii), the power-benefit points $\{(P_i, B_i) \mid i = 0, \ldots, NS_j(k)\}$ are computed as follows:

(a) The initial power-benefit points are set equal to $P_0 = 0$ and $B_0 = \mathrm{B}_{j,k}(0)$;

(b) The width of each power segment is set equal to

$$PSW_j(k) = \frac{P_j^{\max}(k)}{NS_j(k)} \tag{9.33}$$

(c) For each $i = 1, \ldots, NS_j(k)$, the power point P_i is set equal to $P_0 + iPSW_j(k)$, and the benefit point B_i is set equal to $\mathrm{B}_{j,k}(P_i)$.

CHAPTER 10

THE LINKED SWING-CONTRACT MARKET DESIGN

10.1 CHAPTER PREVIEW

This chapter proposes a *linked* collection of ISO-managed swing-contract markets with look-ahead horizons that can range in duration from multiple years to minutes.[1] The linkage is to be achieved by having reserve offers and price-sensitive reserve bids cleared in earlier markets be carried forward on the books of the ISO as a portfolio of contracts that can be adaptively updated in subsequent markets.

As illustrated in Figure 10.1, this linked collection of markets can include one or more *long-term forward markets (LFMs)* with look-ahead horizons ranging from multiple days to years to facilitate the efficient entry and exit of existing reserve providers as well as needed investment in new reserve providers. It can also include one or more *short-term forward markets (SFMs)* with look-ahead horizons measured in hours, such as a day-ahead market (DAM), to facilitate advance reserve contracting for real-time operations. Finally, it can include one or more *very short-term forward markets (VSFMs)* with look-ahead horizons of one hour or less, such as a real-time market (RTM), in order to address any remaining reserve needs for net load balancing in near real-time.

The portfolio of preexisting and newly cleared reserve contracts carried forward by the ISO permits a two-way linkage among these forward markets. Each current forward market is conditioned not only on all still-active reserve contracts cleared in past forward markets but also on opportunities to procure additional reserve contracts in future forward markets. In addition, each current forward market is conditioned on current system operating conditions.

[1] Recall from Chapter 6 that the look-ahead horizon LAH(T) for a swing-contract market M(T) is the time interval between the close of M(T) and the start of a future operating period T; see Figure 6.1.

A New Swing-Contract Design for Wholesale Power Markets, First Edition. Leigh Tesfatsion.

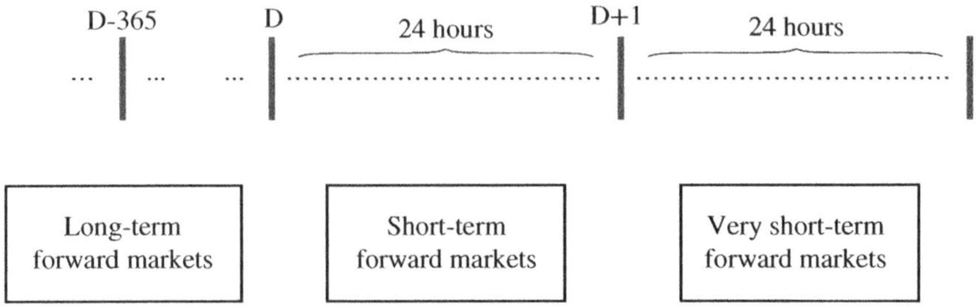

Figure 10.1 Linked collection of swing-contract markets permitting the advance procurement of reserve (dispatchable power-paths) for an operating day D+1.

The purpose of this *Linked Swing-Contract Market Design* is to permit the ISO to procure flexible dependable reserve (dispatchable power-paths) for real-time operations through an efficient centralized market-based process, not to monopolize power trading. Participants in this centralized market process are free to engage in other types of power transactions.

For example, participants are free to hedge their price and/or volumetric risk by securing financial instruments through privately negotiated bilateral trades or through financial exchange transactions. Moreover, as discussed at some length in Section 5.7, they are free to engage in privately negotiated physically covered bilateral contracts for power as long as the resulting power outcomes are appropriately self-scheduled in a swing-contract market to secure transmission support.

The key advantages of the Linked Swing-Contract Market Design as a mechanism for reserve procurement are as follows:

- *Increased transparency of market operations*: Having a centralized conceptually coherent market process for reserve procurement would provide all market participants with a clearer understanding of this process and would thus reduce opportunities for the exercise of market power.
- *Efficiency gains*: This centralized market process should result in more efficient reserve procurement than bilateral contracting. Dispatchable resources can offer their varied capabilities for providing reserve (dispatchable power-paths), with open up-front specification of all required availability and performance compensation. This should result in improved valuations and reduced transaction costs, hence less resource wastage.
- *Reliability gains*: The ability of cleared reserve providers to ensure full coverage of their avoidable costs should facilitate efficient entry and exit of existing reserve sources as well as efficient investment in needed new reserve sources.

A critical issue that arises for any linked market design is "time inconsistency." This issue is addressed for the Linked Swing-Contract Market Design in the next two sections of this chapter. The remaining sections of the chapter then discuss the envisioned operations of an LFM, SFM, and VSFM within this design.

10.2 MULTISTAGE OPTIMIZATION AND TIME INCONSISTENCY

A *multistage optimization problem (MOP)* is a decision problem for which decisions are made at successive stages (points in time) in an attempt to achieve a single stated objective. A MOP is said to be *time-inconsistent* if re-optimization at a later stage results in deviations from the optimal solution determined at an initial stage.

More precisely, consider a MOP for which a decision must be taken at each successive stage t in a decision interval $[t_0, t_1]$. Let s_0^{opt} denote the optimal solution (i.e. decision sequence) for $[t_0, t_1]$ calculated at stage t_0. Let s_τ^{reopt} denote the *reoptimized* solution for $[\tau, t_1]$ calculated at any later stage $\tau \in (t_0, t_1]$. The MOP is said to be *time inconsistent* if there exists a later stage $\tau^* \in (t_0, t_1]$ such that the reoptimized solution $s_{\tau^*}^{\text{reopt}}$ for $[\tau^*, t_1]$ deviates from the optimal solution for $[\tau^*, t_1]$ determined by s_0^{opt} at the initial stage t_0.

In general, unless the objective function and constraints for a MOP satisfy stringent regularity conditions permitting a recursive dynamic programming representation, the MOP will exhibit time inconsistency [29], [147, Chapter 5], [174, 179, 195]. These regularity conditions pertain to the structure of a MOP, not to the information state of the decision-maker. While uncertainty regarding future events can result in time inconsistency for a MOP, the absence of uncertainty is not sufficient to ensure the MOP is time consistent.

For example, in economic MOP studies focusing on intertemporal competitive market equilibrium over a decision horizon $[t_0, t_1]$, the following three regularity conditions are commonly imposed in order to ensure time consistency. First, the objective is the maximization of utility (benefit) for a single "representative" consumer subject to system constraints. Second, from each vantage point $\tau \in [t_0, t_1]$ the consumer's preference ordering over consumption paths $\mathbf{c} = \{c(t) \mid t_0 \leq t \leq t_1\}$ can be represented by a utility functional taking form

$$\mathbf{U}_\tau(\mathbf{c}) = \int_{t_0}^{t_1} \beta(t - \tau) u(c(t)) dt \tag{10.1}$$

where $u(c(t))$ measures the utility (benefit) attained by the consumer from consumption $c(t)$ at time t, and the consumer's discount function $\beta(t - \tau)$ takes the exponential form[2]

$$\beta(t - \tau) = \exp^{-\rho[t-\tau]} \tag{10.2}$$

[2]If the discount function $\beta(t - \tau)$ deviates from the exponential form (10.2), the consumer's preference ordering over consumption paths can systematically change with each change in vantage point τ, resulting in time-inconsistent choice behavior [174]. For example, the consumer might exhibit *hyperbolic* (impatient) choice behavior in the following sense: At each successive time τ, the consumer chooses an action with a high immediate benefit at the expense of longer run benefits with the intention (never realized) of being more cautious at each future time $t > \tau$; see, e.g. [163].

for some nonnegative discount factor ρ. Third, the system constraints (production technology, budget restrictions, market clearing, ...) are such that the system has a Markov chain form, i.e. the transition of the system from a current state to a next state depends only on the current state, not on previous history.

How do these observations on time inconsistency relate to the context at hand: namely, RTO/ISO-managed wholesale power markets? As discussed at some length in Section 2.3, these markets are currently anchored by a DAM and RTM operating in tandem over time. On each day D, a DAM SCUC/SCED optimization results in LMPs and scheduled dispatch solutions for all 24 hours of day D+1. RTM SCED optimizations are then conducted at frequent intervals during day D+1 to resolve discrepancies between scheduled dispatch solutions and the actual dispatch levels needed to ensure market clearing (net load balancing) on day D+1.

On a practical level, RTO/ISOs have implemented market rules designed to encourage the convergence of DAM and RTM LMP solutions. For example, as carefully explained in [14], traders are permitted to submit virtual (physically uncovered) supply offers into a DAM on day D for delivery on day D+1 whose quantity positions must be liquidated (reduced to zero) on day D+1 based on RTM prices. In addition, RTO/ISOs and power system researchers are exploring methods for improving the accuracy of net load forecasts.

Conceptually, industry practitioners and researchers are exploring various MOP alternatives to existing DAM/RTM formulations. One proposed alternative is a two-stage stochastic SCUC/SCED formulation; see, e.g. [16, 156, 191]. Binary unit commitment decisions for next-day operations are determined and announced on day D, conditional on a constructed set of possible net load scenarios for day D+1 with assigned probabilities of occurrence. In addition, scenario-conditioned price and dispatch solutions are determined for day D+1. A specific price and dispatch solution is then implemented during day D+1 in approximate accordance with the net load scenario realized during day D+1.

The practical implementation of two-stage stochastic SCUC/SCED formulations faces two major hurdles. First, the constructed set of net load scenarios must adequately cover the space of possible real-time net load outcomes while still maintaining computational feasibility. Second, the probabilities of occurrence assigned to these scenarios must be empirically credible.

Other proposed MOP approaches include multiple-interval RTMs [85]; multiple-day markets [91, Sec. 1.2]; and multistage SCUC/SCED robust optimization approaches (e.g. [11, 121, 201, 202, 204]) for which net load uncertainties are represented by bounded uncertainty sets rather than by scenarios with associated probability assessments. To reduce computational complexity and/or to exploit the successive reduction of uncertainty, many of these MOP studies incorporate or recommend reoptimization at the start of each successive decision stage.

Whether in existing or proposed forms, SCUC/SCED optimizations generally include intertemporal system constraints to capture basic physical restrictions. Examples include minimum down/up time constraints, ramping constraints, and storage capacity constraints. Intertemporal system constraints can prevent the representation of these optimizations in a recursive time-consistent form.[3]

More precisely, as carefully demonstrated in [179], the crux of the problem is that the optimal decision for each interior stage τ of a MOP with intertemporal system constraints is conditioned on anticipated decision choices for future stages $t > \tau$. Consequently, the first-order necessary conditions for optimality at stage τ impose *anticipatory restrictions* on future decision choices once decision choices for stage τ are irrevocably implemented. However, these anticipatory restrictions do not explicitly appear among the system constraints for later re-optimizations. Subsequent reoptimized decision choices can thus violate these anticipatory restrictions, resulting in time inconsistency, unless the MOP satisfies stringent recursive regularity conditions implying that these anticipatory restrictions are either redundant for later reoptimizations or trivially satisfied.

Why not simply impose time consistency as a constraint on market operations? As illustrated in [179], the time-consistent solution that results from the imposition of any such constraint will typically be suboptimal in comparison with the time-inconsistent solution when evaluated from an ex post vantage point. As succinctly stated by Strotz [174, pp. 170, 171]: "To continue to obey a fixed (plan) just because it was optimal when viewed at an earlier date is not rational if that plan is not the optimal one at the present date."

10.3 SETTLEMENT TIME-CONSISTENCY OF SWING-CONTRACT MARKETS

As seen in Chapters 6 and 7, swing-contract markets include intertemporal system constraints; hence, time inconsistency is also an issue for the Linked Swing-Contract Market Design. However, six special characteristics of this design mitigate these time inconsistency concerns.

- First, the reserve offers submitted into an ISO-managed swing-contract market M(T) consist of portfolios of swing contracts in firm or option form. The exercise times for submitted option swing contracts provide additional flexibility to the ISO with regard to resource commitment for operating period T.
- Second, the only outcome the ISO is required to announce to market participants at the close of M(T) is which reserve offers and price-sensitive reserve bids have been cleared for period T.
- Third, the only settlements the ISO implements for M(T) in advance of period T are offer-price payments to dispatchable resources whose reserve offers submitted into M(T) have been cleared. The purpose of these payments is to ensure

[3]For example, Hua et al. [85] demonstrate how time inconsistency can arise for multi-interval real-time wholesale power markets with ramping restrictions when approximately solved by means of a rolling horizon method.

coverage of availability costs. These payments can be amortized to reflect the availability costs actually incurred between the closing time for M(T) and the start-time for period T.

- Fourth, all performance compensation and power-usage charges for period T are based on actual period-T outcomes; they are not assessed and paid until after the close of period T. Consequently, these payments do not require adjustments of any previous settlements or settlement obligations.
- Fifth, all performance compensation paid ex post to dispatchable resources with cleared M(T) reserve offers must be in conformity with the performance payment methods that these resources included within their reserve offers. Consequently, all such payments must be structurally consistent with the performance payments anticipated by the ISO and market participants at the close of M(T).
- Sixth, all power-usage payments charged ex post to LSEs with cleared M(T) price-sensitive reserve bids must be in conformity with the willingness to pay for power-path deliveries as expressed in these reserve bids. Consequently, all such charges must be structurally consistent with the power-usage charges anticipated by the ISO and market participants at the close of M(T).

These six characteristics imply that the Linked Swing-Contract Market Design exhibits *settlement time-consistency*. This feature sharply distinguishes this design from the design of current RTO/ISO-managed wholesale power markets. The latter markets do not exhibit settlement time-consistency because they do not adhere to the principle that no performance compensation or power-usage charges should be assessed in advance of real-time operations.

Specifically, the latter markets are two-settlement systems. DAM operations during each day D result in DAM LMP settlements for generators and LSEs at the end of day D; and RTM operations during each day D result in RTM LMP settlements for generators and LSEs at the end of day D. The RTM settlements on day D are adjustments to DAM LMP settlements determined on day D−1; these adjustments are undertaken to correct for deviations between DAM-scheduled dispatch levels for day D and actual dispatch levels realized during day D.

In addition, DAM and RTM LMP settlements can be subject to further adjustments based on ex post LMPs calculated on the basis of actual realized loads. Moreover, further out-of-market (OOM) settlement adjustments are typically undertaken ex post to ensure coverage of actual resource commitment costs.

The lack of settlement time-consistency in current US RTO/ISO-managed wholesale power markets tremendously complicates their operation. This is reflected in the extraordinary number, length, and complexity of the business practice manuals provided by each RTO/ISO to explain market operations in its regional footprint. See, for example, the documentation [49, 50] provided by ERCOT to explain ERCOT's DAM/RTM operations.

10.4 SWING-CONTRACT LONG-TERM FORWARD MARKETS

As illustrated in Figure 10.1, the Linked Swing-Contract Market Design can include one or more LFMs with look-ahead horizons ranging from multiple days to years. The ability to incorporate system constraints with specificity will presumably be low for LFMs, given their relatively long look-ahead horizons. Nevertheless, LFMs incorporating roughly anticipated long-term reserve requirements could still play an important role in encouraging appropriate entry of new resources with long investment lead-times.

This section describes, in general terms, the envisioned operation of an ISO-managed swing-contract market LFM(T) that takes place well in advance (one or more years) of a future operating period T. This market is organized as an ISO-managed exchange for reserve (dispatchable power-paths) for period T. The participants in this market submit offers and bids for reserve for period T, subject to system constraints.

The reserve offers submitted into LFM(T) consist of portfolios of swing contracts, in firm or option form, that offer collections of dispatchable power-paths for possible dispatched delivery during period T. These reserve offers can be submitted by entities with existing power facilities, or by entities planning to develop such facilities in advance of period T.

The offer price associated with each swing contract offered into LFM(T) consists of a requested amortized payment schedule for coverage of avoidable fixed cost between the close of LFM(T) and the start of period T. This avoidable fixed cost is the cost that must be incurred by the offering entity to ensure the availability of its offered power-paths for possible dispatched delivery during period T. Examples include permitting and licensing fees, legal fees, capital investment cost, rental cost, and opportunity cost. As will be discussed more carefully in Section 15.4, the ability of entities contemplating long-term investment in power facilities to submit reserve offers into LFMs that permit the recovery of verified avoidable fixed costs through amortized offer prices could greatly alleviate revenue insufficiency problems in current US RTO/ISO-managed wholesale power markets [168, Sec. III.A].

Reserve bids submitted into LFM(T) consist of fixed or price-sensitive demands for power-path deliveries during period T. Reserve bids can be submitted by Load-Serving Entities (LSEs), but they can also be submitted by the ISO, itself, based on a long-term assessment of period-T reserve needs.

Following the conclusion of the offer/bid submission process for LFM(T), the ISO determines which reserve offers and price-sensitive reserve bids to clear for LFM(T) to achieve its goal of maximizing the total net benefit of market participants subject to system constraints. The portfolio of cleared reserve contracts resulting from LFM(T) can be adaptively updated in subsequent LFMs, in subsequent SFMs held hours in advance of T, in subsequent VSFMs such as might be held during reliability assessment periods immediately prior to the start of period T, and in real-time operations during period T.

10.5 SWING-CONTRACT SHORT-TERM FORWARD MARKETS

No matter how carefully entities plan to meet future power needs through long-term contracts, unexpected events will arise that necessitate adjustments in these plans as operating periods draw near. For the ISO, updated system data could affect anticipated system reserve requirements. For generators, changes in fuel costs and needed maintenance repairs could affect their reserve offers. For LSEs, improved weather data could affect their forecasts for the future power usage of their customers, hence their reserve bids.

The need for a SFM is currently recognized in all seven US RTO/ISO-managed energy regions. Each region supports the operation of a DAM that permits market participants to prepare in advance for next-day real-time operations. As detailed in [76, App., Table 1], six energy regions (CAISO, ERCOT, MISO, NYISO, PJM, SPP) have co-optimized DAMs that determine unit commitments, cleared hourly location-specific amounts of "energy" and "reserve" for next-day operations, and hourly LMPs and *market clearing prices (MCPs)*[4] for these cleared amounts. In the remaining energy region (ISO-NE), the DAM determines cleared hourly location-specific dispatch levels and LMPs for next-day operations; reserve is separately secured in a forward reserve market.

The Linked Swing-Contract Market Design envisions the inclusion of one or more ISO-managed SFMs, at least one of which is a DAM. However, the sole focus of these SFMs is on the procurement of flexible dependable reserve (dispatchable power-paths) for near-term operations.

For example, let SFM(D+1) denote an ISO-managed swing-contract DAM held on day D. The goal of the ISO is to ensure the availability of an adequate amount of reserve to meet net load balancing needs on day D+1, consistent with the maximization of total net benefit for market participants.

The design envisions the ISO as a clearinghouse that enters into SFM(D+1) with an existing portfolio of cleared reserve offers and bids already recorded on its books. The ISO wheels this existing portfolio through SFM(D+1), permitting it to be augmented[5] with additional reserve offers and bids through a contract-clearing optimization process as outlined and illustrated in Chapters 6 and 7.

---

[4]Hourly MCPs for reserve are typically calculated as the dual variables for hourly zone-specific reserve requirement constraints at SCED optimal solution points.

[5]RTO/ISOs managing current US wholesale power markets generally try to avoid de-commitment of previously committed units, as a matter of fairness for their market participants. Similarly, contract breakage by an ISO managing a Linked Swing-Contract Market Design should only be undertaken in emergency conditions. However, it might be desirable to permit a previously cleared contract to be amended prior to real-time operations if all parties to the contract agree to this amendment; for example, this is the practice in ERCOT [50, Sec. 6.4.7]. One possible downside would be increased opportunity for strategic manipulation of the contract submission process. Another possible downside would be the loss of settlement time-consistency. This is a planned topic for future study.

The ISO managing SFM(D+1) is thus able to draw on three different sources to ensure the availability of flexible dependable reserve for operations on day D+1: (i) firm and/or option swing contracts cleared for day D+1 in swing-contract markets held prior to day D: (ii) firm and/or option swing contracts for day D+1 that are newly offered into SFM(D+1) on day D; and (iii) firm and/or option swing contracts for day D+1 that could be cleared in subsequent short-term or very short-term swing-contract markets to be held prior to or during day D+1.

Market SFM(D+1) operates similarly to a standard DAM with regard to the physical obligations imposed on market participants. A dispatchable resource whose reserve offer is cleared for day D+1 must be able to provide dispatched delivery of any one of its offered power-paths during day D+1. An LSE whose price-sensitive reserve bid is cleared for day D+1 must be able to communicate price or power-usage set points to its retail customers during day D+1 to ensure power withdrawals in accordance with any one of the power-usage paths included in this bid.

However, as carefully explained in Section 10.3, the settlement process for the swing-contract market SFM(D+1) differs substantially from the settlement process for a standard DAM. The only new settlements triggered by the optimal clearing of SFM(D+1) are offer price payments for newly cleared reserve offers. Moreover, this optimal clearing does not require adjustments of any previous settlements or settlement obligations.

10.6 SWING-CONTRACT VERY SHORT-TERM FORWARD MARKETS

The Linked Swing-Contract Market Design also envisions the inclusion of at least one ISO-managed VSFM. The purpose of this VSFM is to ensure the availability of sufficient reserve (dispatchable power-paths) for very near-term operations.

Let VSFM(T) denote an ISO-managed swing-contract market conducted in advance of a very near-term operating period T. The goal of the ISO is to maximize the total net benefit of market participants, subject to system constraints that require the procurement of adequate reserve to ensure net load balancing needs are met during period T. Market participants include dispatchable resources able to submit newly issued reserve offers for period T and LSEs able to submit newly issued price-sensitive reserve bids for period T.[6]

As for the SFMs discussed in Section 10.5, the ISO enters into VSFM(T) with an existing portfolio of cleared reserve offers and cleared price-sensitive reserve bids already recorded on its books. The ISO wheels this existing portfolio

[6]Existing US RTO/ISO-managed wholesale power markets do not permit LSEs to submit fixed (nonprice sensitive) demand bids into very short-term planning operations, such as RTMs or reliability assessment commitment (RAC) processes taking place immediately in advance of an operating period. ISO load forecasts are used in place of LSE fixed demand bids. This feature is retained for the linked swing-contract market design.

through VSFM(T), permitting it to be augmented with additional reserve offers and price-sensitive reserve bids through a contract-clearing optimization process as outlined and illustrated in Chapters 6 and 7.

Market VSFM(T) operates similarly to a standard RTM with regard to the physical obligations imposed on market participants. A dispatchable resource whose submitted reserve offer is cleared for period T must be able to provide dispatched delivery during period T of any one of its offered power-paths. An LSE whose price-sensitive reserve bid is cleared for period T must be able to communicate price or power-usage set points to its retail customers during period T to ensure power withdrawals in accordance with any one of the power-usage paths included in this bid.

However, the settlement process for VSFM(T) differs substantially from the settlement process for a standard RTM. As carefully explained in Section 10.3, the only new settlements triggered by the optimal clearing of VSFM(D+1) are offer price payments for newly cleared reserve offers. Moreover, this optimal clearing does not require adjustments of any previous settlements or settlement obligations.

10.7 SWING-CONTRACT DEPLOYMENT IN REAL-TIME OPERATIONS

This section provides a general description of the envisioned real-time operations of the Linked Swing-Contract Market Design.

For simplicity of exposition, the description ignores critically important network analysis, *supervisory control and data acquisition (SCADA)*, and weather forecasting activities that must be undertaken during real-time operations to maintain up-to-date estimates for system operating conditions and up-to-date forecasts for must-service loads and nondispatched generation across the transmission grid.[7] The description also assumes that the transmission grid is a stand-alone electrical infrastructure that is not seamed with grids from other energy regions.

The goal of the ISO managing the Linked Swing-Contract Market Design is to procure and deploy dispatchable resources in an efficient manner to ensure continuous power balance across the transmission grid, subject to system constraints. This power balance requires power injection to match power withdrawal (extraction and losses) at each point in time. The availability of dispatchable power injections and withdrawals is determined by the existing portfolio of cleared reserve offers and cleared price-sensitive reserve bids. *Dispatch efficiency* is measured in terms of resource nonwastage in two senses: physical nonwastage; and value nonwastage.[8]

In pursuit of this goal for a real-time operating period T, the ISO conducts a final very short-term forward market VSFM(T) in advance of T; see Figure 10.2.

[7] See, for example, the discussion of ERCOT real-time activities provided in [50, Sec. 6.5.7.1].

[8] In economics, value nonwastage is measured by "Pareto efficiency." For the case at hand, a feasible dispatch of resources is *Pareto efficient* if there exists no other feasible dispatch of these resources such that all market participants are at least as well off and at least one market participant is strictly better off.

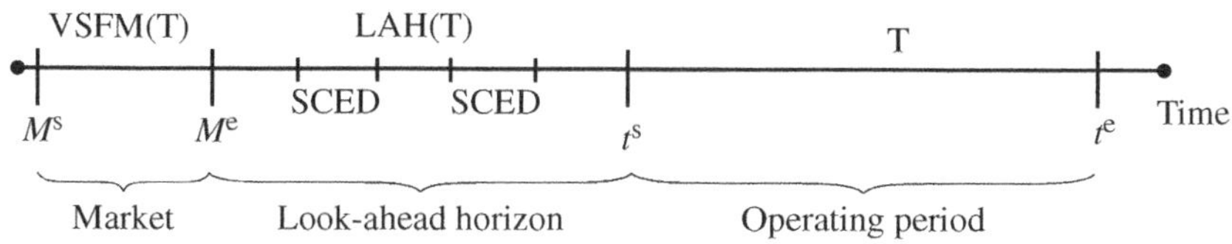

Figure 10.2 Real-time operations for the Linked Swing-Contract Market Design.

The formulation of this market includes the ISO's current estimates for system operating conditions at the start of T and for net must-service load and power losses during T.

The optimal solution for VSFM(T) determines the final portfolio ISOPort(T) of reserve offers and price-sensitive reserve bids to be cleared for period T through market operations. It also determines a *SCED optimal solution* for period T that includes an anticipated period-T dispatch schedule for cleared reserve offers and an anticipated period-T power-usage schedule for cleared price-sensitive reserve bids.

The ISO then conducts one or more SCED optimizations during the look-ahead horizon LAH(T) for VSFM(T), conditional on ISOPort(T). These SCED optimizations incorporate successively improved estimates for system operating conditions at the start of period T and for net must-service load and power losses during period T. These SCED optimizations determine optimal solutions for period T, conditional on ISOPort(T) and these estimates.

Let SCED(T) denote the final SCED conducted by the ISO during LAH(T), and let OptSCED(T) denote the optimal solution for SCED(T). At the conclusion of SCED(T), the ISO implements OptSCED(T) in approximate form by means of set points communicated to the dispatchable resources and LSEs whose cleared reserve offers and cleared price-sensitive reserve bids are included in ISOPort(T). Any remaining power imbalance during period T is handled by primary frequency control at the local level.

The physical implementation of real-time operations for the Linked Swing-Contract Market Design is similar to the physical implementation of real-time operations in current US RTO/ISO-managed wholesale power markets. Both depend on real-time telemetry to implement an ultimately determined SCED solution for an operating period T, conditional on available dispatchable resources.

Where significant advantage occurs is in the real-time settlement process. For the Linked Swing-Contract Market Design, the only settlements pertaining to an operating period T that occur prior to period T are offer price payments to the dispatchable resources with cleared reserve offers included in ISOPort(T). The only settlements pertaining to period T that occur subsequent to period T are as follows:

- Performance compensation for power-path deliveries, paid ex post to dispatchable resources with cleared reserve offers in ISOPort(T) in accordance with the performance payment methods these resources have included within their offers;

- Demand charges for power usage, paid ex post by LSEs with cleared price-sensitive reserve bids in ISOPort(T) in accordance with the willingness-to-pay information these LSEs have included within their bids;
- ISO allocation of net reserve cost to net must-service loads, and transmission service cost to transmission-using market participants, as described in Section 6.7.

In contrast, the real-time settlement processes for current US RTO/ISO-managed wholesale power markets are extraordinarily complicated. For example, ERCOT's real-time settlements include: various types of LMPs and LMP-based settlement prices calculated ex ante and ex post for buses, zones, and hubs; payments for variously defined ancillary services; and OOM payments to compensate for various types of required reliability services (e.g. reliability must-run obligations). The description of these real-time settlements provided in Sections 6.6 and 6.7 of the ERCOT manual [50, pp. 119-183] is 64 pages long.

CHAPTER 11

ILLUSTRATION: LINKED DAY-AHEAD AND HOUR-AHEAD SWING-CONTRACT MARKETS

11.1 CHAPTER PREVIEW

The previous chapter describes in general terms how the Linked Swing-Contract Market Design can facilitate efficient net load balancing in real-time operations. The linkages among ISO-managed swing-contract markets with look-ahead horizons ranging from multiple years to minutes are to be achieved by having the ISO carry forward on its books an adaptively updated portfolio of all reserve offers and bids cleared in earlier swing-contract markets.

This chapter presents a numerical example, adapted from [76, 77], that illustrates this linkage for the special case of a day-ahead swing-contract market and an hour-ahead swing-contract market whose linked timing configurations are depicted in Figures 11.1 and 11.2. An additional purpose of this numerical example is to illustrate how the reserve offers submitted into swing-contract markets by dispatchable resources can consist of swing-contract *portfolios*.

11.2 HOUR-AHEAD MARKET WITH RESERVE OFFERS CONSISTING OF SWING-CONTRACT PORTFOLIOS

For simplicity of exposition, suppose the ISO managing the hour-ahead swing-contract market VSFM(H) anticipates that transmission congestion and power losses will be negligible during hour H. Suppose, also, that there are no LSE price-sensitive reserve bids cleared for H in any previous swing-contract market, including the day-ahead swing-contract market SFM(D+1).[1]

[1] The presence of transmission congestion, power losses, and/or LSE price-sensitive reserve bids is not critical for the purposes of this chapter.

A New Swing-Contract Design for Wholesale Power Markets, First Edition, Leigh Tesfatsion.

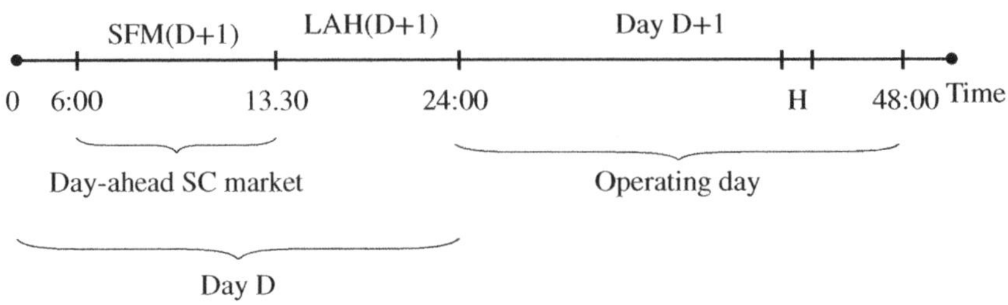

Figure 11.1 A day-ahead swing-contract market SFM(D+1) linked to operating hour H on day D+1.

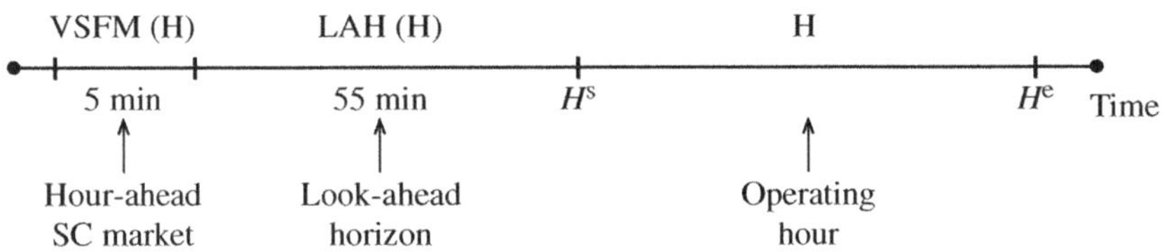

Figure 11.2 An hour-ahead swing-contract market VSFM(H) for operating hour H on day D+1.

The participants in VSFM(H) include (i) dispatchable generators that submit reserve offers for H and (ii) non-dispatchable variable energy resources whose *non-dispatchable generation (NDG)* produced during H is fully injected into the transmission grid. This NDG for H is treated as negative must-service[2] load. The ISO's *forecasted net must-service load* for H, formulated at the start of VSFM(H), consists of forecasted must-service load minus forecasted NDG. This forecasted net must-service load is depicted in Figure 11.3.

The ISO's objective is to ensure that its *uncertainty set* U(H) for net must-service load during hour H can feasibly be covered by a suitable combination of existing and newly cleared reserve offers at minimum possible cost. The uncertainty set U(H) is a corridor around the ISO's hour-H forecasted net must-service load profile depicted in Figure 11.3. This corridor is characterized by system-wide down/up reserve requirements RRD(H) and RRU(H) for hour H, expressed as decimal percentages of forecasted net must-service load.

More precisely, let $\widehat{NL}^f(\mathrm{M})$ denote the ISO's forecasted net must-service load (MW) at the start of minute M during hour H. Then, $\mathrm{U(H)} = \{\mathrm{U(M)} \mid \mathrm{M} \in \mathrm{H}\}$, where the minute-M uncertainty set U(M) consists of the range of power levels $p(\mathrm{M})$ satisfying

$$[1 - \mathrm{RRD(H)}] \cdot \widehat{NL}^f(\mathrm{M}) \leq p(\mathrm{M}) \leq [1 + \mathrm{RRD(H)}] \cdot \widehat{NL}^f(\mathrm{M}) \qquad (11.1)$$

[2]Recall that must-service load, while not price sensitive, can be both uncertain and variable.

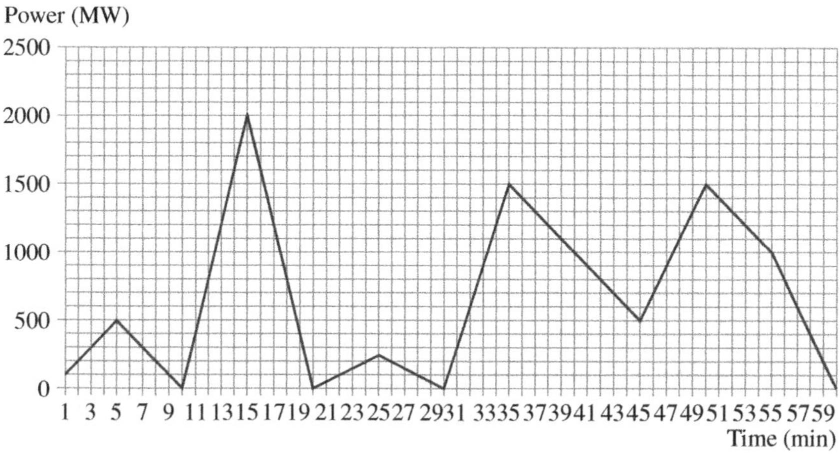

Figure 11.3 ISO-forecasted net must-service load profile for hour H of day D+1 at the start of VSFM(H), expressed at the granularity of minutes. Source: [77].

To achieve this objective, the ISO undertakes a SCED optimization for VSFM(H), expressed at the granularity of minutes. This optimization is conditioned on: (i) system operating conditions at the start of VSFM(H); (ii) all reserve offers submitted into VSFM(H) that were cleared for H in previous swing-contract markets; (iii) all newly issued reserve offers submitted into VSFM(H); and (iv) ISO-forecasted net must-service load for H, as depicted in Figure 11.3. The system constraints for each $M \in H$ include a system-wide power balance constraint requiring total anticipated power dispatch to equal total forecasted net must-service load at the start of M; and the system-wide down/up reserve requirement constraint (11.1) for M.

The dispatchable generators participating in VSFM(H) consist of three *Generation Companies (GenCos)*, denoted by G1, G2, and G3. These three GenCos have the following down/up ramp-rate limits and min/max power capacity limits:

$$\text{G1}: r_1^D = r_1^U = 120 \text{ MW/min}, \ P_1^{\text{min}} = 0 \text{ MW}, \ P_1^{\text{max}} = 600 \text{ MW}$$
$$\text{G2}: r_2^D = r_2^U = 200 \text{ MW/min}, \ P_2^{\text{min}} = 0 \text{ MW}, \ P_2^{\text{max}} = 700 \text{ MW}$$
$$\text{G3}: r_3^D = r_3^U = 300 \text{ MW/min}, \ P_3^{\text{min}} = 0 \text{ MW}, \ P_3^{\text{max}} = 900 \text{ MW}$$

The reserve offer submitted by each GenCo Gi into VSFM(H) consists of a finite number of swing-contract portfolios, called *GenPorts*. The jth GenPort submitted by GenCo Gi, denoted by $\text{GenPort}_{i,j}$, includes a non-negative offer price $\alpha_{i,j}$ together with a finite number of swing contracts for hour H.[3] The ISO can clear at most one of the GenPorts included in GenCo Gi's reserve offer. GenCo Gi must then

[3]If $\text{GenPort}_{i,j}$ consists of only one swing contract, its offer price $\alpha_{i,j}$ reduces to the usual swing-contract offer price. Otherwise, if $\text{GenPort}_{i,j}$ consists of multiple swing contracts $\text{SC}_{i,j,u}$, a single offer price $\alpha_{i,j}$

be able to simultaneously fulfill the terms of all of the swing contracts included within this cleared GenPort.

For simplicity of exposition, the swing contracts comprising each submitted GenPort are assumed to take a firm form[4] with a single offered time-of-use performance price ϕ (\$/MWh). The delivery start-time t^s and end-time t^e for each swing contract are expressed at the granularity of minutes $\{1, \ldots, 60\}$ for hour H. Omitting references to delivery locations[5] and exercise time, the reserve offers submitted into VSFM(H) by GenCos G1, G2, and G3 for hour H of day D+1 are as follows[6]:

G1's reserve offer consists of a single GenPort previously cleared in SFM(D+1):

$$\begin{aligned} \text{GenPort}_{1,1} &= \{\alpha_{1,1}, \text{SC}_1\}, \text{ where} \\ \text{SC}_1 &= [t^s = 0,\ t^e = 60,\ |r| \le 100,\ 0 \le p \le 500,\ \phi = 100] \end{aligned} \tag{11.2}$$

G2's reserve offer consists of three newly issued GenPorts, where each GenPort includes multiple swing contracts:

$$\begin{aligned} \text{GenPort}_{2,1} &= \{\alpha_{2,1}, \text{SC}_{2,1,1}, \text{SC}_{2,1,2}\}, \text{ where} \\ \text{SC}_{2,1,1} &= [t^s = 10,\ t^e = 20,\ |r| \le 200,\ 0 \le p \le 600,\ \phi = 135] \\ \text{SC}_{2,1,2} &= [t^s = 30,\ t^e = 60,\ |r| \le 200,\ 0 \le p \le 600,\ \phi = 130] \end{aligned} \tag{11.3}$$

$$\begin{aligned} \text{GenPort}_{2,2} &= \{\alpha_{2,2}, \text{SC}_{2,2,1}, \text{SC}_{2,2,2}, \text{SC}_{2,2,3}\}, \text{ where} \\ \text{SC}_{2,2,1} &= [t^s = 0,\ t^e = 10,\ |r| \le 100,\ 0 \le p \le 100, \phi = 105] \\ \text{SC}_{2,2,2} &= [t^s = 10,\ t^e = 20,\ |r| \le 200,\ 0 \le p \le 600, \phi = 135] \\ \text{SC}_{2,2,3} &= [t^s = 30,\ t^e = 60,\ |r| \le 200,\ 0 \le p \le 600, \phi = 130] \end{aligned} \tag{11.4}$$

$$\begin{aligned} \text{GenPort}_{2,3} &= \{\alpha_{2,3}, \text{SC}_{2,3,1}, \text{SC}_{2,3,2}, \text{SC}_{2,3,3}\}, \text{ where} \\ \text{SC}_{2,3,1} &= [t^s = 0,\ t^e = 10,\ |r| \le 100,\ 0 \le p \le 100,\ \phi = 105] \\ \text{SC}_{2,3,2} &= [t^s = 10,\ t^e = 20,\ |r| \le 200,\ 0 \le p \le 700,\ \phi = 140] \\ \text{SC}_{2,3,3} &= [t^s = 30,\ t^e = 60,\ |r| \le 200,\ 0 \le p \le 700,\ \phi = 135] \end{aligned} \tag{11.5}$$

---

appears in place of individual swing-contract offer prices $\alpha_{i,j,u}$. This single offer price can be used by GenCo Gi to ensure coverage of all avoidable fixed cost that Gi would have to incur to guarantee the *joint* availability of all net load balancing services offered for hour H by the multiple swing contracts $\text{SC}_{i,j,u}$ included in $\text{GenPort}_{i,j}$.

[4]Recall from Section 4.3 that a swing contract submitted into a swing-contract market is called *firm* if it has a single exercise time that coincides with the close of this market.

[5]Since absence of transmission congestion is assumed for operating hour H, the precise grid locations for anticipated power deliveries are irrelevant.

[6]Note that the terms of the swing contracts included within each submitted GenPort can be simultaneously fulfilled because their offered hour-H delivery subintervals $[t^s, t^e)$ are disjoint.

G3's reserve offer consists of two newly-issued GenPorts, where each GenPort includes three swing contracts:

$$
\begin{aligned}
\text{GenPort}_{3,1} = &\{\alpha_{3,1}, \text{SC}_{3,1,1}, \text{SC}_{3,1,2}, \text{SC}_{3,1,3}\}, \text{ where} &(11.6)\\
\text{SC}_{3,1,1} &= [t^s = 10,\ t^e = 20,\ |r| \leq 300,\ 0 \leq p \leq 900,\ \phi = 175]\\
\text{SC}_{3,1,2} &= [t^s = 33,\ t^e = 39,\ |r| \leq 200,\ 0 \leq p \leq 400,\ \phi = 155]\\
\text{SC}_{3,1,3} &= [t^s = 48,\ t^e = 54,\ |r| \leq 200,\ 0 \leq p \leq 400,\ \phi = 155]\\
\text{GenPort}_{3,2} = &\{\alpha_{3,2}, \text{SC}_{3,2,1}, \text{SC}_{3,2,2}, \text{SC}_{3,2,3}\}, \text{ where} &(11.7)\\
\text{SC}_{3,2,1} &= [t^s = 10,\ t^e = 20,\ |r| \leq 300,\ 0 \leq p \leq 900,\ \phi = 175]\\
\text{SC}_{3,2,2} &= [t^s = 30,\ t^e = 39,\ |r| \leq 200,\ 0 \leq p \leq 400,\ \phi = 150]\\
\text{SC}_{3,2,3} &= [t^s = 44,\ t^e = 54,\ |r| \leq 200,\ 0 \leq p \leq 400,\ \phi = 150]
\end{aligned}
$$

As previously noted, the non-negative offer price $\alpha_{i,j}$ for each $\text{GenPort}_{i,j}$ is the payment requested by GenCo Gi for guaranteeing the hour-H availability of the dispatchable power-paths characterized by the terms of the swing contracts comprising $\text{GenPort}_{i,j}$. If the ISO clears $\text{GenPort}_{i,j}$, GenCo Gi must be paid the offer price $\alpha_{i,j}$.

In addition, if $\text{GenPort}_{i,j}$ is cleared, GenCo Gi must be compensated ex post for any ISO-instructed delivery of power during hour H in accordance with the contractual terms of the swing contracts comprising $\text{GenPort}_{i,j}$. This compensation is determined by the performance payment methods ϕ included among the contractual terms of the swing contracts in $\text{GenPort}_{i,j}$. For the case at hand, each of these performance payment methods takes the form of a pre-specified price (\$/MWh) for delivered power during hour H.[7]

The ISO is permitted to clear at most one GenPort from each GenCo participating in the SCED optimization for VSFM(H). The resulting cleared GenPorts can thus be represented in the form of an *optimal ISO Portfolio (ISOPort)*, as follows:

$$
\text{ISOPort}^* = \{\text{GenPort}_1^*, \text{GenPort}_2^*, \text{GenPort}_3^*\} \tag{11.8}
$$

The ISO is obligated to clear $\text{GenPort}_{1,1}$ submitted by GenCo G1, since $\text{GenPort}_{1,1}$ was previously cleared for hour H in SFM(D+1); thus, $\text{GenPort}_1^* = \text{GenPort}_{1,1}$. However, the ISO can decide not to clear any GenPort from GenCo G2 and/or GenCo G3, i.e. $\text{GenPort}_k^* = \text{None}$ is possible for $k = 2$ and/or $k = 3$.

[7]For example, each $\text{SC}_{i,j,u}$ in $\text{GenPort}_{i,j}$ could correspond to a distinct generation unit u owned by Gi, and the performance price $\phi_{i,j,u}$ included in $\text{SC}_{i,j,u}$ could be the anticipated marginal generation cost (\$/MWh) for unit u during hour H.

11.3 SCED SOLUTION FOR HOUR-AHEAD SWING-CONTRACT MARKET

11.3.1 Overview

Given a specific numerical value for the offer price $\alpha_{i,j}$ of each submitted $\text{GenPort}_{i,j}$, a SCED solution for VSFM(H) can be computationally determined using the SCED portion of the analytical SCUC/SCED optimization formulation developed in Chapter 7 for the base-case swing-contract market M(T) with operating period T set equal to H. Here, instead, the SCED solution is characterized in graphical terms for increased intuitive understanding.

11.3.2 Power Balance

The form of the SCED optimization for VSFM(H) described in Section 11.2 assures that any optimal ISOPort cleared by the ISO for VSFM(H) must permit the achievement of a *zero balance gap (ZBG)* for operating hour H. That is, this optimal ISOPort must permit an exact balancing of cleared dispatchable generation against the ISO's forecasted net must-service load profile depicted in Figure 11.3.

For example, Figures 11.4 and 11.5 show how each of the following two ISOPorts enables the achievement of a ZBG:

$$\text{ISOPort}_1 = \{\text{GenPort}_{1,1}, \text{GenPort}_{2,2}, \text{GenPort}_{3,1}\} \tag{11.9}$$

$$\text{ISOPort}_2 = \{\text{GenPort}_{1,1}, \text{GenPort}_{2,3}, \text{GenPort}_{3,1}\} \tag{11.10}$$

These figures indicate the dispatch of generation from particular GenPorts for particular GenCos.

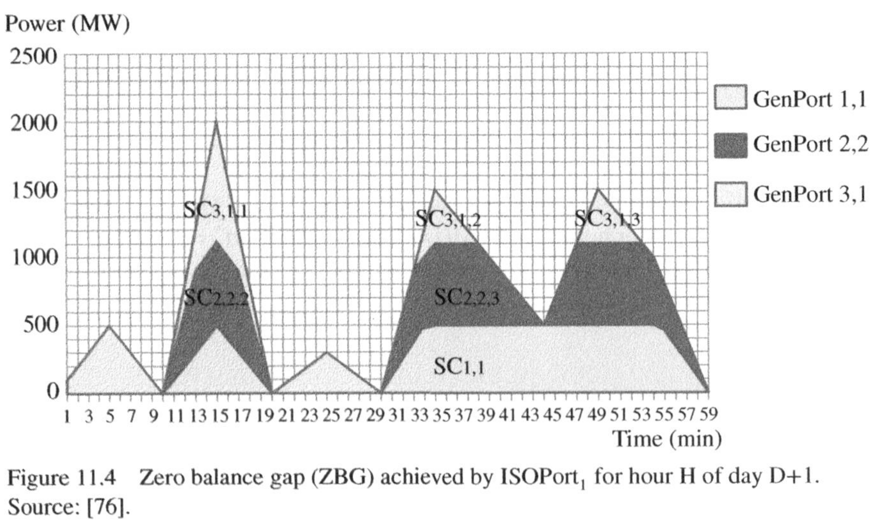

Figure 11.4 Zero balance gap (ZBG) achieved by ISOPort_1 for hour H of day D+1. Source: [76].

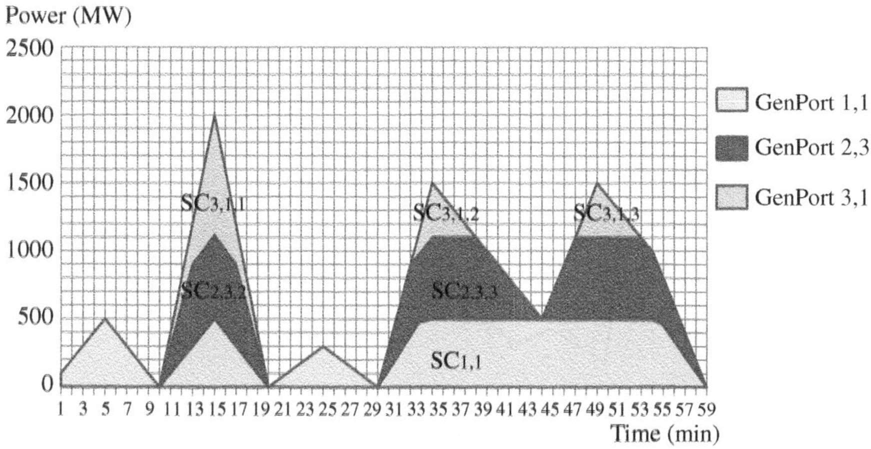

Figure 11.5 Zero balance gap (ZBG) achieved by ISOPort_2 for hour H of day D+1. Source: [77].

11.3.3 Coverage of the ISO's Uncertainty Set

Suppose an ISOPort submitted into VSFM(H) achieves a ZBG for H, conditional on the ISO's hour-H forecasted net must-service load profile depicted in Figure 11.3. Since the swing contracts constituting this ISOPort offer swing (flexibility) in their attributes, the ISOPort can also achieve a ZBG for a range of hour-H net must-service load profiles that deviate from this ISO forecast. Hereafter, this range will be called the *inherent reserve range (IRR)* of the ISOPort for hour H.

The IRR of an ISOPort with swing in its contractual terms is a robust-control device for net load balancing, eliminating the need for the ISO to consider detailed net load scenarios and scenario probabilities. The optimal ISOPort's IRR must cover the ISO's uncertainty set U(T) for operating period T, by construction, since U(T) is characterized by reserve requirement constraints appearing among the system constraints for the SCED optimization for VSFM(H). However, the exact form of the optimal ISOPort's IRR depends in a complicated manner on the particular attribute specifications of the swing contracts that constitute this ISOPort as well as on the minute-by-minute operating state of the dispatchable resources that have offered these swing contracts. See, e.g. the analytically expressed run-time form (7.30) of the IRR for the base-case swing-contract market M(T) presented in Chapter 7.[8]

Consequently, prior to obtaining the SCED-optimal solution for VSFM(H), the IRR associated with each possible ISOPort for VSFM(H) can only be approximated.

[8]For this base-case swing-contract market M(T), each submitted reserve offer consists of a single swing contract submitted by a dispatchable resource for operating period T; hence, a cleared ISOPort for M(T) consists of at most one submitted swing contract from each participating dispatchable resource.

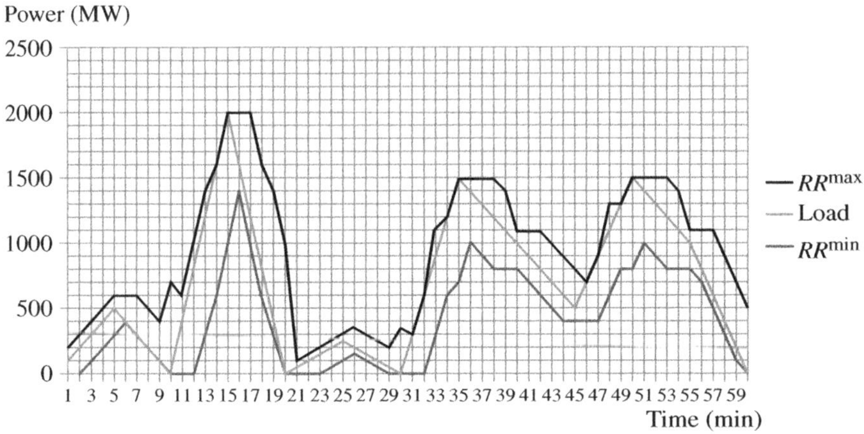

Figure 11.6 Approximate inherent reserve range of ISOPort_1 for hour H of day D+1. Source: [76].

Figures 11.6 and 11.7 plot approximate IRRs for ISOPort_1 and ISOPort_2 in (11.9) and (11.10) by assuming that each GenCo at the start of each minute M during hour H is as its ZBG-generation level. The depicted approximate IRRs in Figures 11.6 and 11.7 for ISOPort_1 and ISOPort_2 are conditional on the ISO's forecasted hour-H net must-service load profile shown in Figure 11.3 and on the ISO's hour-H ZBG implementations for ISOPort_1 and ISOPort_2 as shown in Figures 11.4 and 11.5.

More precisely, the approximate IRR depicted in Figure 11.7 for ISOPort_2 was derived by means of the following steps, applicable for any ZBG ISOPort. At the start of each minute M of hour H, calculate the minimum and maximum power levels

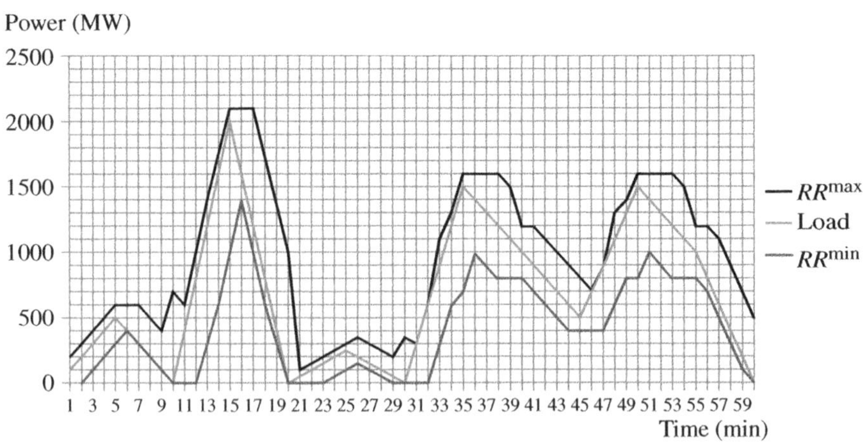

Figure 11.7 Approximate inherent reserve range of ISOPort_2 for hour H of day D+1. Source: [77].

$RR_M^{\min}$ and $RR_M^{\max}$ that could be attained at the end of minute M. These minimum and maximum power levels depend on (i) the contractual terms of the swing contracts constituting ISOPort_2; (ii) the particular ZBG implementation of ISOPort_2 for hour H; and (iii) the ZBG operating state at the start of each minute M of hour H for each GenCo with a reserve offer included in ISOPort_2.

Specifically, for each GenCo supplier Gi, and for each minute M during the operating hour H, let $\text{Gen}_{i,M}$ denote the ZBG generation level (MW) of Gi at the start of M. Also, let $r^D_{i,M}/r^U_{i,M}$ denote the down/up ramp-rate limits (MW/min) for Gi during M, and let $P^{\min}_{i,M}/P^{\max}_{i,M}$ denote the min/max power limits (MW) for Gi at the end of M. Then, the lower and upper bounds on the power levels that could be delivered by Gi at the end of M, conditional on its ZBG state at the start of M, are given by

$$PI^L_{i,M} = \max\ \{\text{Gen}_{i,M} - r^D_{i,M},\ P^{\min}_{i,M}\} \tag{11.11}$$

$$PI^U_{i,M} = \min\ \{\text{Gen}_{i,M} + r^U_{i,M},\ P^{\max}_{i,M}\} \tag{11.12}$$

The minimum power level $RR_M^{\min}$ attainable for the system as a whole at the end of minute M can be approximated by summing the lower power bounds (11.11) across the GenCo set $\text{G}_H = \{\text{G1}, \text{G2}, \text{G3}\}$ for hour H. Similarly, the maximum power level $RR_M^{\max}$ attainable for the system as a whole at the end of minute M can be approximated by summing the upper power bounds (11.12) across the GenCo set G_H for hour H. The inherent reserve range IRR(M) at the end of minute M is then approximately given by the power-level interval between these summed lower and upper bounds:

$$\text{IRR(M)} = [RR_M^{\min},\ RR_M^{\max}] = \left[\sum_{Gi \in \text{G}_H} PI^L_{i,M},\ \sum_{Gi \in \text{G}_H} PI^U_{i,M}\right] \tag{11.13}$$

Finally, the IRR over the entire hour H, expressed at the granularity of minutes, is approximately given by

$$\text{IRR(H)} = \{\text{IRR(M)} \mid \text{M} \in \text{H}\} \tag{11.14}$$

This is the method used to derive the plots in Figures 11.6 and 11.7 for the approximate hour-H IRRs for ISOPort_1 and ISOPort_2 described in (11.9) and (11.10).

The GenCos can seek compensation for the IRR characteristics of their offered GenPorts through their GenPort offer prices and through the performance prices ϕ appearing among the contractual terms of their GenPorts. This includes, in particular, compensation for any down/up power needed to balance deviations between ISO-forecasted and actual net must-service loads for hour H.

11.3.4 Constrained Minimization of Expected Cost

The expected cost of any ISOPort = (GenPort_1, GenPort_2, GenPort_3) submitted into VSFM(H) that achieves a ZBG for operating hour H and covers the ISO's uncertainty set U(H) consists of an availability payment for ensuring the availability of dispatchable power during H and an expected performance payment for power-path delivery during H.

The availability payment for the ISOPort is determined by the portfolio offer prices $\{\alpha_1, \alpha_2, \alpha_3\}$ that must be paid to GenCos G1, G2, and G3 for the procurement of GenPort_1, GenPort_2, and GenPort_3. The expected payment for power-path delivery calculated for the ISOPort depends on the ISO's net must-service load forecast for operating hour H and on the performance payment methods included among the swing contracts that comprise the GenPorts included in ISOPort.

For example, to calculate the expected performance payment implied by the ZBG implementation of ISOPort_2 depicted in Figure 11.5, one could first measure the energy (MWh) for each of the areas in Figure 11.5 with a distinct color shading; each such area corresponds to a distinct swing-contract implementation. Next, multiply each of these energy amounts by the performance price ϕ (\$/MWh) included among the contractual terms of the corresponding swing contract. Finally, add up all of these amounts.

11.4 LINKED DAY-AHEAD AND HOUR-AHEAD MARKETS

This section considers more carefully the prior operation of the day-ahead swing-contract market SFM(D+1) depicted in Figure 11.1 in relation to the operation of the hour-ahead swing-contract market VSFM(H) depicted in Figure 11.2. For notational simplicity, SFM(D+1) will hereafter be referred to as the *day-ahead market (DAM)* and VSFM(H) will hereafter be referred to as the *real-time market (RTM)*.

RTM operations conditioned on a single previously cleared GenPort were illustrated in Sections 11.2 and 11.3. This RTM illustration will now be extended to show more carefully how RTM operations are linked to previous DAM operations. For simplicity of exposition, the following two assumptions are maintained for the DAM. First, the reserve bid submitted into the DAM by each LSE j takes a fixed block-energy form; i.e. it consists of a single constant power-usage level $p_j^f(H)$ (MW) for the entire hour H. Second, the ISO managing the DAM anticipates that transmission congestion and power losses will be negligible during hour H.

Let $L_H^{NF,\text{DAM}}$ denote the DAM net must-service load forecast for hour H, given by aggregate DAM LSE reserve bids for hour H minus the ISO's day-ahead forecast for NDG during hour H. Suppose the DAM reserve requirement constraints for hour H take form (11.1) with 10% system-wide down/up reserve requirements; i.e. RRD(H) = RRU(H) = 0.10. The DAM uncertainty set U(H) for hour H thus consists of a corridor around $L_H^{NF,\text{DAM}}$ whose lower and upper bounds are determined as 90% and 110% of $L_H^{NF,\text{DAM}}$, respectively. This corridor, hereafter referred to as the *DAM 10% power corridor*, is depicted in Figure 11.8.

As illustrated in Figure 11.9, the DAM net must-service load forecast $L_H^{NF,\text{DAM}}$ will typically differ from the net must-service load forecast $L_H^{NF,\text{RTM}}$ subsequently

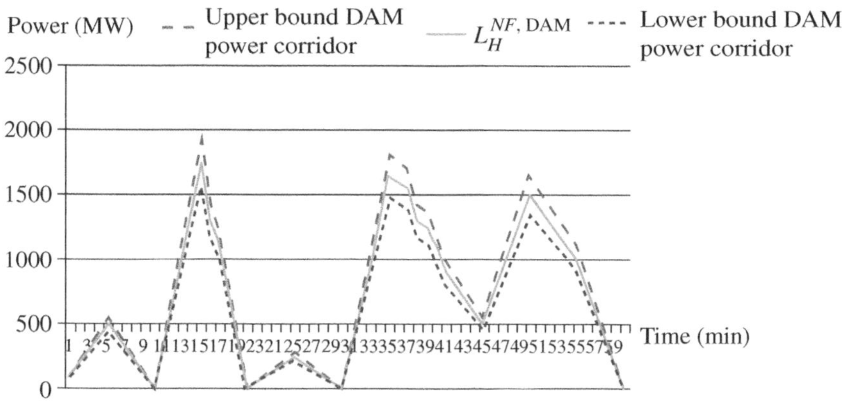

Figure 11.8 The DAM 10% power corridor for hour H of day D+1, conditional on the DAM net must-service load forecast $L_H^{NF,DAM}$ for hour H of day D+1. Source: [76].

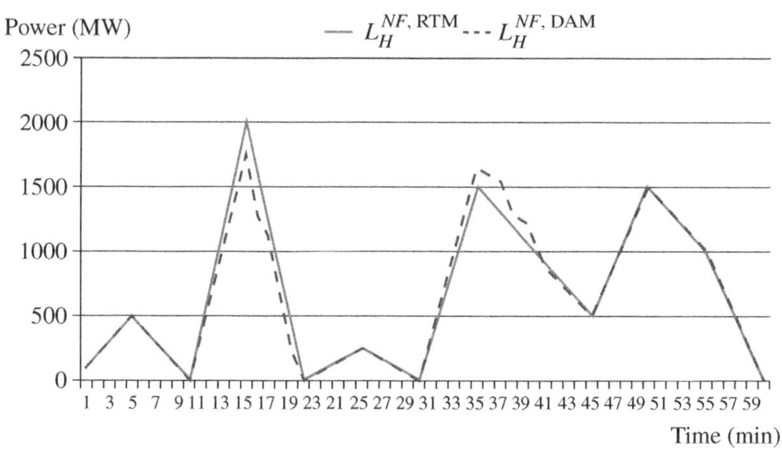

Figure 11.9 RTM forecasted net must-service load $L_H^{NF,RTM}$ vs. DAM forecasted net must-service load $L_H^{NF,DAM}$ for hour H of day D+1. Source: [76].

formulated by the ISO at the start of the RTM.[9] For example, net must-service load could be affected by uncertain weather conditions, and the ISO could have improved information about these weather conditions at the time of the RTM relative to the information available to the ISO at the time of the DAM.

[9]Note that the RTM net must-service load profile $L_H^{NF,RTM}$ in Figure 11.9 coincides with the RTM net must-service load profile earlier depicted in Figure 11.3.

The ISO's RTM objective is to ensure a least-cost ZBG for hour H of day D+1, conditional on system operating conditions, reserve requirement constraints, and the ISO's updated net load forecast $L_H^{NF,\text{RTM}}$ for H. Suppose the RTM reserve requirement constraints take form (11.1) with RRD(H) = RRU(H) = 0.05.[10] The ISO must then ensure that reserve offers are cleared with sufficient swing (flexibility) in their offered attributes that they are capable of covering a corridor of potential net must-service load profiles around the ISO's forecasted net must-service load profile $L_H^{NF,\text{RTM}}$ whose lower and upper bounds are determined as 95% and 105% of $L_H^{F,\text{RTM}}$, respectively. Call this corridor the *RTM 5% power corridor*.

The gap $G_M^L = [PC_M^{L,\text{RTM}} - PC_M^{L,\text{DAM}}]$ between the lower bound $PC_M^{L,\text{RTM}}$ of the RTM 5% power corridor and the lower bound $PC_M^{L,\text{DAM}}$ of the DAM 10% power corridor for minute M of hour H determines the down-power amount P_M^{down} that the ISO anticipates it will need to procure in the RTM in order to ensure power balance for minute M of hour H. Specifically, $P_M^{\text{down}} = \min\{G_M^L, 0\}$.

Similarly, the gap $G_M^U = [PC_M^{U,\text{RTM}} - PC_M^{U,\text{DAM}}]$ between the upper bound $PC_M^{U,\text{RTM}}$ of the RTM 5% power corridor and the upper-bound $PC_M^{U,\text{DAM}}$ of the DAM 10% power corridor for minute M of hour H determines the up-power amount P_M^{up} that the ISO anticipates it will need to procure in the RTM in order to ensure power balance for minute M of hour H. Specifically, $P_M^{\text{up}} = \max\{G_M^U, 0\}$.

Figure 11.10 depicts the RTM down/up power requirements P_M^{down} and P_M^{up} implied by the lower and upper bounds $PC_M^{L,\text{RTM}}$, $PC_M^{L,\text{DAM}}$, $PC_M^{U,\text{RTM}}$, and $PC_M^{U,\text{DAM}}$

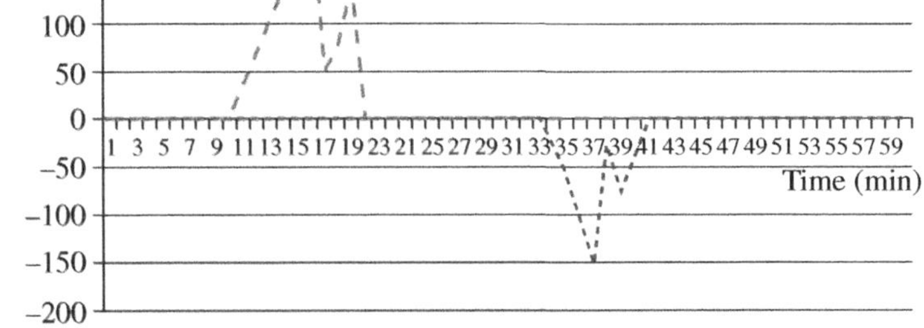

Figure 11.10 Anticipated RTM down/up power procurement needed to balance net load during hour H of day D+1, given an RTM 5% power corridor and a DAM 10% power corridor. Source: [76].

---

[10]Net load forecast errors for the RTM can be expected to be smaller than net load forecast errors for the DAM, since the RTM forecasts are hour-ahead rather than day-ahead. This presumption is reflected in the specification of smaller RTM down/up reserve requirements RRD(H) and RRU(H).

for each minute M of hour H. Note that no RTM down-power procurement is needed during minutes 10–20 of hour H because $PC_M^{L,DAM} < PC_M^{L,RTM} \leq 0$ over this time interval. On the other hand, RTM up-power procurement is needed during minutes 10–20 of hour H because $PC_M^{U,RTM} > PC_M^{U,DAM} \geq 0$ during this time interval.

Thus, the RTM is intrinsically linked to the previous operation of the DAM. The reserve offers cleared in the DAM for hour H ensure coverage of a DAM uncertainty set expressed as a DAM 10% power corridor around a DAM net must-service load forecast for H. These cleared reserve offers might not be sufficient to provide complete coverage of the subsequent RTM uncertainty set, expressed as an RTM 5% power corridor around an updated RTM net must-service load forecast for hour H. The clearing of any newly offered reserve offers in the RTM is then conditioned on the availability of these previously cleared reserve offers.

More generally, a very short-term swing-contract market VSFM(H) for an operating hour H is conditioned on a contract inventory consisting of all reserve offers and price-sensitive reserve bids cleared for H in previously conducted swing-contract markets. VSFM(H) is a balancing mechanism permitting this contract inventory to be supplemented as needed with newly cleared reserve offers and price-sensitive reserve bids to ensure net load balancing for hour H. As illustrated in Figure 11.10, the size of the trade volume in VSFM(H) can be very small; it will tend to vary inversely with the amount of swing in the contract inventory.

CHAPTER 12

STANDARD MODELING OF A COMPETITIVE MARKET

12.1 CHAPTER PREVIEW

This chapter discusses the standard economic modeling of a competitive market, stressing aspects relevant for the operations of current US RTO/ISO-managed wholesale power markets. For this purpose, some common economic presumptions are relaxed. For example, demands and supplies are expressed as step-function schedules rather than as continuously differentiable functions. The material in this chapter is adapted from Tesfatsion [180].

12.2 KEY DEFINITIONS

In standard economic usage, a *good* is an exchangeable physical item able to benefit an entity that consumes or uses it; and a *service* is an action that can be taken by an entity to provide benefit to one or more other entities. Examples of goods include apples, tractors, and paintings. Examples of services include haircuts, medical examinations, and food catering.

An *asset* is anything of durable value, that is, anything that acts as a means to store value over time. *Real assets* are assets in physical form (e.g. land, equipment, houses), including "human capital" assets embodied in people (e.g. natural abilities, learned skills, knowledge). *Financial assets* are claims against real assets, either directly (e.g. company stock shares) or indirectly (e.g. money holdings, bonds).

A *commodity* is an asset with a standard unit of measurement that permits the substitution of one unit for another with no change in valuation. A commodity can thus be valued and traded in bulk multi-unit form. Examples of commodities include standardized agricultural products such as No. 1 Hard Red Winter Wheat (measured in metric tons), standardized industrial products such as Light Crude Oil (measured in barrels), and standardized financial products such Microsoft Corporation (MSFT) common stock (measured in number of shares).

A New Swing-Contract Design for Wholesale Power Markets, First Edition, Leigh Tesfatsion.

A *market* is any public context in which buyers and sellers come together to participate in the purchase and sale of goods, services, or assets. An *auction market* is a market that is centrally managed by an *auctioneer* (or "central clearing house") tasked with facilitating trades between buyers and sellers. Market quantity and price outcomes are determined by the particular trade and settlement rules enforced by the auctioneer. A *double-auction market* [98, 126] is a two-sided auction market in which buyers submit demand bids, sellers submit supply offers, and the auctioneer matches demand to supply by an appropriate setting of one or more market prices.

Finally, a double-auction market is said to be a *competitive market* if the buyer and seller participants condition their individual bids and offers on market prices, taken as given. A traditional assumption used to motivate this price-taking behavior is that the buyer and seller participants are so numerous that a change in any individual bid or offer has only a negligible effect on market prices, all else equal.

12.3 STANDARD COMPETITIVE MARKET ASSUMPTIONS

The standard economic model of a competitive market is a spot-market model. Buyers and sellers are assumed to be competing in a double-auction market at a particular point in time to buy and sell a single commodity Q at a market-determined price. All seller fixed costs are assumed to be sunk, and seller supply curves for Q are assumed to coincide with the rising portion of their marginal cost curves.

Typically, the unit of measurement for the commodity Q is assumed to be infinitely divisible. This permits demand and supply schedules for Q to be represented analytically as continuous functions that can be plotted as continuous demand and supply curves in quantity-price space. In addition, a typical competitive market analysis assumes that demand curves are strictly downward sloping and supply curves are strictly upward sloping.

Examples include a spot-market for natural gas (Henry Hub) for which gas quantities q are measured in MMBtu and gas prices π are expressed in US dollars (USD) per MMBtu; a spot-market for crude oil (Western Texas Intermediate) for which oil quantities q are measured in barrels and oil prices π are expressed in USD per barrel; and a spot-market for gourmet apples (Honeycrisp) for which apple quantities q are measured in bushels and apple prices π are expressed in USD per bushel.

12.4 LAW OF ONE PRICE FOR COMMODITIES

The *Law of One Price* for a commodity Q states that, in the absence of trade frictions, each unit of Q available at a given time should sell at the same price. Expressed more generally, in the absence of trade frictions, each unit of Q should sell at the same present value price at any given present time. The Law of One Price follows immediately from the definition of a commodity.

Examples of trade frictions that can prevent units of Q from selling at the same price at a given time include transport costs, tariffs, and trade barriers (e.g. import restrictions). These types of trade frictions can affect the valuation of units of Q available at different locations.

Standard competitive market theory assumes away all trade frictions. The typical goal of a competitive market analysis for a commodity Q is to determine the unique equilibrium quantity q^* and price π^* for Q, expressed analytically as the unique intersection point (q^*, π^*) in a quantity-price plane of a strictly downward sloping demand function for Q and a strictly upward sloping supply function for Q.

12.5 COMPETITIVE MARKET: BASIC FORMULATION

Consider a collection of buyers and sellers participating in a competitive market at a particular time t. The buyers and sellers are competing to buy and sell quantities q of a commodity Q at bid/offer prices π.

An *(inverse true) demand schedule* for a buyer j is a schedule giving the maximum price π that buyer j would be willing to pay for each successive quantity unit demanded. These successive maximum purchase prices are referred to as buyer j's *purchase reservation values*. Conversely, an *(inverse true) supply schedule* for a seller i is a schedule giving the minimum price π that seller i would be willing to accept in payment for each successive quantity unit supplied. These successive minimum sale prices are referred to as seller i's *sale reservation values*.

Inverse demand and supply schedules give reservation values for each additional quantity unit, whereas *ordinary* demand and supply schedules give maximum quantities that buyers and sellers are willing to purchase and sell for each successively higher price. Inverse forms are used in this chapter to facilitate the graphical representation of demand and supply schedules in individual and aggregate forms. The qualifier "true" indicates that a demand or supply schedule accurately reflects purchase or sale reservation values, a maintained assumption in this chapter unless otherwise noted.[1] For ease of exposition, the qualifiers "inverse" and "true" for demand and supply schedules will hereafter be omitted.

A simple illustration of a competitive market for gourmet apples is given in Table 12.1. Apple demands and supplies are measured in bushels. The table entries give purchase and sale reservation values measured in \$/bushel. For example, the table entries for buyer B2 indicate that his purchase reservation value is \$70 for a first bushel, \$50 for a second bushel, \$25 for a third bushel, and \$0 for any additional bushels. Similarly, the table entries for seller S1 indicate that his sale reservation value is \$10 for a first bushel, \$50 for a second bushel, \$65 for a third bushel, and \$70 for a fourth bushel. Four bushels are his maximum supply capacity, so his sale reservation value for any additional bushels is effectively infinite.

[1] Section 12.9 considers whether buyers and sellers can profitably exploit the rules of a competitive market by strategically reporting bids and offers that do not reflect their true purchase and sale reservation values.

TABLE 12.1 Apple market example: Supply and demand schedules for two sellers and three buyers

Bushels	S1	S2	B1	B2	B3
1	\$10.00	\$10.00	\$70.00	\$70.00	\$25.00
2	\$50.00	\$50.00	\$70.00	\$50.00	0
3	\$65.00	\$70.00	\$65.00	\$25.00	0
4	\$70.00	\$70.00	\$65.00	0	0
5	∞	∞	0	0	0
6	∞	∞	0	0	0
7	∞	∞	0	0	0

The *aggregate demand schedule* gives the maximum purchase reservation value π for each successive quantity unit demanded, where the maximum is calculated across all individual buyers. The aggregate demand schedule is constructed from the individual buyers' demand schedules by arranging the buyers' reservation values for successive quantity units in *descending* order. This aggregate demand schedule can be represented as a nonincreasing relationship $\pi = \mathrm{D}(q)$, where π denotes the maximum purchase reservation value for the last quantity unit demanded at q.

The *aggregate supply schedule* gives the minimum sale reservation value π for each successive quantity unit supplied, where the minimum is calculated across all individual sellers. The aggregate supply schedule is constructed from the individual sellers' supply schedules by arranging the sellers' reservation values for successive quantity units in *ascending* order. This aggregate supply schedule can be represented as a nondecreasing relationship $\pi = \mathrm{S}(q)$, where π denotes the minimum sale reservation value for the last quantity unit supplied at q.

Figure 12.1 plots the aggregate demand and supply schedules for the apple market with three buyers and two sellers whose individual demand and supply schedules are given in Table 12.1. Note that these plotted aggregate demand and supply schedules depict the maximum purchase reservation value (across all buyers) and the minimum sale reservation value (across all sellers) for each successive bushel of apples.

Figure 12.2 plots the aggregate demand and supply schedules for a different apple market with two buyers and three sellers. The maximum purchase reservation value (MaxBuyPrice) and the minimum sale reservation value (MinSellPrice) for each successive bushel of apples, calculated across all buyers and sellers respectively, are given on the right-hand side of the figure.

A quantity-price combination $e^* = (q^*, \pi^*)$ for a competitive market is called a *competitive market clearing (CMC) point* for this market if plots of the aggregate demand schedule $\mathrm{D}(q)$ and aggregate supply schedule $\mathrm{S}(q)$ (with flat segments included) intersect at e^*; that is, if $\pi^* = \mathrm{S}(q^*) = \mathrm{D}(q^*)$. The quantity q^* is called a *CMC output level*, and the price π^* is called a *CMC price level*.

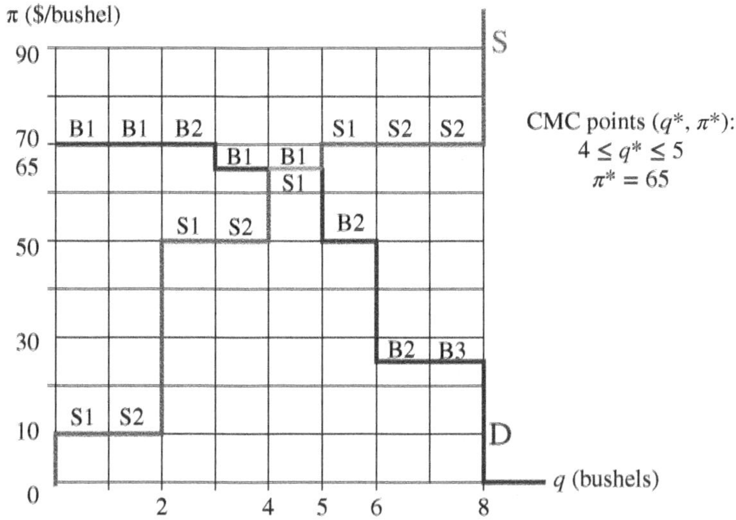

Figure 12.1 Plots of the aggregate demand and supply schedules D and S corresponding to the individual demand and supply schedules reported in Table 12.1. The competitive market clearing points (q^*, π^*) are given by $4 \leq q^* \leq 5$ and $\pi^* = 65$.

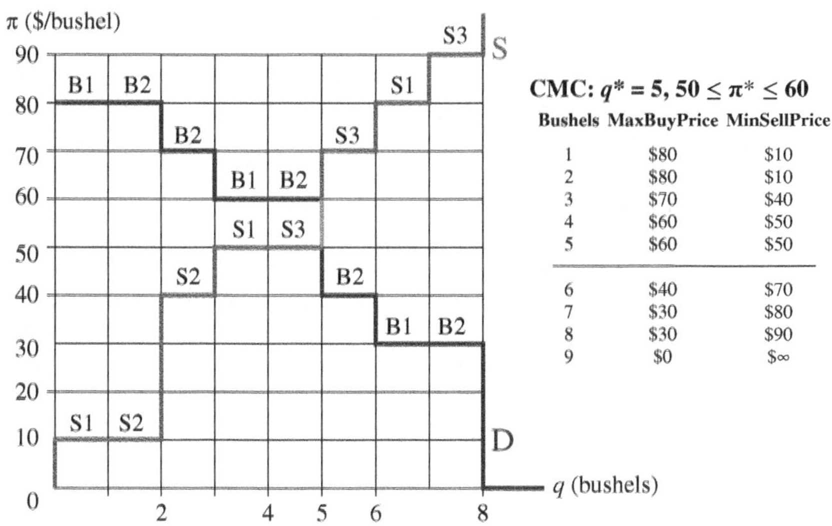

Figure 12.2 A second illustration of aggregate supply and demand schedules for a competitive market. The competitive market clearing points (q^*, π^*) are given by $q^* = 5$ and $50 \leq \pi^* \leq 60$.

The CMC points for a competitive market can be multiple, unique, or non-existent at any given point in time. Multiplicity of CMC points for a given competitive market CM at a given time t is only possible if the aggregate demand and supply schedules for CM at t have coincident horizontal or vertical flat segments.[2] See, for example, Figures 12.1 and 12.2.

12.6 NET SURPLUS EXTRACTION

To convey with clarity the efficiency implications of CMC points for a competitive market, the concept of "net surplus" must first be explained. The *net buyer surplus on a quantity unit bought* is the difference between the buyer's reservation value for this unit and the actual price paid by the buyer for this unit. The *net seller surplus on a quantity unit sold* is the difference between the actual sale price received by the seller for this unit and the seller's reservation value for this unit.

By construction, $e^* = (q^*, \pi^*)$ is a CMC point for a competitive market if and only if the following two conditions hold at e^*: (a) Taking π^* as the given market price, no buyer perceives an opportunity to *strictly* increase its extracted net buyer surplus by changing its quantity bid; and (b) taking π^* as the given market price, no seller perceives an opportunity to *strictly* increase its extracted net seller surplus by changing its quantity offer.

Thus, if all buyers and sellers are net-surplus-seeking entities *that take prices as given* at a CMC point, they will perceive no reason to unilaterally deviate from their current quantity demands and supplies at this CMC point. Alternatively, these buyers and sellers will perceive no reason to unilaterally deviate if they are so numerous that they cannot, in fact, affect market prices through their own individual actions.

The *net buyer surplus* at a CMC point $e^* = (q^*, \pi^*)$ is the area between the aggregate demand schedule and the horizontal line at the price level π^*, up to q^*. The net buyer surplus at e^* measures buyer welfare at this point. It is the difference between the maximum payment that buyers are willing to make for q^* units and the amount $\pi^* q^*$ that buyers would actually pay for the purchase of q^* units at price π^*.

The *net seller surplus* at a CMC point $e^* = (q^*, \pi^*)$ is the area between the horizontal line at the price level π^* and the aggregate supply schedule, up to q^*. The net seller surplus at e^* measures seller welfare at this point. It is the difference between the amount $\pi^* q^*$ that sellers would actually receive for the sale of q^* units at price p^* and the minimum payment they are willing to receive for the sale of q^* units.

[2]Competitive market models in economic textbooks typically assume a strictly decreasing aggregate demand function $\pi = D(q)$ and a strictly increasing aggregate supply function $\pi = S(q)$ that have a unique intersection point (q^*, π^*) in the (q, π)-plane. In contrast, demand bids and supply offers in US RTO/ISO-managed wholesale power markets are typically required to take the form of step or piecewise-linear functions in the (MW, \$/MWh) plane that can result in coincident flat segments in aggregate demand and supply functions. See, for example, the demand bid and supply offer forms required by MISO [131, Exhibits 4-8, 4-30].

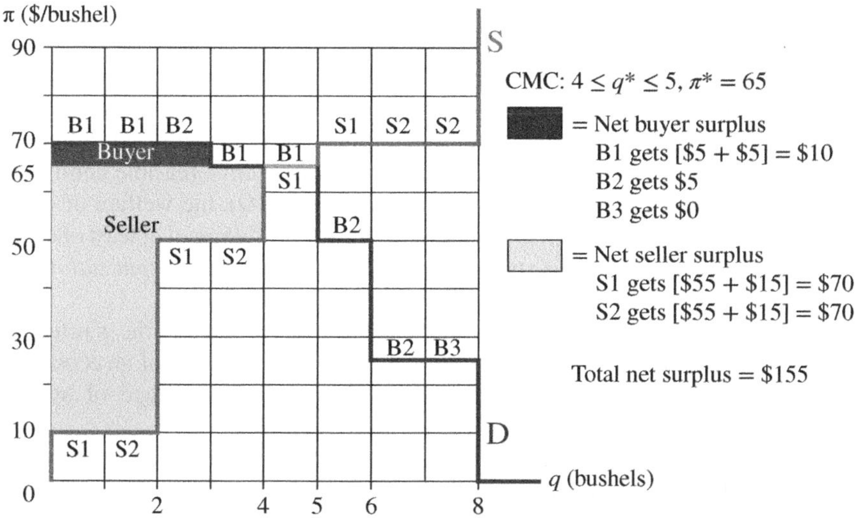

Figure 12.3 Net buyer surplus, net seller surplus, and total net surplus calculations and depictions for the aggregate demand and supply schedules presented in Table 12.1 and plotted in Figure 12.1.

The *total net surplus (TNS)* at a CMC point $e^* = (q^*, \pi^*)$ is the sum of net buyer surplus and net seller surplus at e^*. By construction of e^*, this TNS gives the *maximum* amount of net surplus that can feasibly be extracted from the market; i.e. there is no way to arrange trades so that TNS is greater than at e^*. This follows from the arrangement of buyer and seller reservation values in descending and ascending orders in the construction of the aggregate demand and supply schedules whose intersection determines e^*.

The net buyer surplus, net seller surplus, and TNS corresponding to any one of the multiple CMC points depicted in Figure 12.1 for an illustrative competitive market are calculated and plotted in Figure 12.3. Note that TNS is invariant across the multiple CMC points. This TNS invariance holds generally for competitive markets with multiple CMC points, by the constructive definition of a CMC point.

12.7 MARKET EFFICIENCY METRIC

In economics, the *efficiency* of a system refers to the non-wastage of the system's resources. Here "resources" refers to physical assets (e.g. equipment, workers) that can be employed as inputs for the production of outputs that are valued by the people participating in the system.

Economists identify two types of wastage: (i) *physical wastage*, in the sense that some valued units of resource remain unused; and (ii) *wastage of value*, in the sense that some units of resource are not being used by those who value them most.

A system with no physical wastage is said to exhibit "productive efficiency" and a system with no wastage of value is said to exhibit "Pareto efficiency."

More precisely, a particular configuration (I, O) of inputs and outputs is said to be *productively efficient* if O is as large as possible, given I, and I is as small as possible for the given O. A particular configuration (I, O) of inputs and outputs is said to be *Pareto efficient* if there does *not* exist any *alternative* feasible configuration (I', O') of inputs and outputs for which, relative to (I, O), the welfare of every person participating in the system is at least as high and the welfare of at least one person participating in the system is strictly higher. In the absence of *output satiation*,[3] productive efficiency is a necessary condition for Pareto efficiency.

Short-run efficiency of an economic system at a particular time t refers to non-wastage of *existing* system resources at t. *Longer-run efficiency* of an economic system over some designated time interval T refers to non-wastage of system resources during T, given initial conditions at the start of T.

Let CM(t) denote a competitive market at a particular time t, as described in Section 12.5. The standard metric used by economists to measure the "market efficiency" of CM(t) addresses both productive and Pareto efficiency.

Let MaxTNS(t) denote the maximum possible total net surplus that buyers and sellers could feasibly extract from CM(t). If a CMC point exists, MaxTNS(t) coincides with the TNS extracted at this CMC point. Also, let ActualTNS(t) denote the *actual* total net surplus that buyers and sellers manage to extract from CM(t). Then, the *market efficiency* of CM(t) is measured by the following metric:

$$\mathrm{ME}(t) = \left(\frac{\mathrm{ActualTNS}(t)}{\mathrm{MaxTNS}(t)} \right) \times 100\% \tag{12.1}$$

By construction, ME(t) $\leq 100\%$. Moreover, if buyers never purchase quantity units at prices above their purchase reservation values and sellers never sell quantity units at prices below their sale reservation values, then $0\% \leq$ ME(t).

The quantity units lying to the left of the largest CMC quantity level are called *inframarginal*, and all other quantity units are called *extramarginal*. ME(t) will be 100% for CM(t) if all inframarginal units trade, no extramarginal units trade, and the price paid by the buyer equals the price paid to the seller for each traded unit. Conversely, ME(t) can fall below 100% for CM(t) only if some inframarginal unit fails to trade, or some extramarginal unit is traded, or the price paid by a buyer in some trade is not equal to the price received by the seller in this trade.

For example, consider the competitive market depicted in Table 12.1, Figures 12.1 and 12.3 for which the unique CMC price is $\pi^* = \$65$/bushel. Suppose sellers S1 and S2 sell their five inframarginal bushels to buyers B1 and B2 at successive sales price $\pi_1, \ldots, \pi_5$ such that π_k lies between the maximum purchase reservation value and minimum sale reservation value for each quantity

[3]A person is said to be *satiated* with respect to an output at time t if the receipt of an additional unit of this output at t would *not* increase the welfare of that person.

unit $k = 1, \ldots, 5$. Then, the extracted TNS will be MaxTNS = \$155 (i.e. the sum of the five purchase reservation values minus the sum of the five sale reservation values) regardless of the precise sales prices or the precise buyer–seller match-ups for each quantity unit sold.

Conversely, suppose S1 instead sells his first (\$10) inframarginal bushel to the extramarginal buyer B3, whose purchase reservation value for this bushel is only \$25. Then the actual TNS extracted in this market will be no greater than \$115, regardless of remaining trades, implying a deadweight loss[4] of at least \$40 and a value for ME(t) that is strictly below 100%.

12.8 MARKET EFFICIENCY AND PRICING RULES

Economists adhere to the market efficiency metric (12.1) for a competitive market because they are concerned that efficiency be achieved in both the short and longer run. Their short-run concern is to ensure maximum TNS extraction using *existing* resources. Their longer-run concern is to provide proper incentives to encourage investment in *new* resources for producing and extracting more TNS in the future. In particular, buyers with new ideas for gaining more benefit per unit of output purchased should have an incentive to introduce these ideas into a market to compete against current buyers for the extraction of buyer surplus. Similarly, sellers with new ideas for supplying output at lower unit cost should have an incentive to introduce these ideas into a market to compete against current sellers for the extraction of seller surplus.

It is instructive to consider how the market efficiency of a competitive market is potentially enhanced in both the short run and longer run under a uniform CMC pricing rule. As seen in Section 12.7, the TNS extracted at any CMC point for a competitive market at a particular time t is the maximum TNS that can be extracted from this market at t. Thus, short-run efficiency is achieved for this market at t if the market price for all quantity units is set at a CMC price level.

However, the extraction of MaxTNS at a particular time t for a competitive market does not *require* a uniform CMC pricing rule. Rather, as noted in Section 12.7, MaxTNS is extracted at t if the following three conditions hold: (i) Traded output is at a CMC level; (ii) only inframarginal units are traded; and (iii) for each traded unit, the price paid by the buyer equals the price paid to the seller.

For example, these three conditions are consistent with a *k-discriminatory pricing rule* ($k \in [0, 1]$) defined as follows: For any matched buyer–seller pair with purchase reservation value π^b and sale reservation value π^s satisfying $\pi^b \geq \pi^s$, set the *strike price* for this pair at the weighted-average level $\pi^k \equiv k\pi^b + [1 - k]\pi^s$ lying between their reservation values. This pricing rule simply distributes the net surplus generated by a trade between a buyer–seller pair in a ratio determined by k without affecting the amount of this net surplus.

---

[4] In economics, a *deadweight loss* is defined to be possible net surplus that is not extracted either by buyers or by sellers, thus resulting in a reduction in TNS.

In particular, ignoring possible deleterious effects on incentives for truthful revelation of costs and capacities, as well as serious concerns related to recovery of non-variable costs, short-run efficiency could alternatively be achieved by using a 0-discriminatory pricing rule that awards all generated net surplus to buyers. This is essentially the approach advocated for wholesale power markets in a series of studies on "payment cost minimization"; see, e.g. Yan et al. [198].

On the other hand, from a longer-run efficiency point of view, setting market price at a uniform CMC price level encourages both buyers and sellers to innovate. In contrast, use of a k-discriminatory pricing rule, particularly with an extreme setting of $k = 1$ or $k = 0$, would tend to discourage innovation by reducing or even eliminating the gains to innovation by buyers or sellers.

For example, a seller investing in new technology in order to lower its production costs, and hence its sale reservation values, would experience lower sale prices under a k-discriminatory pricing rule with $k < 1$. Conversely, a CMC uniform pricing rule permits sellers of inframarginal units to keep any additional net seller surplus (net earnings) garnered from a lowering of their costs and hence a lowering of their sale reservation values. This provides sellers with an incentive to find new ways to supply output using fewer inputs or to supply more output from the same inputs.

Similarly, setting market price at a uniform CMC price level permits buyers of inframarginal units to keep any additional net buyer surplus (net earnings) garnered from the development of new ways of achieving more benefit from each purchased unit of output and hence an increase in their purchase reservation values. This provides buyers with an incentive to pursue such developments.

The above analysis presumes that the auctioneer for a competitive market knows (or can determine) the true reservation values of buyers and sellers for successive units of Q. In fact, however, it is not necessarily to the advantage of buyers and sellers to reveal these private attributes to the auctioneer in their reported supply and demand schedules. This issue is taken up in the following section.

12.9 STRATEGIC TRADE BEHAVIOR AND TRADER MARKET POWER

Let M denote a double-auction market for a commodity Q. Suppose buyer demand bids and seller supply offers for Q take the form of *reported demand schedules* and *reported supply schedules*, respectively. Suppose, also, that the goal of each buyer and seller is attainment of the highest possible extraction of net surplus for itself. All else equal, then, buyers are better off the lower the prices they pay for the quantity units they purchase, and sellers are better off the higher the prices they are paid for the quantity units they sell.

Buyer market power refers to the ability of a buyer to profitably lower the purchase price(s) it pays for the quantity units it purchases, relative to CMC conditions. *Seller market power* refers to the ability of a seller to profitably raise the sale price(s) it is paid for the quantity units it sells, relative to CMC conditions.

Buyers and sellers participating in M can attempt to exercise market power through their reported demand and supply schedules in two basic ways. First, they can misrepresent their true reservation values, a practice referred to as *economic capacity withholding*. Alternatively, they can reduce the number of units they bid to buy or offer to sell below their feasible purchase and sale amounts, a practice referred to as *physical capacity withholding*.

For example, suppose M is an apple market comprising two buyers and three sellers whose true purchase and sale reservation values at time t are as depicted in Figure 12.2. Suppose buyers and sellers at time t report their true purchase and sale reservation values to the auctioneer in their reported demand and supply schedules, and the auctioneer announces that the market-clearing CMC point for M at t is $e^* = (q^*, \pi^*)$ = (5 bushels, \$50/bushel). At e^*, seller S1 sells two bushels at a price of \$50/bushel and attains a net seller surplus equal to \$40 = [\$50 − \$10] + [\$50 − \$50].

Now suppose, instead, that seller S1 omits from his reported demand schedule the availability of his second inframarginal bushel at \$50/bushel, thus engaging in physical capacity withholding. All other aspects of the reported demand and supply schedules remain the same. The upper portion of the aggregate supply schedule calculated by the auctioneer is then shifted to the left by one bushel, resulting in what the auctioneer perceives to be a unique "CMC" point $e' = (q', \pi')$ = (4 bushels, \$60/bushel). At e', seller S1 attains a net seller surplus equal to \$50 = [\$60 − \$10] from the sale of one bushel, a higher net seller surplus than the \$40 it would have attained at e^* had it truthfully reported its supply schedule. Indeed, *all three* sellers attain higher net seller surplus at e' than at e^*; that is, all three sellers benefit from the exercise of market power by seller S1.

Recall from Section 12.5 that buyers and sellers in a *competitive* market have no incentive to individually change their quantity purchases or sales at CMC points because, *by assumption*, they *either* take the CMC market price as given *or* they are too numerous to have any effect on the market price through their individual actions. The important point to observe about the aggregate demand and supply schedules depicted in Figure 12.2 is that neither of these "competitive" conditions can reasonably be assumed to hold. Rather, seller S1 can easily calculate that he has the ability to increase the market price to his own advantage by withdrawing one of his quantity units from sale; that is, seller S1 has seller market power.

The next chapter describes, in simplified form, the flow of daily ISO activities in US RTO/ISO-managed day-ahead markets, which are organized as double-auction markets. As will be seen, the detection of market power is particularly problematic in these markets due to the effects of grid congestion on market prices.

CHAPTER 13

US RTO/ISO-MANAGED MARKETS: EFFICIENCY AND MARKET POWER

13.1 CHAPTER PREVIEW

The exact market optimization formulations currently in use in US RTO/ISO-managed wholesale power markets are proprietary. However, summary descriptions of these formulations can be found in RTO/ISO business practice manuals, training manuals, technical reports, conference papers, and journal articles; see, e.g. [24, 25, 37, 48, 75, 90, 123, 124, 130, 140, 148, 170, 177].

Section 13.2 describes, in simplified form, the logical flow of daily day-ahead market (DAM) and real-time market (RTM) operations in current US RTO/ISO-managed wholesale power markets. Section 13.3 next develops an illustrative analytical DAM model. Employing net surplus concepts introduced and motivated in Chapter 12, Section 13.4 uses this DAM model to examine how total net surplus (TNS) is currently extracted in US RTO/ISO-managed DAMs.

More precisely, the analytical and simulation findings reported in Section 13.4 show that TNS extraction in RTO/ISO-managed DAMs differs in two fundamental ways from TNS extraction in the simple competitive market model analyzed in Chapter 12. First, the RTO/ISO "auctioneer" itself extracts positive net surplus from the DAM whenever there is grid congestion; and the amount of this ISO net surplus extraction increases dramatically as the percentage of *price-sensitive* load-serving entity (LSE) demand is decreased from 100% to 0%, all else equal. Second, the RTO/ISO has to approximate LSE (buyer) and GenCo (seller) net surplus in the DAM objective function using *reported* LSE demand bids and GenCo supply offers that can differ from true LSE and GenCo reservation values.

Sections 13.5–13.8 use these TNS extraction findings, together with RTO/ISO net surplus allocation precepts, to examine the relationship between seller market power and market inefficiency in current US RTO/ISO-managed DAMs. Section 13.9 discusses how the outcomes of privately negotiated physical bilateral contracts are scheduled in these markets.

A New Swing-Contract Design for Wholesale Power Markets, First Edition. Leigh Tesfatsion.

For ease of exposition, this chapter hereafter refers to US RTO/ISO-managed wholesale power markets in abbreviated form, as *ISO markets*, and the system operators for these markets are simply referred to as ISOs.

13.2 DAILY MARKET OPERATIONS

As described in general terms in Section 2.3, an ISO market is organized as a two-settlement system consisting of a daily DAM operating in tandem with daily RTM processes over a high-voltage AC transmission grid. As depicted in Figure 2.2, the scheduled generation dispatch levels determined in the DAM and RTM are separately settled in accordance with DAM-determined and RTM-determined LMPs.

The participants in an ISO market include a collection of LSEs and GenCos distributed across the buses of the transmission grid. The ISO's goal is to ensure over time the reliable and efficient operation of the ISO market. The goal of each LSE is to secure for itself over time the highest possible net earnings through the daily purchase of electric power in the ISO market and the resale of this electric power to its retail customers. The goal of each GenCo is to secure for itself over time the highest possible net earnings through the daily sale of electric power in the ISO market.

A more detailed description of the daily ISO market operations depicted in Figure 2.2 is as follows:

- During the morning of day D, each LSE chooses a *demand bid* to submit to the day-D DAM based on its forecasted loads for day D+1. Each demand bid consists of two parts: a *fixed demand bid* (i.e. a 24-hour price-insensitive load profile); and 24 *price-sensitive demand bids* (one for each hour), each consisting of a demand function defined over a purchase capacity interval.
- During the morning of day D, each GenCo chooses a *supply offer* to submit to the ISO for use in all 24 hours of the day-D DAM. Each supply offer consists of a reported marginal cost function defined over a reported power capacity interval.
- After receiving LSE demand bids, GenCo supply offers, and physical bilateral trade schedules[1] during the morning of day D, the ISO conducts a DAM SCUC/SCED optimization to determine unit commitments, scheduled generation dispatch levels, and LMPs for each hour H of day D+1.
- The ISO resolves any differences that arise during day D between actual day-D power usage and the generation dispatch levels scheduled in the day-(D−1) DAM by solving appropriate RTM SCED optimizations during day D conditional on continually updated net must-service load forecasts and system state estimates.

[1]As will be clarified in Section 13.9, these schedules are the physical outcomes of physically covered trades negotiated between two power traders at privately negotiated prices. These physical outcomes need to be incorporated into the ISO's DAM optimization to the extent that their delivery requires use of the transmission grid.

- The day-D RTM dispatch solutions are communicated to GenCos as real-time dispatch set points by means of an automatic generation control (AGC) system.
- The ISO uses day-D DAM LMPs to settle the power demands and supplies cleared in the day-D DAM for day D+1.
- The ISO uses day-D RTM LMPs to settle any needed adjustments to the generation dispatch levels scheduled in the day-(D−1) DAM for delivery on day D.
- The ISO calculates ex-post LMPs on the basis of actual day-D loads and actual day-D generator responses to dispatch instructions, as determined by telemetry.
- The ISO uses these ex-post LMPs, as well as other administratively determined charges, to settle any additional needed adjustments in power injections and withdrawals on day D to ensure net load balancing on day D. Make-whole payments to ensure coverage of unit commitment costs might also be made.

The above simplified description of daily ISO market operations omits consideration of reserve procurement. As detailed in [46], various forms of reserve are procured in current US ISO markets either through separate reserve markets or through DAM/RTM co-optimization of energy and reserve. In a DAM/RTM co-optimization, the procurement of reserve is undertaken by incorporating ISO-specified demand bids for reserve into the objective function and/or ISO-specified system-wide and zonal reserve requirements into the constraints.

A second complication omitted from the above simplified description is that DAMs permit the submission of *virtual* offers/bids in an attempt to ensure convergence between DAM and RTM LMPs.[2] Virtual offers/bids are offers/bids to sell/buy a block of energy at a designated price and location that are not backed by physical generation or physical load. Virtual offers/bids must be appropriately covered to ensure their quantity positions are liquidated (reduced to zero) at any scheduled delivery point. For example, a virtual supply offer submitted into a day-D DAM for delivery on day D+1 could be settled for its liquidated position based on RTM prices for day D+1.

A third complication omitted from the above simplified description is that additional ISO-managed SCUC optimizations are typically conducted subsequent to a DAM to commit additional resources for next-day operations. These additional processes are undertaken if the ISO concludes that the resources committed in the DAM based on LSE demand bids are in fact insufficient (in amount and/or in location) to ensure the balancing of next-day net loads as forecasted by the ISO. Any resource procurement costs incurred in these additional processes are passed through to market participants ex post, taking into account the extent to which their actual real-time load or generation has deviated from scheduled DAM amounts. These additional processes are variously referred to in the different ISO markets, e.g. as *Residual Unit Commitment* processes in CAISO, as *Reliability Unit Commitment*

[2] Evidence presented in [14] for MISO indicates that virtual offers/bids do not necessarily achieve this price-arbitrage goal.

processes in ERCOT, and as *Reliability Assessment Commitment* processes in MISO.

Moreover, the above simplified description does not explain the manner in which DAM SCUC/SCED optimal solutions are obtained. For computational tractability, some larger US ISO markets have resorted to the use of a two-stage optimization process. A DAM SCUC optimization formulated as a MILP problem is first solved for all 24 hours of the following day, subject to unit commitment constraints and a limited number of system constraints. The resulting binary 0-1 unit commitment solutions for next-day operations are then inserted into hourly DAM SCED optimizations formulated as linear programming problems subject to detailed system constraints, but with unit commitment constraints omitted, to obtain scheduled dispatch, reserve, and LMP solutions for next-day hourly operations. See, for example, MISO [124]. On the other hand, for a sufficiently small ISO market, DAM unit commitments, scheduled dispatch levels, reserve supplies, and LMP solutions can all be jointly obtained from a single comprehensive DAM SCUC/SCED optimization; see, e.g. CAISO [24].

As computational methods continue to improve, presumably all ISO markets will have an incentive to transition to the implementation of single comprehensive DAM SCUC/SCED optimizations.

13.3 ILLUSTRATIVE ANALYTICAL DAM FORMULATION

This section provides an illustrative analytical modeling for daily DAM operations in current ISO markets, as depicted in Figure 2.2. This modeling includes analytical formulations for the LSE demand bids and GenCo supply offers submitted into the DAM during each day D.

For simplicity of exposition, attention is focused on a SCED optimization for an ISO-managed DAM, conditional on SCUC-optimal unit commitments. The goal of the ISO undertaking this SCED optimization is to maximize TNS subject to power balance constraints, transmission line capacity limits, and GenCo power capacity constraints.[3] The participants in this ISO-managed DAM consist of LSEs $j = 1, \dots, J$ and GenCos $i = 1, \dots, I$.

Each LSE j services the power usage demands of its retail customers by aggregating these demands into demand bids for submission into the DAM. More precisely, each LSE j services two types of retail customers: (i) customers with fixed-price contracts that buy power p from LSE j at a regulated rate r ($/MWh) and (ii) customers with dynamic-price contracts that buy power p from LSE j at retail prices π^{Ret} ($/MWh) set by LSE j.

For each day D, LSE j's *demand bid* submitted into the day-D DAM for hour H of day D+1 thus consists of two parts: (i) a fixed (non-price-sensitive) demand bid,

[3] As will be seen below in (13.17), when all demand takes the form of must-service (non-price-sensitive) load compensated at a predetermined regulated rate r, the maximization of TNS is equivalent to the minimization of GenCo total avoidable cost.

p^F_{Lj}(H,D+1) (MW), giving the forecasted amount of power to be sold downstream at rate r during hour H of day D+1 to its retail customers with fixed-price contracts; and (ii) a *price-sensitive demand function* F_j ($/MWh) given by

$$\pi = \mathrm{F}_j(p^S_{Lj}, \mathrm{H}, \mathrm{D}{+}1) = e_j(\mathrm{H}, \mathrm{D}{+}1) - 2f_j(\mathrm{H}, \mathrm{D}{+}1)p^S_{Lj} \tag{13.1}$$

with positive coefficients e_j and f_j, where (13.1) is defined over a *purchase interval*

$$0 \le p^S_{Lj} \le \mathrm{Pur}^{\mathrm{U}}_j(\mathrm{H}, \mathrm{D}{+}1) \equiv \frac{e_j(\mathrm{H,D}{+}1)}{2f_j(\mathrm{H,D}{+}1)} \tag{13.2}$$

The demand function F_j in (13.1) gives LSE j's *purchase reservation value* for power evaluated at p^S_{Lj}, i.e. the maximum price that LSE j is willing to pay for an incremental increase in its power usage set point at the start of hour H, given that its current power usage set point at the start of hour H is p^S_{Lj}. This purchase reservation value is assumed to be derived from the purchase reservation values of its retail customers with dynamic-price contracts.

For each day D, GenCo i's *supply offer* submitted into the day-D DAM for each hour H of day D+1 is its *marginal cost function* MC_i ($/MWh) given by

$$\pi = \mathrm{MC}_i(p_{Gi}) = b_i + 2c_ip_{Gi} \tag{13.3}$$

with positive coefficients b_i and c_i, where (13.3) is defined over a *power capacity interval*

$$0 \le p_{Gi} \le \mathrm{Cap}^{\mathrm{U}}_i \tag{13.4}$$

The marginal cost function MC_i in (13.3) gives GenCo i's *sale reservation value* for power, evaluated at p_{Gi}, i.e. the minimum price that GenCo i is willing to accept in payment for an incremental increase in its dispatch set point at the start of hour H, given that its current dispatch set point at the start of hour H is p_{Gi}.

Finally, to simplify the exposition, it is assumed that GenCos do not incur start-up, shut-down, no-load, or opportunity costs, and that all of their fixed costs are *sunk*, i.e. non-avoidable. Consequently, the *avoidable cost function* ($/h) for each GenCo i for any hour H of day D+1 is given by

$$\mathrm{GCost}_i(p_{Gi}) = \int_0^{p_{Gi}} \mathrm{MC}_i(p)dp = b_ip_{Gi} + c_i[p_{Gi}]^2 \tag{13.5}$$

where p_{Gi} satisfies (13.4).

13.4 NET SURPLUS EXTRACTION IN THE ILLUSTRATIVE DAM

Suppose LSE j, located at bus $n(j)$, is cleared in the day-D DAM at a total power demand level p_{Lj}(H,D+1) = [p^F_{Lj}(H,D+1) + p^S_{Lj}(H,D+1)] at price $\mathrm{LMP}_{n(j)}$(H,D+1) for each hour H of day D+1. The resulting cost ($) assessed to LSE j for all 24 hours of day D+1, settled at the end of day D, is then given by the summation of its LMP payment obligations:

$$\text{LCost}_j(\text{D+1}) = \sum_{\text{H}=00}^{23} [\text{LMP}_{n(j)}(\text{H,D+1})p_{Lj}(\text{H,D+1})] \cdot 1 \text{ hour} \tag{13.6}$$

The net surplus accruing to the "last" MW of power sold by LSE j to its dynamic-price retail customers, evaluated at any total sale quantity p, is given by $[\text{F}_j(p,\text{H,D+1}) - \pi^{\text{Ret}}] + [\pi^{\text{Ret}} - \text{LMP}_{n(j)}(\text{H,D+1})]$, where π^{Ret} denotes the retail price charged by LSE j for this last MW. The first bracketed term is the net surplus portion accruing to the retail customers and the second bracketed term is the net surplus portion accruing to LSE j. For simplicity, it will hereafter be supposed that LSE j is able to extract all net surplus from its dynamic-price retail customers by charging these customers their maximum willingness-to-pay for each MW increment, i.e. by setting $\pi^{\text{Ret}} = F_j(p,\text{H,D+1})$ at each power level p.[4]

The gross surplus extracted by LSE j from the day-D DAM based on its forecasted retail sales for day D+1 is then given by its revenue ($), calculated as

$$\text{LRev}_j(\text{D+1}) = \sum_{\text{H}=00}^{23} \left[r \cdot p_{Lj}^F(\text{H,D+1}) + \int_0^{p_{Lj}^S(\text{H,D+1})} \text{F}_j(p,\text{H, D+1})dp \right] \cdot 1 \text{ hour} \tag{13.7}$$

LSE Net Surplus ($), the aggregation of all net surplus extracted from the day-D DAM by LSEs $j = 1, \ldots, J$ based on forecasted retail sales for day D+1, is thus given by

$$\text{LNS(D+1)} = \text{LRev(D+1)} - \text{LCost(D+1)} \tag{13.8}$$

where

$$\text{LRev(D+1)} = \sum_{j=1}^{J} [\text{LRev}_j(\text{D+1})] \tag{13.9}$$

$$\text{LCost(D+1)} = \sum_{j=1}^{J} [\text{LCost}_j(\text{D+1})] \tag{13.10}$$

Suppose GenCo i, located at bus $n(i)$, is dispatched in the day-D DAM at level $p_{Gi}(\text{H,D+1})$ at price $\text{LMP}_{n(i)}(\text{H,D+1})$ for each hour H of day D+1. The *revenue* ($) due to GenCo i for all 24 hours of day D+1, to be settled at the end of day D, is then given by

$$\text{GRev}_i(\text{D+1}) = \sum_{\text{H}=00}^{23} [\text{LMP}_{n(i)}(\text{H,D+1})p_{Gi}(\text{H,D+1})] \cdot 1 \text{ hour} \tag{13.11}$$

[4]At the other extreme, if the LSE were to set $\pi^{\text{Ret}} = \text{LMP}_{n(j)}(\text{H,D+1})$, all of the net surplus extracted by the LSE would be awarded to its dynamic-price retail customers.

Let the avoidable cost (\$/h) that would be incurred by GenCo i for any hour H of day D+1, should it carry out its day-D DAM dispatch p_{Gi}(H,D+1), be denoted by GCost_i(H,D+1). Then, GenCo i's avoidable cost (\$) for its entire scheduled dispatch on day D+1 is given by

$$\text{GCost}_i(\text{D+1}) = \sum_{\text{H}=00}^{23} [\text{GCost}_i(\text{H,D+1})] \cdot 1 \text{ hour} \tag{13.12}$$

It follows that the *GenCo Net Surplus* (\$) extracted by all of the GenCos from the day-D DAM for their scheduled dispatch on day D+1 is given by

$$\text{GenNS(D+1)} = [\text{GRev(D+1)} - \text{GCost(D+1)}] \tag{13.13}$$

where

$$\text{GRev(D+1)} = \sum_{i=1}^{I} \text{GRev}_i(\text{D+1}) \tag{13.14}$$

$$\text{GCost(D+1)} = \sum_{i=1}^{I} \text{GCost}_i(\text{D+1}) \tag{13.15}$$

The *ISO Net Surplus* (\$) extracted from the day-D DAM is defined to be the difference between LSE cost (LMP payments) and GenCo revenues (LMP receipts), based on the power demands and supplies submitted into the day-D DAM for day D+1. This ISO Net Surplus is given by

$$\text{ISONS(D+1)} = [\text{LCost(D+1)} - \text{GRev(D+1)}] \tag{13.16}$$

Suppose a standard DC-OPF SCED optimization is used for the DAM, which assumes zero power losses. Then, the ISO Net Surplus (13.16) must necessarily be nonnegative; see Ref. [3, Proposition 2.1]. Moreover, in the absence of transmission congestion, this ISO Net Surplus must be zero. The latter result holds because absence of congestion implies absence of LMP separation across the grid buses; hence, every LSE LMP payment is offset by a GenCo LMP receipt. Conversely, congestion arising anywhere on the grid necessarily results in the separation of LMPs at two or more bus locations [144].

Finally, the TNS (\$) extracted from the day-D DAM is defined to be the summation of all of the net surplus extracted by the LSEs, the GenCos, and the ISO. For the illustrative DAM at hand, this TNS is given by

$$\begin{aligned}\text{TNS(D+1)} &= \text{LNS(D+1)} + \text{GenNS(D+1)} + \text{ISONS(D+1)} \\ &= [\text{LRev(D+1)} - \text{GCost(D+1)}]\end{aligned} \tag{13.17}$$

Figure 13.1 provides a simple example of LSE, GenCo, and ISO net surplus extraction from a day-D DAM for an operating hour H during day D+1. The DAM

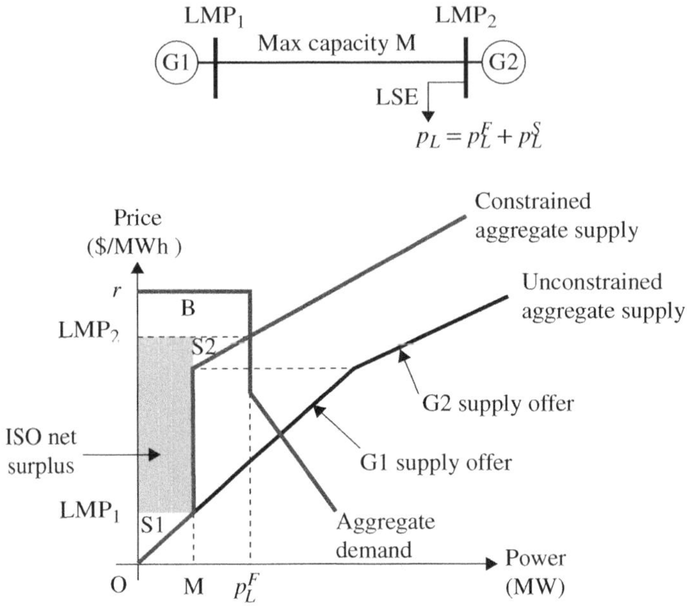

Figure 13.1 An example of net surplus extraction in a DAM for a 2-bus system. A transmission capacity limit M prevents a cheaper GenCo G1 at bus 1 from fully servicing LSE demand at bus 2. Source: Adapted from [157].

operates over a 2-bus transmission grid populated by an LSE at bus 2 and GenCos G1 and G2 at buses 1 and 2, respectively. The power line connecting the two buses has a maximum transmission capacity M, which prevents the cheaper GenCo G1 from fully servicing the LSE's power demand p_L at bus 2.

More precisely, all power demand is located at bus 2, and is serviced by the LSE. The aggregate demand for power at bus 2 consists of two parts: (i) a fixed demand p_L^F (priced at a regulated retail rate r), where $p_L^F > \mathrm{M}$ and (ii) a demand function for price-sensitive demand p_L^S defined over a purchase interval $[0, \mathrm{Pur}_L^{\mathrm{U}}]$. The resulting aggregate demand curve is labeled *aggregate demand* in Figure 13.1.

The supply offer of GenCo G1 consists of a strictly increasing linear marginal cost function $\mathrm{MC}_1(p)$ defined over a power capacity interval $[0, \mathrm{Cap}_1^{\mathrm{U}}]$, and the supply offer of GenCo 2 consists of a strictly increasing linear marginal cost function $\mathrm{MC}_2(p)$ defined over a power capacity interval $[0, \mathrm{Cap}_2^{\mathrm{U}}]$. The aggregation of these supply offers, without consideration of transmission constraints, will be referred to as the *unconstrained aggregate supply curve*.

For visual simplification, Figure 13.1 plots the unconstrained aggregate supply curve for the case in which MC_1 is *more* steeply sloped than MC_2 and there exists a power level $p^{\mathrm{kink}} > 0$ satisfying $\mathrm{MC}_1(p_1^{\mathrm{kink}}) = \mathrm{MC}_2(0)$. In this case, the unconstrained aggregate supply curve is a concave continuous piece-wise linear curve with a kink point at p^{kink}. Merit-order dispatch would require the

ISO to switch from G1 power generation to G2 power generation at power levels exceeding p^{kink}.[5]

In the absence of transmission congestion, market clearing would take place at the point where the aggregate demand curve intersects the unconstrained aggregate supply curve. At this point, the cheaper GenCo G1 would be the sole supplier for all cleared demand, and this cleared demand would include both fixed and price-sensitive portions. However, as indicated in Figure 13.1, this would require G1 to deliver power to bus 2 at a level that exceeds the transmission capacity limit M, which is not feasible. Once G1's power level hits M, the ISO is in fact forced to switch from G1 to G2 to service the remaining demand at bus 2.

Thus, the *effective* aggregate supply curve is the kinked curve in Figure 13.1 labeled *constrained aggregate supply*; and market clearing (subject to system constraints) takes place at the point where the aggregate demand curve intersects this constrained aggregate supply curve. At this constrained market clearing point, the LSE at bus 2 is charged LMP_2 for each MW of its cleared fixed demand $p_L^F > \text{M}$. The LSE's cost (LMP payment) is therefore given by $\text{LMP}_2 p_L^F$. A portion M of this fixed demand is supplied by GenCo G1 at bus 1, who is paid a price LMP_1 for this supply. The remaining portion $[p_L^F - \text{M}]$ of this fixed demand is supplied by GenCo G2 at bus 2, who is paid a price $\text{LMP}_2 > \text{LMP}_1$ for this supply.

The net surplus extractions depicted in Figure 13.1 for operating hour H are measured in \$/h units. The buyer net surplus extracted by the LSE is given by the area labeled B. The seller net surplus extracted by GenCo G1 is given by the triangular area labeled S1, and the seller net surplus extracted by GenCo G2 is given by the triangular area labeled S2. The ISO net surplus [LCost − GRev], given by

$$[\text{LMP}_2 \cdot p_L^F] - [\text{LMP}_1 \cdot \text{M}] - \text{LMP}_2 \cdot [p_L^F - \text{M}] = [\text{LMP}_2 - \text{LMP}_1]\text{M} \tag{13.18}$$

is depicted as a shaded area.

The TNS for operating hour H, measured in \$/h units, is the summation of these LSE, GenCo, and ISO net surplus extractions. That is, TNS is the summation of B, S1, S2, and the shaded area depicted in Figure 13.1. This summation can alternatively be expressed as the area between the aggregate demand curve and the constrained aggregate supply curve, starting from the power level $p = 0$ and ending at the power level p_L^F at which these two curves intersect.

Figure 13.1 illustrates several important general properties of ISO net surplus extraction in ISO markets that were briefly noted earlier in this section. In the absence of a binding transmission grid limit M, the ISO Net Surplus would reduce to zero and LMP_1 would equal LMP_2. However, given M is binding, the ISO Net Surplus is strictly positive and LMP_1 separates from LMP_2.

[5] Another possibility is that $\text{MC}_1(\text{Cap}_1^{\text{U}}) \leq \text{MC}_2(0)$ with MC_1 *less* steeply sloped than MC_2. In this case, in the absence of transmission grid congestion, G1's upper power capacity limit (not merit-order dispatch) would force a switch from G1 to G2 to meet any power demand in excess of Cap_1^{U}. Thus, if $\text{MC}_1(\text{Cap}_1^{\text{U}}) = \text{MC}_2(0)$, the unconstrained aggregate supply curve is a *convex* continuous piecewise-linear curve with a kink at Cap_1^{U}. Alternatively, if $\text{MC}_1(\text{Cap}_1^{\text{U}}) < \text{MC}_2(0)$, the unconstrained aggregate supply curve is a piece-wise linear curve that jumps discontinuously at Cap_1^{U} from MC_1 to MC_2.

13.5 MARKET POWER IN THE ILLUSTRATIVE DAM: TYPE-I ERROR

As discussed in Section 12.9, *buyer market power* refers to the ability of a buyer to profitably lower the purchase price(s) it pays for the quantity units it purchases, relative to competitive market clearing (CMC) conditions. Conversely, *seller market power* refers to the ability of a seller to profitably raise the sale price(s) it is paid for the quantity units it sells, relative to CMC conditions. Buyer market power is not considered problematic in current ISO DAMs since most demand still takes a must-service (price-insensitive) form. Consequently, only seller market power is considered in this section.

Seller market power takes two main forms. First, a seller can engage in *economic withholding of capacity* by reporting higher-than-true marginal costs. Second, a seller can engage in *physical withholding of capacity* by reporting a lower-than-true maximum power capacity. Seller market power in either form can hinder the efficient and fair use of existing resources as well as the proper signaling of future transmission and generation investment needs.

For example, economic withholding of capacity can induce higher prices for cleared supply as well as *out-of-merit-order dispatch*, i.e. more expensive generation dispatched in place of less expensive generation. This results in inefficient (and politically important) transfers of net surplus away from LSEs and their retail customers and toward GenCos.

Physical withholding of capacity can induce higher prices for remaining offered capacity and hence higher net surplus for GenCos that withhold only a portion of their capacities. It can also result in out-of-merit-order dispatch. In addition, physical withholding of capacity increases the chances of *inadequacy events* in which offered capacity is insufficient to meet total fixed demand. Such events force ISOs to take potentially costly special actions, such as additional reserve commitments or load shedding, to avoid the breakdown of power balance on the grid.

The exercise of seller market power in ISO DAMs can thus have substantial adverse effects on the efficiency, reliability, and fairness of market operations. Nevertheless, it is difficult to construct measures for its reliable detection [167, 192].

On the one hand, the complexity of the rules and regulations governing ISO DAM operations creates opportunities for GenCos to exercise seller market power, either individually or in coalitions, while still complying with all rules and regulations. On the other hand, the physical laws governing power flow through transmission grids cause these grids to be strongly connected networks. Injections or withdrawals of power at one location on the grid can have substantial near-immediate effects on prices and power flows at distant locations, creating the appearance of seller market power even if no seller market power has been deliberately exercised.

Regarding the latter issue, consider the impact of grid network effects on ISO DAM prices. The LMP determined by means of a day-D ISO DAM SCED optimization at a particular grid bus *b* for a particular hour H of day D+1 is typically calculated as a *right-hand system marginal cost*: namely, the least cost *to the system* of servicing an *additional* increment of fixed demand at bus *b* during hour H of day D+1, subject to system constraints.[6] This additional increment of fixed demand can be serviced in three possible ways: (i) increasing generation somewhere in the system; (ii) cutting back on cleared price-sensitive demand (if any) somewhere in the system to free up generation for servicing the additional fixed demand; or (iii) some combination of the two.

A GenCo *i* is said to be *right-hand marginal* at a system operating point if it is operating strictly below its upper power capacity limit Cap_i^{U} at this point. Suppose LMPs are calculated as right-hand system marginal costs. Then the LMP at any bus where there is a right-hand marginal GenCo *i* is given by *i*'s *right-hand marginal cost*, i.e. by the change in *i*'s avoidable cost given an incremental *increase* in its power generation.[7] Moreover, the LMP at any bus with no right-hand marginal GenCo is determined as a linear combination of the right-hand marginal costs of the right-hand marginal GenCos at other buses.

If the injection of power at a particular grid location leads to grid congestion, this will cause at least some separation of LMPs across the grid. Moreover, if price-sensitive demand then has to be curtailed, this grid congestion will also result in a failure to fully extract all buyer and seller net surplus that would be possible to extract based on demand and supply conditions alone, a situation economists refer to as deadweight loss.

Consider, for example, the benchmark (no-learning) 5-bus DAM test case reported in [112] that was simulated by means of the AMES Wholesale Power Market Test Bed [185]. Figure 13.2 depicts the 5-bus transmission grid for this test case. Figure 13.3 depicts the true GenCo marginal cost curves and power capacity limits.

Figure 13.4 displays the simulated hourly aggregate demand and unconstrained aggregate supply curves for the peak-demand hour 17. The initial vertical portion of the aggregate demand curve corresponds to LSE fixed demand and the remaining portion corresponds to LSE price-sensitive demand. The ratio of fixed to price-sensitive demand is calibrated to demand data from MISO. The intersection of these aggregate demand and supply curves occurs at (1245 MW, \$32/MWh). Note that the cleared demand at this CMC point includes both fixed and price-sensitive demand.

---

[6]If the optimized DAM SCED objective function is a differentiable function of the constraint constant cc appearing in the power balance constraint for bus *b* during hour H of day D+1, then the LMP is uniquely determined as this derivative; the qualifier "right hand" is not needed. However, if the optimized SCED objective function is left-differentiable and right-differentiable with respect to cc, but not differentiable with respect to cc, then the *sign* of the incremental change in cc will matter for the determination of the resulting change in the optimized SCED objective function. Note that LMPs are *not* conceptually equivalent to the CMC price levels defined in Section 12.5. See [120] for a careful definition and analytical derivation of LMPs as possibly non-unique dual variable solutions for power balance constraints at SCED optimization points; and see [69, 196] for a variety of surprising and confounding properties of LMPs.

[7]For later purposes, note that the right-hand marginal cost of a GenCo *i* operating at its upper power capacity limit Cap_i^{U} is effectively infinite because its marginal cost curve is vertical at Cap_i^{U}.

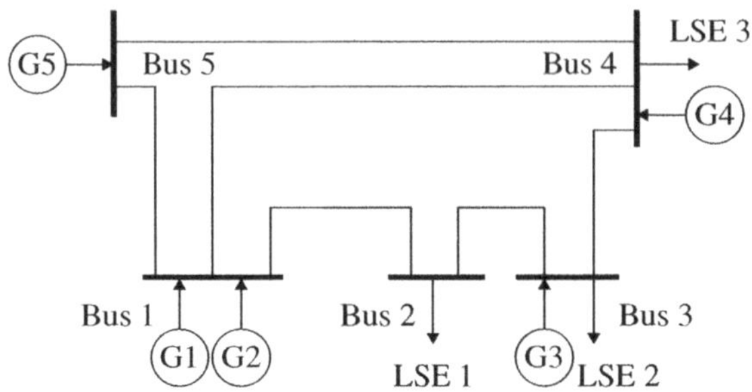

Figure 13.2 5-Bus grid [108] used for the 5-bus DAM test cases reported in [112].

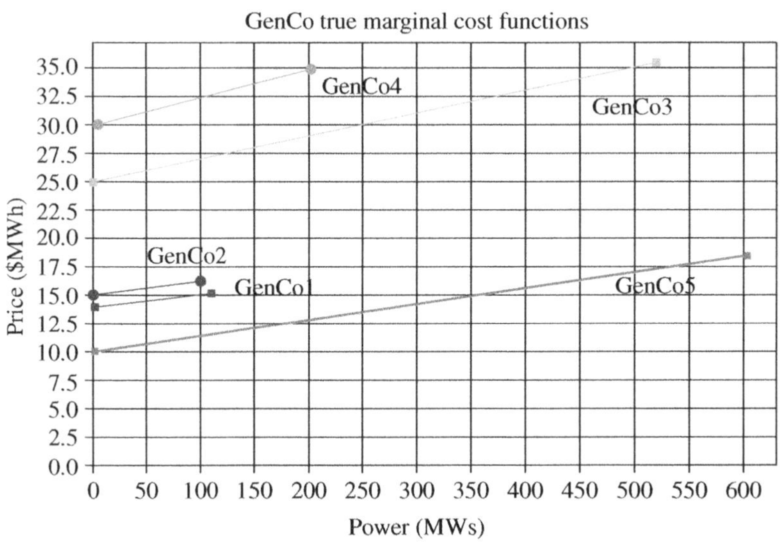

Figure 13.3 True GenCo marginal cost curves and power capacity limits for the 5-bus DAM test cases reported in [112].

However, as shown in [112], due to congestion on the transmission line between bus 1 and bus 2, only fixed demand (1153.58 MW) is actually cleared; no price-sensitive demand is actually cleared, indicating a deadweight loss. Moreover, the congestion causes strong LMP separation across the grid. The LMP at bus 1 is \$14.02/MWh, the LMP at bus 2 is \$78.24/MWh, the LMP at bus 3 is \$66.07/MWh, the LMP at bus 4 is \$32.61/MWh, and the LMP at bus 5 is \$17.32/MWh. The average of these actual LMP values is \$41.65/MWh, much higher than the CMC price level \$32/MWh depicted in Figure 13.4.

As is well understood, the DAM LMP paid to each capacity-constrained (non-marginal) GenCo for each MW it is dispatched to sell can strictly exceed its *left-hand* marginal cost. The MWs supplied by these capacity-constrained GenCos constitute inframarginal quantity units, similar to the quantity units to the left of $q^* = 5$ for the competitive market model depicted in Figure 12.1. What is not as

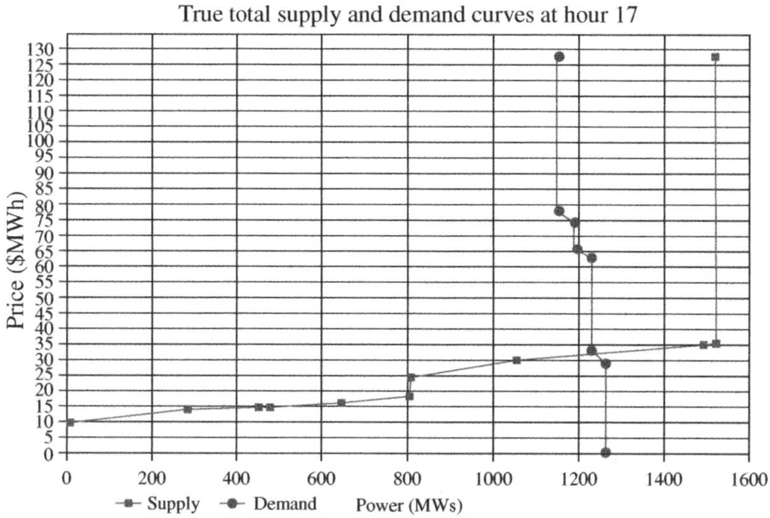

Figure 13.4 Aggregate demand and unconstrained aggregate supply curves during peak-demand hour 17 for the benchmark (no-learning) 5-bus DAM test case reported in [112].

well understood, however, is that an LMP at some bus n can strictly exceed the *right-hand* marginal cost of all *right-hand* marginal GenCos located at buses $n' \neq n$ if grid congestion requires out-of-merit-order dispatch.

For example, to service an additional 1 MW of fixed demand at some bus n during some hour H in the presence of grid congestion might require that less expensive generation at some second bus n' be backed down, e.g. by 2 MW during hour H at \$20/MWh, and that more expensive generation at some third bus n'' be brought up, e.g. by 3 MW during hour H at \$30/MWh, in order to avoid overloading an already constrained transmission line. In this case, the right-hand system marginal cost of servicing an additional 1 MW of fixed demand at bus n during hour H – i.e. the LMP at bus n for hour H – is \$50/MWh = [3· \$30 – 2·\$20]/MWh. If the GenCo at n'' supplying 3 MW during hour H at \$30/MWh has additional power capacity available during hour H at a marginal cost less than \$50/MWh, then the LMP at bus n strictly exceeds the right-hand marginal cost of this right-hand marginal GenCo at bus n''.

Indeed, this scenario occurs during the peak-demand hour 17 depicted in Figure 13.4 for the 5-bus DAM test case reported in [112]. In this test case every GenCo submits its true marginal cost function as its DAM supply offer. Thus, no GenCo exercises seller market power by engaging either in economic capacity withholding or in physical capacity holding. Nevertheless, at the resulting DAM solution point the LMP at bus 2 strictly exceeds the right-hand marginal costs of the three right-hand marginal GenCos G1, G4, and G5 located at buses 1, 4, and 5.

The above example indicates that care is needed when attempting to detect seller market power in ISO markets by means of metrics that rely solely on comparisons between price and marginal cost. One such measure is the well-known *Lerner*

Index $[\pi^* - \text{MC}(q^*)]/\pi^*$, where π^* denotes a market price for a commodity Q and q^* denotes the quantity of Q that is cleared at this price; see [167]. A finding that $\pi^* > \text{MC}(q^*)$ is commonly interpreted to mean that seller market power is being exercised. However, as shown by the above example, such conclusions could be completely erroneous.

13.6 MARKET POWER IN THE ILLUSTRATIVE DAM: TYPE-II ERROR

Section 13.5 discusses how a market monitor using standard types of seller market power metrics could *erroneously* conclude that generators in ISO markets *are* exercising seller market power. This section uses 5-bus DAM test case outcomes from Li and Tesfatsion [112] to illustrate the reverse issue: namely, that market monitors using standard types of seller market power metrics could *erroneously* conclude that generators in ISO markets *are not* exercising seller market power.

The 5-bus transmission grid and true GenCo marginal cost curves for these 5-bus DAM test cases are depicted in Figures 13.2 and 13.3. A key treatment factor for these test cases is the absence or presence of GenCo learning capabilities. In the no-learning treatment, each GenCo submits its true marginal cost function (13.3) as its hourly DAM supply offer. In the learning treatment, each GenCo uses reinforcement learning in an attempt to improve its net earnings over time through economic capacity withholding. Specifically, each GenCo i experiments with alternative settings $b_i^{\text{R}} \geq b_i$ and $c_i^{\text{R}} \geq c_i$ for the marginal cost function $\text{MC}_i^{\text{R}}(p) = b_i^{\text{R}} + 2c_i^{\text{R}}p$ that it reports to the ISO as its hourly DAM supply offer.

For both the no-learning and learning treatments, congestion persistently occurs on the transmission line connecting bus 1 to bus 2, hereafter referred to as *line 1–2*. This congestion results in LMP separation across all five buses.

The ratio of potential price-sensitive demand to potential total demand (both fixed and price sensitive) is measured in these 5-bus DAM test cases by a metric called the *R ratio*. As seen in Figure 13.5, a switch from the no-learning treatment to the learning treatment results in an increase in average hourly LMP levels for each given R value; and this increase is substantial for small R values (i.e. for cases in which demand is largely fixed). This LMP increase occurs because *each* of the five GenCos in the learning treatment succeeds in learning over time how to advantageously exercise economic capacity withholding by reporting higher-than-true marginal costs.

This exercise of economic capacity withholding by the five GenCos in the learning treatment also has dramatic effects on ISO and GenCo net surplus extractions. These dramatic effects are graphically depicted in Figures 13.6 and 13.7.

Consider, for example, the $R = 0.0$ (100% fixed demand) daily data presented for the no-learning treatment in Figure 13.6 and for the learning treatment in Figure 13.7. Average LSE cost (LMP payments) on day 1000 for the learning treatment are $5,040,530.89, an approximately 6.7-fold increase relative to the no-learning treatment. However, average ISO net surplus on day 1000 for the learning

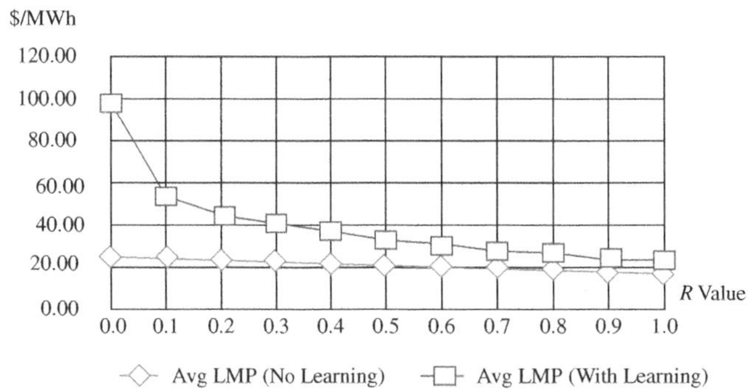

Figure 13.5 Average hourly LMP outcomes on day 1000 for the 5-bus DAM test cases reported in [112], both with and without GenCo learning, as the portion R of price-sensitive LSE demand varies from $R = 0.0$ (100% fixed) to $R = 1.0$ (100% price sensitive).

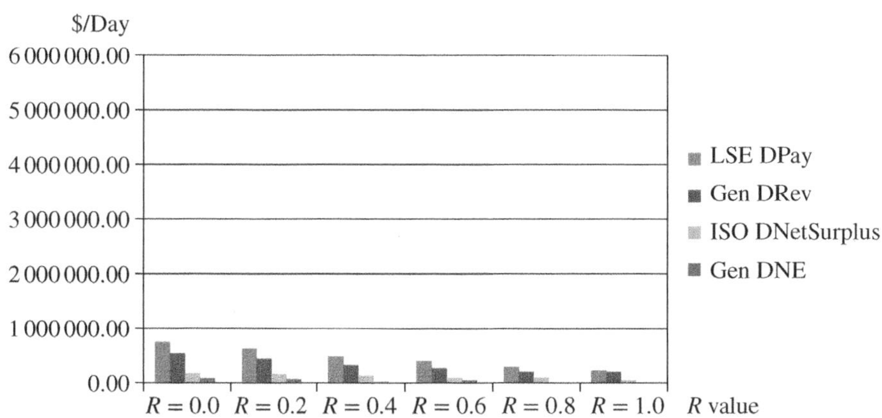

Figure 13.6 *No-Learning Treatment*: Average LSE cost (DPay), GenCo revenue (DRev), ISO net surplus (DNetSurplus), and GenCo net surplus (DNE) during a typical day D for the 5-bus DAM test cases reported in [112] for which GenCos *do not* have learning capabilities. The portion of price-sensitive LSE demand varies from $R = 0.0$ (100% fixed) to $R = 1.0$ (100% price sensitive).

treatment is then $2,097,620.96, an almost 10-fold increase relative to the no-learning treatment. Indeed, average ISO net surplus under learning is similar in magnitude to average GenCo net surplus under learning ($2,441,646.71).

Since total demand for $R = 0.0$ is the same under learning and no learning, the 10-fold increase in average ISO net surplus extraction under learning implies that the average LMP paid by the LSEs is substantially higher than the average LMP received by the GenCos. This is due to the approximately sixfold increase under learning in the average LMP for bus 2, which has the largest load (LSE 1) and no generation, and to the much smaller increases under learning in the average LMPs for buses 1 and 5, which have generation but no load.

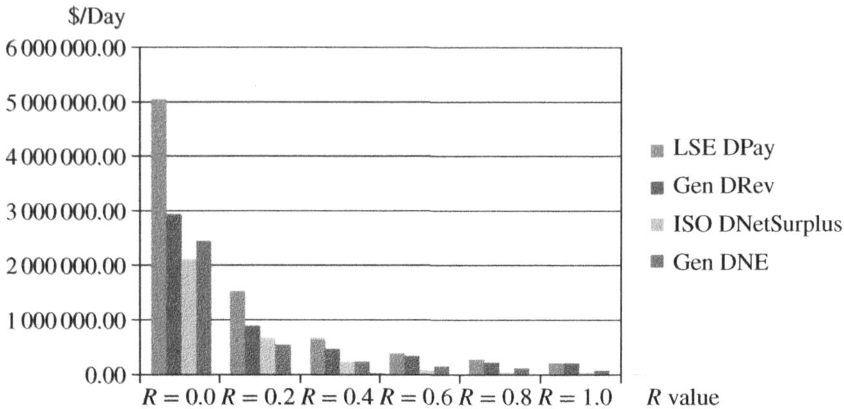

Figure 13.7 *Learning Treatment*: Average LSE cost (DPay), GenCo revenue (DRev), ISO net surplus (DNetSurplus), and GenCo net surplus (DNE) during day 1000 for the 5-bus DAM test cases reported in [112] for which GenCos *do* have learning capabilities. The portion of price-sensitive LSE demand varies from $R = 0.0$ (100% fixed) to $R = 1.0$ (100% price sensitive).

Another regularity observed in Figures 13.6 and 13.7 is that average LSE cost (LMP payments), ISO net surplus, and GenCo net surplus each exhibit a marked monotonic decrease as R increases from $R = 0.0$ (100% fixed demand) to $R = 1.0$ (100% price-sensitive demand). The explanation for this monotonic decrease is as follows.

Consider, first, the no-learning treatment outcomes reported in Figure 13.6. Given low R values, the LSEs have very low price resistance; their fixed demands constitute the bulk of their total demands. Around the peak-demand hour 17, due in part to congestion on line 1–2, the ISO must dispatch the most expensive GenCos 3, 4, and 5 to meet the large LSE fixed demand, i.e. these GenCos are pivotal suppliers for hour 17. This results in relatively high LMPs. As R increases, however, the LSEs are increasingly able to resist high prices through demand withholding. This results in lower LMPs, lower total demand, and lower avoidable costs of production. Average GenCo revenues and LSE payments are thus lower, and average GenCo net surplus outcomes are also lower because the decrease in GenCo avoidable costs is more than offset by the decrease in GenCo revenues. Similarly, average ISO net surplus is lower because the decrease in average GenCo revenues is more than offset by the decrease in average LSE payments.

Next consider the learning-treatment outcomes in Figure 13.7. For $R = 0.0$ the average LSE cost (LMP payments), ISO net surplus, and GenCo net surplus attained under the learning treatment are substantially greater than their corresponding levels attained under the no-learning treatment in Figure 13.6. As R increases, however, these average learning levels all eventually drop below their corresponding no-learning levels, the switch point occurring at about $R = 0.6$.

The explanation for these switch-point findings can be deduced from detailed LMP and total demand findings for the no-learning and learning treatments. When GenCos are learners, low R values (implying large fixed demands) provide pivotal suppliers with a substantial opportunity to engage in profitable economic capacity withholding. This dramatically increases LMPs relative to the no-learning treatment, particularly at the load-only bus 2. Since total demand for the learning treatment is only modestly lower than for the no-learning treatment for low R values, the end result is substantially greater average LSE cost (LMP payments), ISO net surplus, and GenCo net surplus.

On the other hand, as R increases and the LSEs acquire an increasing ability to resist high prices through demand withholding, the learning GenCos are increasingly forced to compete with each other for dispatch by lowering their reported marginal costs. This competitive process results in lower LMPs. However, the LMPs resulting under learning remain higher than under no learning for all R values, which in turn induces the LSEs to engage in greater demand withholding under learning.

The end result is that average GenCo revenue, average LSE cost, and average ISO net surplus under learning all fall below their corresponding no-learning values as R approaches 1.0 due to the relatively strong contraction in total demand under learning. As can be verified from the GenCo net earnings data provided in [112, Section IV], the most expensive GenCo 4 is at the greatest disadvantage in this process while the least expensive GenCo 5 is most advantaged.

How easy would it be for market monitors to detect the type of seller market power exhibited by GenCos in these 5-bus DAM test cases under the GenCo learning treatment? Two difficulties can be noted.

First, *each* of the GenCos in these test cases engages in economic capacity withholding; and *each* of these GenCos benefits from this withholding in the sense that its average net surplus extraction increases relative to the no-learning treatment. Average ISO net surplus also increases. The only potential "losers" are the LSEs, whose average LMP payments increase. However, LSEs in current ISO DAMs can protect themselves from higher LMP payments by petitioning state boards to increase their regulated retail rates r. Consequently, the only ultimate losers are retail customers, and these retail customers do not have a direct seat at the ISO DAM table. Thus, a market monitor cannot expect to receive complaints about seller market power from the DAM market participants themselves.

Second, this GenCo economic capacity withholding results in higher hourly LMPs at each bus that are in perfect accordance with the marginal cost functions that the GenCos are reporting to the ISO as their hourly DAM supply offers. Thus, a market monitor will not observe sharp price spikes, widening divergence between LMPs at different buses, or suspicious divergence between prices and reported marginal costs for marginal GenCos. Market monitors that attempt to identify the exercise of seller market power through these types of market observations will all fail.

In short, successful detection of the type of seller market power exercised in these 5-bus DAM test cases with GenCo learning would require a careful auditing of individual GenCos to assess their true marginal costs.

13.7 MARKET INEFFICIENCY IN THE ILLUSTRATIVE DAM

This section uses the ISO DAM formulation presented in Section 13.3 to highlight and expand upon a key point raised in [167]: namely, the use of LMP settlements in current ISO DAMs renders problematic the attainment of market efficiency.

Consider an ISO DAM held on day D for a particular operating hour H of day D+1, modeled as in Section 13.3. For each LSE j, let $n(j)$ denote its bus location. Also, let p^S_{Lj} and p^F_{Lj} denote the quantities of power that LSE j is cleared to buy in the day-D DAM based on its *reported* price-sensitive demand bid $\mathrm{F}^{\mathrm{R}}_j(p)$ and its fixed demand bid, respectively. Finally, let $\mathrm{LMP}_{n(j)}$ denote the LMP determined in the day-D DAM for bus $n(j)$ during hour H of day D+1.

Note that $\mathrm{F}^{\mathrm{R}}_j(p)$ ($/MWh) represents LSE j's *reported* purchase reservation value (maximum willingness to pay) for an incremental change dp in its power delivery during H, evaluated at p. This reported purchase reservation value can differ strategically from LSE_j's true purchase reservation value at p.

Note, also, that LSE j's fixed demand bid is a demand for power delivery at a constant level p^F_{Lj} during hour H, with no accompanying purchase reservation value information.[8] Fixed demand bids submitted into ISO DAMs are treated as must-service power demands except in emergency situations. Consequently, consideration of the benefit accruing to fixed demands does not help to distinguish between the efficiency levels attained under distinct routine market scenarios characterized by the same fixed demand bids because the corresponding fixed-demand benefits simply cancel out when the benefits arising under any two of these scenarios are differenced. For this reason, current ISO DAMs do not solicit purchase reservation values for fixed demand bids; and the benefits accruing to these fixed demand bids are routinely omitted from consideration in ISO DAM objective functions.[9]

The benefit obtained by LSE j from the day-D DAM for hour H of day D+1 is thus assumed to be calculated by the ISO based solely on LSE j's reported price-sensitive demand bid. Specifically, the net surplus ($) for LSE j, calculated as benefit minus cost (LMP payments), takes the form

$$\mathrm{LNS}^{\mathrm{R}}_j = \left[\int_0^{p^S_{Lj}} [\mathrm{F}^{\mathrm{R}}_j(p)]dp - \mathrm{LMP}_{n(j)} \cdot p_{Lj} \right] \cdot 1 \text{ hour} \tag{13.19}$$

where $p_{Lj} = [p^F_{Lj} + p^S_{Lj}]$.

For each GenCo i, let $n(i)$ denote its bus location and let p_{Gi} denote its power level determined in the day-D DAM for scheduled dispatch at bus $n(i)$ during hour H of day D+1. Also, let $\mathrm{LMP}_{n(i)}$ denote the LMP determined in the day-D DAM for

[8] As seen in Section 13.3, if an LSE resells all procured power to retail customers at a regulated flat rate r covering all of its retail marginal costs, then its purchase reservation value (maximum willingness to pay) for this procured power at wholesale would presumably be r.

[9] A special case of this is when all demand bids take a fixed-demand form and the ISO's DAM objective is expressed as the minimization of the cost incurred to service this fixed demand.

bus $n(i)$ during hour H of day D+1. Then, the net surplus ($) extracted by GenCo i from the day-D DAM for hour H of day D+1 is calculated as

$$\mathrm{GenNS}_i^{\mathrm{R}} = \left[\mathrm{LMP}_{n(i)}\, p_{Gi} - \int_0^{p_{Gi}} [\mathrm{MC}_i^{\mathrm{R}}(p)]dp\right] \cdot 1 \text{ hour} \tag{13.20}$$

In (13.20), $\mathrm{MC}_i^{\mathrm{R}}(\mathrm{p})$ ($/MWh) denotes GenCo i's *reported* marginal cost for an incremental change dp in its dispatch set point p at the start of hour H. This reported marginal cost can differ from GenCo i's true marginal cost at p.

The TNS ($) extracted by LSEs and GenCos in the day-D DAM for hour H of day D+1, based on their reported demand bids and supply offers, thus takes the following form:

$$\begin{aligned} \mathrm{TNS}_{\mathrm{L,G}}^{\mathrm{R}} &= \mathrm{LNS}^{\mathrm{R}} + \mathrm{GenNS}^{\mathrm{R}} \\ &\equiv \sum_{j=1}^{J} \mathrm{LNS}_j^{\mathrm{R}} + \sum_{i=1}^{I} \mathrm{GenNS}_i^{\mathrm{R}} \end{aligned} \tag{13.21}$$

Suppose the goal of the ISO in the day-D DAM is to achieve market efficiency by mimicking as closely as possible the actions of an auctioneer managing a standard competitive market, as analyzed in Section 12.5. Thus, the goal of the ISO is to maximize the TNS extracted by buyers (LSEs) and sellers (GenCos) from the day-D DAM, subject to system constraints. However, in attempting to carry out this goal, the ISO must in fact rely on the *reported* demand bids and supply offers of these buyers and sellers.

Consequently, for each hour H of day D+1, the ISO approximates the TNS extracted by LSEs and GenCos in the day-D DAM by the area between the aggregate demand schedule and the unconstrained aggregate supply schedule up to their point of intersection, where these aggregate schedules are constructed from the *reported* price-sensitive demand bids $\mathrm{F}_j^{\mathrm{R}}(p)$ and *reported* supply offers $\mathrm{MC}_i^{\mathrm{R}}(p)$ of the participant LSEs and GenCos. This area is given by

$$\mathrm{Area}_{\mathrm{L,G}}^{\mathrm{R}} = \left(\sum_{j=1}^{J} \int_0^{p_{Lj}^S} [\mathrm{F}_j^{\mathrm{R}}(p)]dp - \sum_{i=1}^{I} \int_0^{p_{Gi}} [\mathrm{MC}_i^{\mathrm{R}}(p)]dp\right) \cdot 1 \text{ hour} \tag{13.22}$$

The LSE cost (LMP payments) and GenCo revenue (LMP receipts) arising from the operations of the day-D DAM are settled through the ISO at the end of day D. As discussed in Section 13.3, the ISO net surplus extracted from the day-D DAM is defined to be LSE cost minus GenCo revenue. For the example at hand, this ISO net surplus ($) is given by

$$\mathrm{ISONS} = \left[\sum_{j=1}^{J} \mathrm{LMP}_{n(j)} p_{Lj} - \sum_{i=1}^{I} \mathrm{LMP}_{n(i)} p_{Gi}\right] \cdot 1 \text{ hour} \tag{13.23}$$

where $p_{Lj} = p_{Lj}^F + p_{Lj}^S$.

Combining (13.20) through (13.23), it is seen that

$$\text{Area}^{\text{R}}_{\text{L,G}} = [\text{TNS}^{\text{R}}_{\text{L,G}} + \text{ISONS}] \tag{13.24}$$

Expression (13.24) shows that two factors can cause $\text{Area}^{\text{R}}_{\text{L,G}}$ to diverge from the *true* TNS $\text{TNS}_{\text{L,G}}$ extracted by LSEs and GenCos from the day-D DAM. First, $\text{TNS}^{\text{R}}_{\text{L,G}}$ can deviate from $\text{TNS}_{\text{L,G}}$ due to strategic reporting of demand bids and supply offers. Second, grid congestion can cause LMPs to separate across the grid, resulting in a strictly positive value for ISONS.

Consequently, the ISO's maximization of $\text{Area}^{\text{R}}_{\text{L,G}}$ does not guarantee the maximization of true total buyer and seller net surplus $\text{TNS}_{\text{L,G}}$. As discussed in Section 12.7, maximum extraction of true total buyer and seller net surplus is the standard economic definition of market efficiency.

With regard to the first difficulty, ISOs generally engage the services of independent market monitors in an attempt to prevent strategic behaviors intended to exploit market power opportunities. However, the findings reported in [110–112], illustrated in Section 13.6, suggest that incentives for the exercise of seller market power through strategic reporting of supply offers will remain a serious concern in ISO markets as long as the bulk of LSE demand is passively bid in as fixed demand.

With regard to the second difficulty, the LMPs calculated for a particular hour H by means of a standard DAM SCED optimization (a DC-OPF problem with zero power losses) collapse to a single uniform LMP level across the grid in the absence of transmission congestion. In this case, the ISO net surplus is zero since the bus power balance constraints for this optimization together imply that total quantity of power bought must equal total quantity of power sold.

On the other hand, Alderete [3, Proposition 2.1] proves for a general DAM DC OPF optimization formulation that ISO net surplus is always strictly positive if there is any transmission congestion. Also, Li and Tesfatsion [112] report substantial positive ISO net surplus outcomes in a suite of computational experiments conducted for 5-bus and 30-bus test cases with transmission congestion when LSE demand is predominately fixed.

In response to this problem, ISOs typically distribute at least part of their ISO net surplus as "congestion rent" payments to participants in markets for *financial transmission rights (FTRs)*.[10] The basic rationale for FTRs is twofold: to provide hedging instruments against LMP volatility caused in large part by transmission congestion; and to encourage appropriate investment in new transmission capacity to alleviate the congestion causing the positive ISO net surplus. However, as discussed more carefully in Section 13.8, it is not clear that FTR markets are currently meeting these goals.

---

[10]As will be explained more carefully in Section 13.9, an FTR obligation is an ISO-managed financial contract that commits holders to receive or make payments based on deviations between LMP outcomes at designated starting and ending buses (nodes) for transmission grid lines. In some energy regions, FTRs are referred to by other names, such as *congestion revenue rights (CRRs)*. For simplicity of exposition, this study uses FTR to refer to all of these financial instruments.

13.8 DAM PERFORMANCE: GENERAL ASSESSMENT

A basic responsibility undertaken by ISOs in ISO DAMs is the maximization of TNS subject to system constraints. An important issue highlighted in earlier sections of this chapter is whether this constrained maximization of TNS properly encourages efficient DAM operations.

As seen in Sections 13.5–13.7, two concerns arise. First, the ISO must rely on *reported* rather than true buyer and seller reservation values when calculating TNS. Second, as illustrated in (13.17), a net surplus dollar accruing to market participants is weighed equally with a net surplus dollar accruing to the ISO itself.

With regard to the first concern, the simulation findings summarized in Sections 13.5–13.7 show cause for worry. The incentives for generators participating in ISO DAMs to misrepresent their marginal costs is high, especially when LSE demand is predominately fixed (insensitive to price).

With regard to the second concern, a number of commentators (e.g. [119]) have compared ISO net surplus with the net surplus of governments in standard welfare economics analyses. However, welfare economists do not assert the *unqualified* desirability of assigning government net-surplus dollars (e.g. tax revenues net of government payments) the same weight as private trader net-surplus dollars in social welfare objective functions. The question always raised when a new tax increase is proposed is what can the government do for the people that they cannot better do for themselves? What is the social benefit or cost of taking a dollar from a private pocket and placing it in the public purse?

Similarly, a key issue for the ISO TNS objective function (13.17) is whether a dollar extracted by the ISO as ISO net surplus is properly treated as having the same social benefit or cost as a dollar flowing to a private market participant. The answer surely depends on how the ISO allocates this ISO net surplus.

As reported in [112], the current practice in many ISO markets is to use ISO net surplus extractions largely as revenue payments to FTR holders. This practice could lead to an overall increase in TNS over the longer run (i.e. to improved longer-run market efficiency) to the extent that it either directly or indirectly incentivizes transmission investments that are socially beneficial because they alleviate congestion.

However, such an outcome is doubtful. The extent to which ISO net surplus payouts to FTR holders actually incentivize new transmission investment is unclear [14, 23], [143, Section 4.2], [158]).[11] Moreover, transmission investment needs can arise for reasons other than congestion, e.g. a need to reach distributed energy resources. Also, congestion might be alleviated more effectively by additional investment in local generation rather than by additional investment in transmission capacity. Consequently, the ability of ISO net surplus extractions to appropriately signal the need for new transmission is questionable.

[11] For example, the CAISO report [23, p. ES-3] reaches the following conclusion: "... the reality has been that the LMP differences have not provided enough incentives to upgrade key facilities even after many types of FTRs and CRRs are provided."

While the social benefits associated with ISO net surplus payouts to FTR holders are unclear, we do have some read on the sizeable nature of the social costs. Benjamin [10, Section V] empirically estimates that FTR market imperfections in ISO-NE, NYISO, and PJM from 2006 through 2008 resulted in hundreds of millions of dollars of additional annual expenses for retail rate-payers in these energy regions, and that this problem appears to be worsening over time.

In addition, as discussed more fully in [112], in some energy regions the residual annual ISO net surplus extractions not needed for FTR payments are used to offset the DAM LMP payments that LSEs are required to make to pay for the power usage of their customers. Here, however, it is important to keep in mind the intended market efficiency rationale for LMP pricing in relation to power demand: namely, to provide an incentive to buyers to bid power demand into the system when and where it can be serviced most cheaply. This incentive could significantly gain in importance if retail markets heed growing pressures to introduce advanced metering and other technologies permitting retail customers to better tailor their power demands to wholesale power prices and to choose their retail suppliers [53]. To the extent that the allocation of ISO net surplus to LSEs dampens their incentive to minimize their DAM LMP payments, it could become an increasing source of social inefficiency in future years.

Another issue also arises. As seen in Section 13.6, ISO and GenCo net surplus extractions dramatically increase when transmission lines are congested, the price-sensitivity of demand is low, and learning GenCos exercise economic capacity withholding. On the other hand, LSE required payments for power also dramatically increase under these conditions. This would appear to provide a desirable incentive to pure LSEs (those without generation ownership) to support congestion reduction measures, increased price-sensitivity of demand, and increased oversight to curtail GenCo exercise of seller market power.

Currently, however, LSE payments for fixed demand at wholesale are typically recovered through the resale of this fixed demand at regulated retail prices. To the extent that LSEs are able to secure timely increases in these regulated retail prices in step with increases in wholesale prices, the LSEs are able to largely insulate themselves from the adverse consequences of higher LMP payments. In this case, no direct participant in the ISO market suffers a loss of net surplus when LMPs increase due to congestion, fixed demand, and/or GenCo capacity withholding. Rather, losses in net surplus are borne by retail customers. Moreover, barriers to entry into transmission, generation, and load servicing could then lead to the persistence over time of socially inefficient wholesale market *rents*, i.e. wholesale market net surplus extractions in excess of the amounts needed to maintain resources in their current productive uses.

Power market researchers recognize that an important goal of market design is to ensure the alignment of market participant objectives with socially desirable outcomes, thus reducing the need for oversight of participant behaviors [158]. Alternatively expressed, power market designs should be incentive-compatible socially beneficial *mechanism designs* [86, 155].

However, the main conclusion drawn from the findings reported in Section 13.7 is that ISO net surplus extractions in current ISO markets are not well-aligned with social welfare. On the one hand, they increase with increases in transmission line congestion, a situation *unfavorable* to efficient resource allocation. On the other hand, the rules currently used to allocate ISO net surplus extractions do not provide effective incentives to market participants or other parties to reduce this congestion.

13.9 SCHEDULING OF BILATERAL CONTRACTS

Measured by energy volume, approximately half of all power trades in the United States currently take the form of privately negotiated physically covered bilateral contracts [74]. This section illustrates how power traders can secure transmission support for these trades in current ISO DAMs.

Consider, for example, the ERCOT DAM operations outlined in Section 5.4. As seen, qualified scheduling entities (QSEs) are permitted to submit energy offers into the ERCOT DAM at Resource Node settlement points[12] that take a three-part supply offer form. In addition, as explained in [47], QSEs are permitted to submit energy bids into the ERCOT DAM at Load Zone settlement points that take the form of proposals to buy energy at a price that decreases with increasing quantity. These energy offers and bids are settled at DAM settlement-point prices.[13]

In order for power traders to self-schedule the power outcomes of privately negotiated bilateral contracts in the ERCOT DAM, they must make use of energy offers and bids that comply with permitted ERCOT DAM forms. As will next be shown, these ERCOT offer/bid forms – augmented with appropriately specified supplementary contracts – permit transmission support for simple bilateral contracts that designate a single "strike price" for the purchase and sale of energy.

Let G denote a dispatchable generator located at a Resource Node consisting of an electrical bus A, and let L denote an LSE located at a Load Zone consisting of a single electrical bus B. Suppose G and L have entered into a simple bilateral contract for hour H of day D+1 that eliminates price risk for both parties. Specifically, this bilateral contract stipulates that, during hour H of day D+1, a specific power level p^* (MW) is to be injected by G at bus A and withdrawn by L at bus B at a specific *strike price* π^* ($/MWh) that is binding on both G and L.

Suppose G and L self-schedule in the day-D ERCOT DAM a maintained power injection p^* at bus A during hour H of day D+1 and a maintained power withdrawal p^* at bus B during hour H of day D+1 in order to secure the transmission services needed to support their bilateral contract agreement. In the subsequent settlement for the ERCOT DAM, generator G will be paid for its maintained power injection p^* in

[12] ERCOT defines three types of settlement points: Resource Nodes (electrical buses where a resource's measured output is settled); Load Zones (groups of electrical buses assigned to the same geographic region); and Hubs (protocol-designated groups of 345 kV buses).

[13] In the ERCOT DAM, the settlement-point price at a Resource Node is the LMP at this node; the settlement-point price for a Load Zone is a load-weighted average of the LMPs at the buses included in this Load Zone; and the settlement-point price for a Hub is a simple average of the LMPs at the buses included in this Hub.

accordance with the LMP determined at bus A for hour H of day D+1; let this LMP be denoted by LMP_A. Conversely, LSE L will be charged for its maintained power withdrawal p^* in accordance with the LMP determined at bus B for hour H of day D+1; let this LMP be denoted by LMP_B.

If at least one transmission line capacity limit is binding for the ERCOT DAM, LMP_A and LMP_B will possibly take on different values. In any case, these LMP values can differ from the bilateral contract strike price π^*. What steps can G and L take in advance of their participation in the ERCOT DAM to ensure that they ultimately end up paying and receiving a single energy price for p^* that coincides with their agreed-upon strike price π^*?

First, generator G can purchase an FTR obligation for a maintained power level p^* from bus A to bus B during hour H of day D+1. This FTR commits G to receive or make a payment $[LMP_B - LMP_A]p^* \cdot 1$ hour for hour H of day D+1. Combining this FTR obligation with its DAM LMP payment $LMP_A p^* \cdot 1$ hour, generator G's combined energy revenue ($) during hour H of day D+1 would be given by

$$LMP_A p^* \cdot 1\text{ hour} + [LMP_B - LMP_A]p^* \cdot 1\text{ hour} = LMP_B p^* \cdot 1\text{ hour} \tag{13.25}$$

Thus, G's FTR obligation effectively changes G's energy sale price from the LMP at bus A to the LMP at bus B.

Second, G and L can augment their bilateral contract with a *contract-for-difference (CFD)* agreement under which: (i) G agrees to "make whole" LSE L if LMP_B exceeds the strike price π^* by paying to L the $ amount

$$[LMP_B - \pi^*]p^* \cdot 1\text{ hour} \tag{13.26}$$

and (ii) L agrees to "make whole" generator G if LMP_B is below the strike price π^* by paying to G the $ amount

$$[\pi^* - LMP_B]p^* \cdot 1\text{ hour} \tag{13.27}$$

Thus, together with G's FTR obligation, this CFD guarantees that G's energy revenue ($) for hour H of day D+1 is given by $\pi^* p^* \cdot 1$ hour and L's energy payment ($) for hour H of day D+1 is given by $\pi^* p^* \cdot 1$ hour, as originally agreed to in their bilateral contract.[14]

In summary, the energy offers and bids permitted by the ERCOT DAM can be used to support the transmission needs of simple physically covered bilateral contract agreements designed to eliminate price risk through designated single strike prices as long as these agreements are appropriately supplemented with FTR obligations and CFD agreements. Securing transmission support for more complicated forms of physically covered bilateral contracts in a manner that preserves the original terms of these bilateral contract would require more complicated types of supplementary financial arrangements.

[14] The procurement price for the FTR obligation also has to be settled between G and L.

CHAPTER 14

COMPARISONS WITH SWING-CONTRACT MARKETS

14.1 CHAPTER PREVIEW

This chapter highlights important conceptual differences between *swing-contract* (SC) markets and current US RTO/ISO-managed wholesale power markets.

Section 14.2 addresses a key issue: What is the essential product transacted in wholesale power markets operating over transmission grids? A concise but careful argument is given that this essential product is reserve, consisting of collections of dispatchable power-paths with diverse attributes, not static amounts of energy (accumulated power). Section 14.3 then stresses an important corollary of this point: namely, that energy in wholesale power markets operating over transmission grids is not a commodity to which the Law of One Price can be applied, even if production and delivery of energy are distinguished by location and time period.

Section 14.4 next argues that the classic arguments in favor of uniform pricing in competitive commodity markets do not apply to SC markets because the product transacted in SC markets is not a commodity, i.e. not an asset whose units are perfect substitutes for each other. Rather, an SC market is a reserve market, where reserve consists of collections of dispatchable power-paths with diverse attributes. Consequently, to achieve efficient resource allocation, resources offering diverse reserve into an SC market must be permitted to seek diverse offer prices for reserve availability and diverse types of performance payments for reserve deployment.

Section 14.5 compares basic attributes and optimization formulations for SC markets and current US RTO/ISO-managed wholesale power markets, focusing on day-ahead markets (DAMs) for concreteness. Three basic advantages of the SC DAM design are highlighted. First, the SC DAM design is a two-part pricing design that permits full separate market-based compensation for reserve availability (insurance) and for actual real-time reserve performance (deployment). Second, no payments for performance are made in advance of actual verified real-time performance. Third, no out-of-market (OOM) make-whole payments are needed to ensure participants receive full compensation for their avoidable costs.

A New Swing-Contract Design for Wholesale Power Markets, First Edition. Leigh Tesfatsion.

As in Chapter 13, this chapter refers to US RTO/ISO-managed wholesale power markets in abbreviated form, as *ISO markets*, and the system operators for these markets are simply referred to as ISOs.

14.2 PRODUCT DEFINITION IN US RTO/ISO-MANAGED MARKETS

Current ISO markets are based on the fiction that the essential product being bought and sold in these markets is energy measured in megawatt-hours (MWh).[1]

Producer variable costs are assumed to be fully expressible as functions of produced energy, conditional on the location and time of this production; and customer benefits are assumed to be fully expressible as functions of delivered energy, conditional on the location and time of this delivery. Other attributes of the injections and withdrawals of power (MW) used to support energy production and delivery, such as reactive power support and down/up ramping, are assumed to be irrelevant for the determination of producer variable costs and customer benefits apart from their effects on overall system reliability.

Consider, for example, an ISO DAM operating over an N-bus transmission grid. The designated pricing locations for this ISO DAM are assumed to be the grid buses $b_1, \ldots, b_N$, with one of these buses (denoted by b_{ref}) arbitrarily selected to be the *reference bus*. The operating day D+1 for the ISO DAM conducted on day D is decomposed into operating hours H = 1, … ,24. The ISO DAM is then treated as a collection of grid-constrained competitive energy markets, one market M(b,H) for each bus-hour combination (b,H).

Suppose the ISO managing the DAM on day D has already conducted a SCUC optimization to determine unit commitments for operating day D+1. The ISO then conducts a security constrained economic dispatch (SCED) optimization to jointly clear the collection of energy markets M(b,H) for day D+1, subject to system constraints. For each bus-hour combination (b,H), this SCED optimization results in a locational marginal price LMP(b,H) ($/MWh) that can be expressed as a summation of energy, transmission, and power-loss components.[2] By construction, for each hour H the LMP determined for the reference bus b_{ref} consists solely of an energy component; and the energy component of the LMP at each other bus for hour H then coincides with this reference-bus energy component.

This ISO DAM design is a conceptually coherent way to determine energy prices, given the presumption that production costs and customer benefits are functions of energy output and energy consumption measured in MWh. In reality,

[1]In accordance with the International System of Units (SI) [138], a *watt* (W) is the standard unit for the measurement of power. A *megawatt* (MW) is one million watts. A *megawatt-hour* (MWh) is the standard unit for the measurement of energy; it is the amount of power accumulated over an hour when the flow of power is maintained at 1 MW during this hour.

[2]Standard DC-OPF SCED formulations often omit consideration of power losses.

however, this presumption is highly problematic; the attributes of power *paths* can matter greatly for the determination of these costs and benefits.

Perhaps the closest analogy in economics is to the modeling of labor markets. Consider the distinction between *labor*, measured in person-hours,[3] and *labor effort* defined to be the work effort exerted by a particular person at a particular point in time. Efficiency wage theorists emphasize the fundamental importance of this distinction; see, e.g. Weiss [194].

Specifically, efficiency wage theorists argue that the classic production function Y(T) = F(L(T)) expressing output Y(T) during a production period T as a function of the person-hours L(T) employed during T, conditional on available capital K(T), is fundamentally flawed. Economists should instead express this production function as a dynamic relationship $y(t) = f(e(t)\mathrm{L(T)})$ for $t \in \mathrm{T}$, where $e(t)$ measures the average amount of labor effort exerted at time t by the workers constituting the period-T labor force L(T). The labor effort $e(t)$ can vary widely over time, affected by worker responses to dynamically changing socio-economic conditions such as relative wage rates (affecting worker morale) and the risk of being fired.

In consequence, given a static amount L(T) of labor (person-hours) hired for a production period T, the total amount of output

$$q(\mathrm{T}) = \int_{\mathrm{T}} y(t)dt = \int_{\mathrm{T}} [f(e(t)\mathrm{L(T)})]dt \tag{14.1}$$

produced during T is a function of the *effective-labor path* given by ELP(T) = $\{e(t)\mathrm{L(T)} \mid t \in T\}$. Producer revenues and customer benefits obtained from the sale and use of output during T are thus functions of ELP(T) as well. In addition, the wage rate paid to a worker is typically assumed to be calculated as the worker's *marginal product of labor*, i.e. the change in output with respect to a change in the worker's labor. Suppose worker wages are calculated and paid as *efficiency wages* (using $f(\cdot)$) rather than as ordinary wages (using $\mathrm{F}(\cdot)$). Then variable production cost during T will also be a function of ELP(T). The bottom line is that the costs and benefits associated with the production and consumption of output during T will typically not be deducible solely from knowledge of L(T); path effects will matter.

Similar to labor, measured in person-hours, energy is measured in megawatt hours (MWhs). Hence, measured energy is inherently a time-dependent concept, the accumulation of power measured in MW over a designated time period T measured in hours h. How power is produced during T can affect its variable production cost, how power is injected into a transmission grid during T can affect its usefulness for net-load balancing, and how power is withdrawn from a transmission grid during T can affect its retail value. Thus, producers, ISOs, and customers might all apply different cost and benefit assessments to a particular quantity of energy produced, generated, or used during T depending on the particular stream of power whose production, injection, or withdrawal during T has resulted in this energy.

[3] A *person-hour* for a category of persons with a specific set of work skills is the work performed by one such person over the course of one hour.

For these reasons, this study argues that reserve (collections of dispatchable power-paths), not energy, is the essential product transacted in wholesale power markets. As stressed and illustrated throughout Chapters 1 through 10, a *power-path* for an operating period T is defined to be the power injected and/or withdrawn *at a single designated grid location* during T.

14.3 WHOLESALE POWER AND THE LAW OF ONE PRICE (NOT)

Recall from Section 12.2 that a *commodity* is an asset with a standard unit of measurement that permits the substitution of one unit for another with no change in valuation. Unit substitutability *with no change in valuation* is essential to ensure that an asset can be valued and traded in bulk multi-unit form. The *Law of One Price* for a commodity Q states that, in the absence of trade frictions, each unit of Q available at a given time should sell at the same price. The Law of One Price is a general precept that follows immediately from the definition of a commodity.

Labor has a standard unit of measurement: namely, a person-hour. Nevertheless, as discussed in Section 14.2, labor fails to be a commodity because person-hours cannot be substituted for each other with no change in valuation. This follows because the labor effort exerted by each worker hired for a particular production period T will typically depend to some extent on personal attributes apart from natural and/or acquired skill levels, such as competitiveness and risk aversion. Consequently, hired workers with identical work skills will typically not produce the same marginal product during T even though, measured in person-hours, each of these workers supplies the same labor during T.

Energy, also, has a standard unit of measurement: namely, a MWh. However, unlike labor, (potential) energy can be packaged in standardized battery form, such as a NusGear portable power generator equipped with a lithium battery pack. When so packaged, energy clearly constitutes a commodity. The standard economic analysis for competitive commodity markets reviewed in Chapter 12 then dictates that the wholesale price π^{W} of a standardized battery should cover its marginal cost of production. In addition, as depicted in Figure 14.1, at any competitive market-clearing (CMC) point the retail price π^{Ret} of a standardized battery must equal π^{W} plus any transportation cost π^{Trans} arising from the need to transport the standardized battery from its wholesale production site to the site of its retail sale.

However, wholesale power markets operating over transmission grids are not energy markets. Power producers generate continuous flows of power that are directly injected into the grid; ISOs procure the availability of power flows as reserve with specific characteristics able to meet grid net-load balancing needs during future operating periods; and power customers use continuous flows of power that are directly withdrawn from the grid. The attributes of these power-paths (amplitude, ramping, duration, ...) have important effects on production costs, reserve values, and customer benefits.

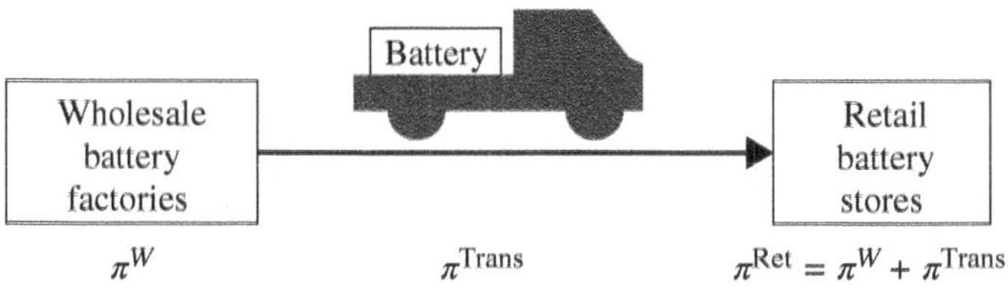

Figure 14.1 Illustration of a competitive energy market. Standardized batteries are produced in wholesale battery factories and then transported to retail battery stores for sale. At any competitive market-clearing point, the retail price π^{Ret} of a standardized battery equals its wholesale price π^{W} plus its transportation cost π^{Trans}.

The Law of One Price therefore does not hold for energy in wholesale power markets operating over transmission grids, even when its measurement is conditioned on location and time period. Tacit admission of this fact is seen in current ISO market efforts to use hedonic pricing to determine separate prices for power-path attributes apart from delivered energy, such as quick-start and fast-ramping capabilities. As discussed in Section 3.2.2, these efforts to compensate resources separately, ex post, for what inherently are joint-product attributes of single power-paths have resulted in complicated and conceptually problematic settlement processes.

It is important to keep in mind the *converse* of the Law of One Price: namely, that assets whose units have differentially valued attributes might reasonably end up being differentially priced as a normal outcome of demand and supply processes. Original paintings sold in art auctions provide a canonical example of this converse.

This study proposes a design for ISO-managed wholesale power markets that is in keeping with this converse: namely, the use of SCs with two-part pricing. The offer prices α for SCs permit explicit recognition, ex ante, of the differential insurance value inherent in available collections of dispatchable power-paths with diverse attributes. The performance payment methods ϕ included in SCs permit explicit recognition, ex post, of the differential performance value of dispatched power-paths with diverse attributes.

14.4 DIFFERENTIAL VS. UNIFORM PRICING

The standard competitive market analysis reviewed in Chapter 12 presumes:

- amounts q of a commodity Q are traded in a market M at a particular time point;
- all buyers and sellers participating in M take π, the price of Q, as given;
- each buyer's demand schedule $\pi = D(q)$ is non-increasing;
- each seller's supply schedule $\pi = S(q)$ is non-decreasing.

Given these four presumptions, Section 12.7 demonstrates that setting the price of Q equal to a uniform CMC price for all buyers and sellers, i.e. a price π^* for which aggregate demand for Q equals aggregate supply of Q, ensures short-run market efficiency, i.e. maximization of the total net surplus (TNS) extracted from M by the buyers and sellers.

In addition, Section 12.8 argues that the setting of a uniform CMC price π^* for Q would also encourage longer-run market efficiency, i.e. non-wastage of opportunities to discover and implement more efficient ways to produce and consume Q in the future. This is the case because price-taking buyers would perceive they have the opportunity to retain any net surplus gains arising from an ability to extract increased benefit from the consumption of inframarginal Q units; and price-taking sellers would perceive they have the opportunity to retain any net surplus gains arising from the ability to produce inframarginal Q units at reduced marginal cost.

The above points are classic arguments made in support of uniform pricing mechanisms; see Helman et al. [75, Section 5.2.3.1] However, these classic arguments only pertain to competitive *commodity* markets; that is, they rely on the assumption that asset units are perfect substitutes for each other. In an asset market for which perfect substitutability of units fails to hold,[4] the achievement of a Pareto efficient allocation of resources will typically require that asset units with diverse attributes be permitted to trade at diverse prices.

As seen in Sections 14.2 and 14.3, wholesale power markets operating over transmission grids are not commodity markets. In particular, energy (MWh) is not a commodity within such a market, even when its production and delivery are conditioned on location and time period. Rather, the essential product transacted in wholesale power markets is collections of dispatchable power-paths with diverse attributes, functioning as reserve. These collections are by no means perfect substitutes for each other.

Clearly, the strategic manipulation of bids and offers through non-truthful reporting of reservation values is an important concern in any market. This concern is surely heightened in non-commodity markets. However, artificial enforcement of uniform prices in such markets does nothing to address this concern. Possible ways to encourage truthful revelation of reservation values in SC markets will be addressed in Section 15.8.

14.5 COMPARISON OF SC AND CURRENT US DAM DESIGNS

Comparisons of basic attributes and optimization formulations for the SC DAM design and current ISO DAM designs are given in Figures 14.2 and 14.3, respectively, under the common simplifying assumption that all LSE reserve bids consist solely of fixed (non-price-sensitive) demand.

As seen in Figure 14.2, several key attributes of the SC DAM design that distinguish it from current ISO DAM designs are as follows. First, the SC DAM

[4] A classic example of such a market, analyzed in many microeconomic textbooks, is a *monopolistically competitive* market whose supply side consists of multiple producers selling multiple products whose units are not perfect substitutes for each other. See, e.g. Kreps [101, pp. 343–346].

		Current DAM	SC DAM
Similarities		• Conducted day-ahead to plan for next-day operations • ISO-managed • Participants include LSEs, dispatchable resources, and VERs • Subject to same system constraints: e.g. transmission, generation, ramping, and power-balance constraints	
Differences	Optimization formulation	SCUC and SCED	Swing contract clearing
	Settlement	Locational marginal prices	Contract-determined prices
	Payment	Payment for next-day power before actual delivery	Payment for availability now and performance ex post
	Out-of-market payments	Make whole-payments (e.g. for unit commitment)	No out-of-market payments
	Info released to participants	Unit commitments, LMPs, and next-day dispatch schedule	Which swing contracts have been cleared

Figure 14.2 SC DAM design compared with current ISO DAM designs.

		Current DAM SCUC	Current DAM SCED	SC DAM optimization
Similarities		• Both SCUC and swing contract (SC) clearing are solved as mixed integer linear programming (MILP) optimization problems subject to system constraints		
Differences	Objective	Min (Start-up/shut-down costs + no-load costs + dispatch costs + reserve cost)	Min (Dispatch costs + reserve costs)	Min (Offer cost + anticipated performance cost + anticipated imbalance cost)
	Unit commitment constraints	Yes	No	Unit commitment constraints are implicit in submitted swing contracts
	Key ISO decision variables	Unit commitments	Energy dispatch and reserve levels	Which swing contracts are cleared
	Settlement	No	LMPs calculated as SCED dual variables	Offer prices paid for cleared swing contracts

Figure 14.3 SC DAM optimization compared with current ISO DAM SCUC/SCED optimizations.

design is a two-part pricing design that permits full separate market-based compensation for service availability (insurance) and actual real-time service performance. Second, the SC DAM design does not require any payments for power-path dispatch to be made in advance of actual dispatch. Third, the SC DAM design does not rely on OOM make-whole payments to market participants to ensure full cost recovery either for power-path availability or for power-path dispatch under normal operating conditions.[5]

[5]As in current ISO DAMs, assurance of reactive power support and voltage control would still have to be undertaken as OOM actions in an SC DAM if a standard DC-OPF approximation is used to eliminate consideration of reactive power and voltages in the system constraints [175, 176]. See [202] for a discussion of generalized DC-OPF formulations whose system constraints include reactive power support and voltage control requirements.

As seen in Figure 14.3, the optimization formulation used for contract clearing in an SC DAM differs from the SCUC/SCED optimization formulations used in current ISO DAMs in four key conceptual regards.

First, the SC DAM objective function permits reserve offers to be cleared in merit order based on a comprehensive accounting of all anticipated costs. These anticipated costs include: offer price payments for cleared reserve offers (covering avoidable fixed cost); performance payments resulting from the real-time dispatch of cleared reserve offers (covering variable cost); and imbalance cost payments (if any) reflecting the possible need to clear additional reserve offers in subsequent SC markets to handle any remaining real-time power imbalance as reflected in non-zero solutions for power-balance slack variables.

In contrast, as illustrated in Section 5.4 for the ERCOT energy region, the supply offers submitted into an ISO DAM SCUC/SCED optimization include a partial reporting of avoidable fixed cost (e.g. start-up cost) as well as a reporting of variable cost in the form of a marginal cost schedule; these costs are then included in the SCUC/SCED objective function. However, subsequent actual compensation consists of OOM make-whole payments together with market settlements for energy and reserve based on dual variable determinations (e.g. LMPs). As discussed and illustrated in Chapter 13, this actual compensation can differ significantly from the costs appearing in the SCUC/SCED objective function.

Second, the SC DAM optimization formulation requires resources to ensure the physical feasibility of their submitted reserve offers. Specifically, resources with cleared reserve offers must be able to ensure: (i) the availability of their offered collections of dispatchable power-paths and (ii) their ability to deliver any one of these dispatchable power-paths in response to real-time dispatch instructions. This implies, in particular, that resources are responsible for ensuring that any unit commitment constraints they face are fully incorporated in their submitted reserve offers.

In contrast, in an ISO DAM SCUC optimization the ISO is responsible for ensuring that unit commitment information is collected in advance from participating dispatchable resources along with their supply offers. The ISO must then incorporate this unit commitment information into the system constraints for the SCUC optimization.

Third, given an SC DAM optimization for an operating period T, the number of binary decision variables (to clear or not) is equal to the number of submitted reserve offers for T. In contrast, in an ISO DAM SCUC optimization for which the operating period T (e.g. a day) is partitioned into NK time-steps k (e.g. 24 hours), the ISO has a separate binary decision variable (to commit or not) for each dispatchable resource for each time-step k. Consequently, all else equal, an SC DAM contract-clearing optimization tends to have a smaller "size" than an ISO DAM SCUC optimization.[6]

Fourth, as explained with care in Section 10.3, SC DAM settlements are time consistent. No performance payments are made in advance of actual performance, and no offer-price or performance payments are adjusted, ex post, as a result of subsequent market or OOM processes. In contrast, ISO DAM settlements are routinely adjusted in subsequent market and OOM processes.

[6] This point will be more carefully addressed in Section 15.9.

CHAPTER 15

ADVANTAGES OF THE LINKED SWING-CONTRACT MARKET DESIGN

15.1 CHAPTER PREVIEW

The Linked Swing-Contract Market Design developed and illustrated in previous chapters of this study has the same basic purpose as current US RTO/ISO-managed wholesale power markets: namely, to ensure efficient continual balancing of net load in real-time operations. Conceptual and practical advantages of the former market design have been noted at various places in these previous chapters.

This chapter provides summary discussions of these advantages. Section 15.2 uses a simple insurance market example to show how the two-part pricing design of a swing-contract (SC) market is analogous to the design of a standard physically covered insurance market. Shorter-term SC markets use two-part pricing to provide appropriate incentives for the entry, retention, and exit of existing dispatchable resources. Longer-term SC markets use two-part pricing to provide appropriate incentives for investment in new dispatchable resources.

Section 15.3 reviews the three basic approaches to long-term reserve procurement currently implemented in US RTO/ISO-managed wholesale power markets: namely, scarcity pricing, centrally managed capacity markets, and load-serving entity (LSE) bilateral contract obligations. These three approaches are contrasted with the two-part pricing approach implemented in longer-term SC markets.

Sections 15.4 and 15.5 then explain how SC markets could facilitate the resolution of the revenue insufficiency and merit-order issues causing problems in current US RTO/ISO-managed wholesale power markets. This resolution would enable better incentives for investment in thermal generation as a backstop power source for uncertain and volatile non-dispatchable generation.

Section 15.6 emphasizes that an SC market M(T) for a future operating period T is an ISO-managed robust-control mechanism for period T. The ISO clears reserve offers and bids for period T, subject to system constraints. These constraints include the requirement that the dispatchable power-paths comprising the cleared reserve

A New Swing-Contract Design for Wholesale Power Markets, First Edition, Leigh Tesfatsion.

offers cover ISO-specified system-wide and zonal uncertainty sets for net load during period T. Detailed specifications of net load scenarios with associated scenario probabilities are not required.

The conceptual coherency of the Linked Swing-Contract Market Design is stressed in Section 15.7. This conceptual coherency reduces the complexity of the design's rules of operation relative to the rules of operation for current US RTO/ISO-managed wholesale power markets. This reduced complexity in turn increases transparency. As discussed in Section 15.8, this increased transparency could help to ensure that SC market participants have neither the incentive nor the opportunity to game market operations for their own advantage at the expense of system efficiency and reliability.

A rough comparison between the size of the contract-clearing optimization for an SC day-ahead market (DAM) and the size of a standard security-constrained unit commitment (SCUC) optimization for a US RTO/ISO-managed DAM is provided in Section 15.9. Size is measured in terms of the number of binary decision variables and the number of system constraints. This size comparison indicates that the SC DAM optimization will tend to be smaller in size.

Finally, Section 15.10 briefly reviews additional desirable features of the Linked Swing-Contract Market Design: namely, open access (fairness); co-optimization of diverse types of reserve; appropriate remuneration for reserve diversity and flexibility; incentives for accurate forecasting and dispatch following; and settlement time-consistency.

15.2 SC MARKETS ARE PHYSICALLY-COVERED INSURANCE MARKETS

Consider the following contract offered to homeowners in a particular geographical region R by a home-fire insurance company:

> C1. Physically covered insurance contract: *During the contract period, our company will collect from you and other homeowners in our region-R insurance pool a relatively small insurance premium in direct or amortized (payment stream) form. If your home suffers fire damage during the contract period, our company will pay ex post for the full repair of your home; your only charge will be a co-payment expressed as a relatively small percentage of the cost of this repair. The purpose of this co-payment is to encourage you and other homeowners to reduce home fire hazards (e.g. overgrown home shrubbery), hence the probability of home fires. Our company depends on insurance premiums and co-payments to cover all of the avoidable cost we must incur over the contract period in order to enter and remain in the home-fire insurance business for region R.*

Contract C1 is a standard type of physically covered insurance contract. The contract is structured to permit risk-averse homeowners to spread their home-fire risk across an insurance pool, thus reducing their personal risk exposure. Should a home-fire occur, the cost of home repair will largely be covered by the insurance company. Homeowners thus gain peace of mind from knowing they are protected against catastrophic loss, whether or not any such loss subsequently occurs. On the other hand, in order to stay in business, the insurance company must be able to set insurance premiums and co-payments for homeowners at levels that guarantee its expected net earnings over the contract period are nonnegative.[1]

Contract C1 is analogous to the way in which an SC market M(T) for a future operating period T ensures avoidable cost coverage for dispatchable resources that are cleared to provide reserve for period T. Recall from Section 6.7 that the ISO managing M(T) allocates all reserve procurement payment obligations across market participants based on their net must-service loads. These reserve procurement payment obligations take two forms: *net availability cost* paid in direct or amortized (payment schedule) form in advance of period T; and *net performance cost* paid ex post, subsequent to verified period-T performance.[2]

The insurance premiums in contract C1 correspond to the ISO's allocation of net availability cost across market participants on the basis of the *anticipated* relative volatility and size of their net must-service loads ("fire hazards") during period T. The co-payments in Contract C1 correspond to the ISO's ex post allocation of net performance cost across market participants on the basis of their *actual* net must-service loads ("fires") during period T.

15.3 LONGER-TERM SC MARKETS SUPPORT NEW INVESTMENT

Current US transmission systems require the continual entry of dispatchable resources for three basic reasons:

- to service increases in power demand over time;
- to provide dispatchable backstop power for the uncertain and volatile power output of increasing numbers of variable energy resources (VERs);

[1]Recall from Section 3.2.7 that net earnings are defined to be earnings minus avoidable cost, where avoidable cost is the sum of avoidable fixed cost and variable cost. Suppose the insurance company is able to ensure its *expected* net earnings over the contract period are nonnegative; and suppose, in addition, that the R-region insurance pool is large, the home-fire risks faced by the participating homeowners are independent of each other, and the (nonnegative) repair costs of homes suffering a fire are identically distributed with a finite expectation and a common finite upper bound. Then the Strong Law of Large Numbers implies that the company's *actual* net earnings over the contract period will be nonnegative with a probability almost surely equal to 1.

[2]Specifically, *net availability cost* consists of net offer-price payment obligations for all reserve offers cleared for period T; and *net performance cost* consists of all ex post performance payments for verified period-T reserve deployment, net of period-T ISO revenues. The latter revenues come from three possible sources: (i) payments received from LSEs for cleared price-sensitive reserve bids; (ii) penalty payments received from LSEs for systematic (correctable) errors in their period-T must-service load forecasts; and (iii) penalty payments received from dispatchable resources for uninstructed deviations from their period-T dispatch signals.

- to offset retirements and de-ratings of thermal generation needed for inertia,[3] and for voltage support through the production and absorption of reactive power.

As discussed in Section 15.2, SC markets are forward reserve markets that function as physically covered insurance markets. Reserve is broadly defined to mean diverse collections of dispatchable power-paths.

The two-part pricing design of the reserve offers submitted by dispatchable resources into SC markets permits full coverage of avoidable costs. For shorter-term SC markets, this cost coverage assurance facilitates the efficient entry, retention, and exit of existing dispatchable resources. For longer-term SC markets, this cost coverage assurance facilitates the efficient financing of needed investments in new dispatchable resources.

In contrast, the seven US RTO/ISO-managed wholesale power markets characterize reserve more narrowly as unencumbered generation capacity (MW). These markets seek to ensure adequate generation capacity for long-term needs in one of three basic ways [17, 95, 96, 133]:

- *Energy-only market (ERCOT)*: The market relies on day-ahead and real-time market prices for energy and ancillary service, supplemental out-of-market (OOM) make-whole payments, and (capped) scarcity prices[4] to ensure adequate generation capacity;
- *Centrally managed capacity market (ISO-NE, MISO, NYISO, PJM)*: The manager of the capacity market specifies system-wide and/or zonal generation capacity demand curves in fixed (vertical) or price-sensitive (sloped) form, and resources compete to supply this generation capacity through the submission of capacity offers;
- *LSE bilateral contract obligations (CAISO, SPP)*: LSEs are obligated to contract in advance with power sources for generation capacity (or self-schedule their own generation capacity) in amounts sufficient to balance the anticipated future power demands of their customers.

---

[3]*Inertia* is the resistance of a physical object to a change in the state of its motion. In current transmission systems, inertia primarily arises from the rotating masses of generators and turbines synchronously connected to the transmission grid. This inertia helps to dampen or offset frequency changes caused by imbalance between power injection and power withdrawal on the grid. Most VERs do not yet contribute to this inertia due to the way in which they are electrically connected to the grid and their lack of energy buffer [190].

[4]A *(capped) scarcity price* π is a price determined at the intersection point of a (price-capped) aggregate demand schedule D and a capacity-constrained aggregate supply schedule S that occurs along the final vertical portion of S. Interpreting the entire aggregate supply curve S (including its final vertical portion) as a system marginal cost curve, the *right-hand* system marginal cost is infinite at this intersection point. When locational marginal prices (LMPs) are calculated as *right-hand* dual variables for power balance constraints, they are not well defined at such intersection points. Nevertheless, the intersection point is a well-defined competitive market clearing (CMC) point as defined in Section 12.5.

More detailed descriptions and assessments for the latter three approaches will next be given in turn.

15.3.1 Energy-Only Market

To help understand how an energy-only market is intended to ensure adequate generation capacity, it is instructive to provide a simple contract analogy. Consider the following contract offered to homeowners in a geographical region R by a home-fire repair company:

> C2. Repair contract*: If your home suffers fire damage during the contract period, our company will repair your home. However, you will be required to pay the full cost of this repair at market-determined prices. Our company depends on these market-determined repair payments, supplemented by OOM government subsidies, to cover all of the avoidable cost we must incur in order to enter and remain in the home-fire repair business for region R.*

The repair contract C2 has three important downsides. First, the home-fire repair company has to depend on government subsidies plus uncertain home-fire repair payments from homeowners experiencing fires in order to cover its avoidable fixed cost as well as its variable cost. Second, homeowners bear all price risk; hence, an individual homeowner could be faced with a catastrophic loss (physical or financial) should a home fire occur. Third, the government subsidies received by the home-fire repair company might not be the best possible use of tax-payer dollars. For example, homeowners might obtain higher net benefit if these subsidy dollars were instead used to provide them with some kind of home-fire insurance plan.

The repair contract C2 captures the basic way in which the energy-only ERCOT market seeks to use market-determined prices for variable cost compensation and supplemental OOM payments for unit commitment cost compensation.

When generation capacity is not scarce, ERCOT relies on DAM and Real-Time Market (RTM) prices to compensate the variable costs of generation capacity suppliers. The start-up and minimum-energy costs of these generation capacity suppliers are separately compensated by means of supplemental OOM make-whole payments. These settlement processes are explained in Section 5.4.

However, when generation capacity is scarce, ERCOT instead makes use of an aggregate demand schedule with an upper price-cap administratively set at an estimated value of lost load (VOLL)[5] to determine the variable cost compensation for cleared generation capacity suppliers. The intersection of the *uncapped* aggregate demand schedule with the capacity-constrained aggregate supply schedule for generation capacity p (MW) determines a "CMC" point (p^*, π^*) for generation capacity.[6]

[5] *VOLL* is a measure of the average willingness of customers to pay for the reliability of their supplied power.

[6] If the point (p^*, π^*) lies along the final vertical portion of the capacity-constrained aggregate supply curve, i.e. if p^* is the maximum possible generation capacity supply, then the price π^* is a scarcity price as defined above in footnote 4.

If the price π^* ($/MWh) does not exceed the VOLL price-cap, then π^* is used to determine the variable cost compensation of cleared generation capacity suppliers. However, if π^* does exceed the VOLL price-cap, the VOLL is used to determine this compensation.[7] As will be discussed in Section 15.4, the use of the VOLL instead of a higher π^* is said to result in "missing money" for the cleared generation capacity suppliers.

Cramton [37, Section VI] and Hogan and Pope [83] are generally supportive of ERCOT's use of scarcity pricing. For example, Cramton (pp. 599–600) states:

> *The higher the scarcity pricing, the stronger is the investment signal. More generation is economically built, which improves reliability. The higher scarcity pricing also motivates demand response to avoid the high price. Indeed, it is the limited amount of demand response that creates the need for scarcity pricing.*

Cramton [37, Section VI] further asserts that scarcity pricing "motivates load to contract for the energy it needs in advance of real time. Forward contracting provides a hedge against volatile real-time prices." Similar arguments in general support of scarcity pricing (and against forward capacity markets) for ensuring adequate generation capacity in wholesale power markets are given by Graves et al. [64].

On the other hand, many commentators have questioned whether an energy-only market design can be successfully implemented in practice, given political and economic realities. See, for example, Joskow [95, 96], Moye and Meyn [133, Section 6.1], and the critical studies reviewed by Bublitz et al. [17, Section 2.3].

One important criticism is the first-noted downside listed above for Contract C2: namely, the energy-only market design does not provide any guarantee to generation capacity suppliers that their market earnings will fully cover their avoidable costs. Although OOM payments can be used to make suppliers whole, reliance on non-market settlements violates the basic principle underlying the energy-only market design: namely, that a wholesale power market should be conducted with minimal reliance on administrative actions.

As stressed by Joskow [96], this problem will be exacerbated as VERs with near-zero marginal costs continue to enter the market, thus driving DAM/RTM energy prices toward zero as fewer dispatchable resources in the merit-order stack are scheduled for dispatch. These revenue insufficiency and merit-order issues will be discussed at greater length in Sections 15.4 and 15.5.

A second related criticism, stressed by Abani et al. [1], Bhagwat et al. [12, p. 12], and Newbery [137], is that the risk-aversion of market participants must be taken into account. Risk-averse generation capacity suppliers in energy-only markets might not be willing to bear the price risk associated with long-term supply and might be unable to find a means to hedge this risk through suitable risk-spreading or risk-pooling arrangements.

---

[7] If the point (p^*,VOLL) lies along the final vertical portion of the capacity-constrained aggregate supply curve, then the VOLL is a capped scarcity price as defined above in footnote 4.

A third criticism, noted by Cramton [37, p. 603], is that relatively high DAM/RTM LMPs and/or scarcity prices during peak-load periods do not provide adequate earnings incentives for long-term investments in new generation capacity because this investment would likely result in a lowering of these prices. That is, earnings projections for planned long-term investments cannot be based on these current prices.

15.3.2 Centrally Managed Capacity Market

In contrast to the energy-only market approach, centrally managed capacity markets express generation capacity requirements by means of RTO/ISO-specified demand curves. Resources compete to supply generation capacity through the submission of capacity offers.

A resource whose capacity offer is cleared in a capacity market receives an immediate capacity payment based on capacity market-clearing prices. In return, the resource is obligated to offer this capacity into subsequent DAM/RTM markets as a fixed (non-price-sensitive) quantity supply. If this capacity is then used to meet energy and/or ancillary service needs, the resource is compensated for this usage at DAM/RTM-determined prices.

Bhagwat et al. [12], Byers et al. [22], and Moye and Meyn [133, Section 6.2] report on the four US electric power regions (ISO-NE, MISO, NYISO, PJM) with centrally managed capacity markets. These four US capacity markets have differing designs that continue to evolve. For example, the capacity markets in ISO-NE and MISO are operated as descending-clock and sealed-bid auctions, respectively, whereas the capacity markets in NYISO and PJM are operated as competitive markets with zonally uniform market-clearing prices. Moreover, ISO-NE, NYISO, and PJM have all transitioned to the use of capped price-sensitive demand curves. However, MISO continues to use a fixed demand curve to express a capacity target based on its planning reserve margin.

Bhagwat et al. [12, Section 3] also report the results of an extensive survey of US capacity market experts. The purpose of this survey was to obtain information about US capacity market performance that could be used to inform the design and operation of capacity markets in the EU. The authors conclude as follows (p. 16): "Based on our survey, US experts generally recommend the use of energy-only markets over capacity markets." The key concern expressed about US capacity markets in this survey was "uncertainty regarding the availability of generation resources that clear the capacity market during scarcity hours." The authors note that this concern arose because "the remuneration from the capacity market is not linked to performance under scarcity conditions, (hence) … some of the generators that have received capacity payments may not be able to honor their commitments during a scarcity event."

However, as detailed in Byers et al. [22, Table 2, p. 67], US capacity markets in recent years have begun to implement direct penalties for non-performance. For example, ISO-NE supplements its capacity market with financial call options to provide financial penalties for under-delivery of generation capacity during shortage events. Generation capacity suppliers must pay the difference between the market-determined energy price and a call option strike price for an operating period T with capacity-constrained supply if they fail to follow through on their capacity supply obligations for period T.

At the end of their study, Byers et al. [22, Section 4, p. 73] conclude: "Where centralized capacity markets exist, our review reveals that there is no clear convergence on how non-performance penalties are defined, how and if these penalties should be applied to renewables, and how qualifying capacity for (variable renewable energy resources) should be calculated." More generally, Bublitz et al. [17] provide a detailed literature review for capacity markets that covers conceptual studies as well as empirical assessments for US and European implementations. Bublitz et al. conclude (p. 1072): "As results from the literature are not always coherent and often only applicable for specific cases, the question of which (capacity market) is most efficient remains open."

Clearly, US centralized-managed capacity markets are still very much a work in progress. It is difficult to compare and contrast the anticipated operation of longer-term SC markets with this moving target. Nevertheless, several critical design differences can be highlighted.

First, capacity payments have some similarity to the offer-price payments received by cleared participants in longer-term SC markets. However, capacity payments are based on market-clearing prices determined by the intersection of incremental supply and demand curves for generation capacity (MW). Consequently, these capacity payments are unable to ensure full coverage of the avoidable fixed costs that investors in new generation capacity would have to incur to guarantee capacity availability during the course of their capacity obligation periods to the extent that these costs include discrete "lumpy" expenditures.

Second, the subsequent DAM/RTM payments received by cleared capacity-market participants for generation capacity provision during emergency conditions or shortage events have some similarity to the performance payments received by cleared participants in SC markets. However, cleared capacity-market participants offer their generation capacity into subsequent DAM/RTM operations as fixed (non-price-sensitive) quantity obligations. Consequently, there is no guarantee that the resulting DAM/RTM payments will cover their actual performance costs.

15.3.3 LSE Bilateral Contract Obligations

As detailed in Chapter 10, the purpose of the Linked Swing-Contract Market Design is to permit reserve (dispatchable power-paths) to be procured through an efficient centralized market-based process. However, participants in linked SC markets can also engage in physically covered bulk-power trades managed by exchanges, dealers, and brokers as long as any resulting physical power outcomes are appropriately self-scheduled in the linked SC markets to secure needed transmission support.

In particular, LSEs participating in linked SC markets can engage in physically covered bilateral contracting to ensure adequate reserve is available to meet the future power demands of their customers. However, bilateral contracting suffers from three well-known problems [4]: greater price risk due to incomplete contract cover; increased opportunity to exercise market power due to market thinness; and higher transactions costs due to lack of standardized contracts. In contrast, important anticipated advantages to be gained by contracting for reserve procurement through linked centrally managed SC markets are as follows:

- *Short-term efficiency gains through increased transparency*: Dispatchable resources can use reserve offers to signal varied capabilities for providing reserve with diverse valued attributes, with open up-front specification of all required compensation for avoidable costs. This should result in improved valuations and reduced transaction costs, hence less wastage of existing resources.
- *Short-term efficiency gains through reduction of risk*: The ability of dispatchable resources with cleared reserve offers to ensure full coverage of their avoidable costs should facilitate the efficient entry, retention, and exit of existing dispatchable resources.
- *Longer-term efficiency gains through improved incentives*: The ability of investors with cleared reserve offers to ensure full coverage of their avoidable costs should facilitate efficient investment in new dispatchable resources.

15.4 SC MARKETS ENSURE REVENUE SUFFICIENCY

Revenue sufficiency refers to a situation in which seller revenues are sufficient to cover seller avoidable costs, thus permitting sellers to stay in business.

As stressed in previous chapters, the Linked Swing-Contract Market Design ensures revenue sufficiency. More precisely, the two-part pricing feature of this design permits dispatchable resources to ensure full coverage of their avoidable costs. This, in turn, encourages the efficient entry, retention, and exit of existing dispatchable resources as well as efficient investment in new dispatchable resources.

In contrast, as discussed in Section 3.2.3, US RTO/ISO-managed markets continue to experience revenue insufficiency problems. Short-term revenue insufficiency arises because DAM/RTM earnings from energy and ancillary service provision do not guarantee coverage of unit commitment costs. RTO/ISOs handle this short-term problem by resorting to OOM make-whole payments.

Longer-term revenue insufficiency arises in these markets because DAM/RTM earnings from energy and ancillary service provision have not been sufficient to encourage adequate investment in new generation capacity. Cramton and Ockenfels [38, Section 2.2] refer to this longer-term revenue insufficiency problem as the *Missing Money Problem*, which they characterize as follows:

> *In 'normal' periods, when there is no shortage of capacity, prices are below the level needed to cover operating and capital costs of new capacity, and in scarcity events, prices are unlikely to accurately reflect the scarcity.*

As detailed in Section 15.3, US RTO/ISO-managed markets are attempting to handle this longer-term Missing Money Problem through scarcity pricing (ERCOT), forward capacity markets (ISO-NE, MISO, NYISO, PJM), and LSE contractual obligations (CAISO, SPP). However, as many commentators have noted,[8] the Missing Money Problem has proved to be difficult to resolve because it arises in part from various OOM administrative rules and practices reflecting political considerations (e.g. fairness to customers), practical cost verification concerns, and perceived system reliability needs. Examples include FERC Order 831 [56], which increases (but does not eliminate) hard upper price-caps on generator energy supply offers, and RTO/ISO-determined *reliability-must-run* (RMR) designations for generators that result in OOM generation dispatch.

15.5 SC MARKETS AMELIORATE MERIT-ORDER CONCERNS

VER participation in current US RTO/ISO-managed DAM/RTM operations is rapidly increasing. VERs tend to have relatively low marginal costs of generation. Consequently, whether VERS are firmed by storage or not, their increased entry is putting downward pressure on DAM/RTM energy and ancillary service prices by decreasing the amount of non-VER dispatchable generation cleared from the merit-order stack. This outcome is referred to in the power systems literature as the *Merit-Order Effect* [162].

On the other hand, the increased participation of VERs that are not fully firmed by storage is increasing the uncertainty and volatility of net load, which in turn is increasing the need for flexible dispatchable generation. Given the current state of US electric energy storage development, this flexible dispatchable generation must largely come from thermal generators.

The conundrum is then as follows. How can US RTO/ISOs ensure they have an adequate amount of flexibly controllable thermal generation to balance the increasingly uncertain and volatile net load resulting from increased VER participation when this increased VER participation is in fact reducing the net earnings of thermal generators, hence their financial viability?

This conundrum can be resolved by permitting thermal generators to obtain appropriate compensation for the full range of valuable services they provide, including the flexibility and controllability of their offered generation. The Linked Swing-Contract Market Design permits this appropriate full compensation.

[8] See, for example, W.W. Hogan [81], M. Hogan [82], and Spence [168, Section III.A].

Specifically, thermal generators participating in an SC market M(T) can submit reserve offers that offer the availability of dispatchable power-paths with diverse attributes for operating period T, thus enabling the ISO managing M(T) to balance uncertain volatile net load during T. In turn, the thermal generators can seek full compensation of their availability cost through their offer prices and full ex post compensation for their performance cost through the performance payment methods they include in their offers.

15.6 SC MARKETS ARE ROBUST-CONTROL MECHANISMS

The Linked Swing-Contract Market Design proposed in the current study is a robust-control approach to reserve procurement for net-load balancing. Reserve is characterized as diverse collections of power-paths, and net load is defined to be power withdrawal net of non-dispatched power injection.

As detailed in Chapters 6–10, the ISO tasked with managing an SC market M(T) for a future operating period T considers how much swing (flexibility) will be needed in cleared reserve offers to ensure coverage of system-wide and zonal net-load uncertainty sets U. Each U is a corridor in a time-power plane for period T, anchored by an expected net-load profile. The upper and lower boundaries for each U are determined by ISO-set reserve requirements. The ISO solves a contract-clearing optimization problem for M(T) that includes coverage requirements for these uncertainty sets U among its system constraints. A detailed specification of net-load scenarios with associated scenario probabilities is not required.

In contrast, the rapidly growing reliance on VERs with uncertain and volatile power outputs has encouraged some researchers to propose the use of stochastic SCUC/SCED formulations to ensure net-load balancing; see, e.g. [132, 145, 193]. These stochastic formulations typically consist of two stages. In the first stage, unit commitments are determined on the basis of *expected* net load for a future operating period T. In the second stage, reserve (unencumbered generation capacity) is procured in an amount sufficient to balance net load during period T for *each* net-load scenario in an ISO-specified set of net-load scenarios for period T.

Consequently, stochastic SCUC/SCED optimization formulations rely on having an accurate modeling of the stochastic data-generating mechanism for net load, with associated net-load probabilities. In addition, to ensure computational tractability, they rely on having a way to express the actual set of possible net-load scenarios in approximate smaller form without any critical loss of information. These challenging requirements have not yet been met.

For example, to be a viable practical option, two-stage stochastic SCUC/SCED formulations must be able to perform better than currently used deterministic SCUC/SCED formulations, which are less computationally demanding. Deterministic SCUC/SCED formulations rely on coverage of net-load uncertainty sets determined by ISO-set reserve requirements; hence, in this particular regard, they are similar to SC market optimizations.

Krishnamurthy et al. [102] and Li and Tesfatsion [114] conducted systematic computational experiments to compare the performance of stochastic and deterministic SCUC/SCED formulations. These experiments use a constructed set $\mathbb{S}^{T}$ of possible "true" net-load scenarios, based on actual ISO-NE thermal generation, wind generation, and load data; each of these net-load scenarios was assigned an equal probability of occurrence. A well-known scenario reduction method was then used to reduce $\mathbb{S}^{T}$ to a smaller approximate net-load scenario set $\mathbb{S}$ with associated probabilities; and an approximate expected net load $\hat{E}[NL]$ was calculated based on $\mathbb{S}$. Performance was measured in terms of "true" expected total energy cost (unit commitment plus dispatch), based on $\mathbb{S}^{T}$.

In the absence of wind generation [102], the deterministic SCUC/SCED formulation (based on $\hat{E}[NL]$) resulted in lower "true" expected total energy cost than the stochastic SCUC/SCED formulation (based on $\mathbb{S}$ with associated probabilities) if the reserve requirements for the deterministic SCUC/SCED formulation were set within "sweet spot" value ranges. In the presence of wind generation [114], the deterministic and stochastic SCUC/SCED formulations resulted in similar "true" expected total energy costs when the reserve requirements for the deterministic SCUC/SCED formulation were set within "sweet spot" value ranges that varied systematically with the level of wind power penetration.

15.7 SC MARKETS REDUCE RULE COMPLEXITY

The Linked Swing-Contract Market Design is a conceptually coherent design for a collection of linked SC markets. Each SC market is an ISO-managed forward reserve market, where reserve consists of diverse collections of dispatchable power-paths. The conceptual design of an SC market M(T) for a future operating period T can be concisely summarized as follows, whatever the length of its look-ahead horizon.

- A reserve offer submitted into M(T) consists of a portfolio of swing contracts.
- Each swing contract is a two-part pricing contract that offers a collection of dispatchable power-paths for possible ISO dispatch during T.
- The offer price of a swing contract can be used to guarantee full compensation for all avoidable fixed cost incurred to ensure reserve availability for T.
- The performance payment method included within a swing contract can be used to guarantee full compensation ex post for any variable cost incurred to deliver an offered power-path during T in accordance with ISO dispatch instructions.
- A reserve bid submitted into M(T) consists of price-sensitive and/or fixed demand for power-path delivery during T.
- The ISO managing M(T) clears reserve offers and price-sensitive reserve bids to maximize total net benefit of market participants subject to system constraints.

- The ISO allocates net availability cost across market participants based on the relative volatility of their net must-service load.
- The ISO allocates net performance cost across market participants based on the relative size of their net must-service load.
- The ISO allocates transmission service cost across market participants based on the relative power imbalance at their grid locations.

In contrast, the rules governing the operations of current US RTO/ISO-managed wholesale power markets are exceedingly complex. Understandably, the rules governing an actual working market will be much more complex than the rules outlined for a proposed market design. However, the key issue is whether there is a clear conceptually coherent set of guiding principles that underlies and justifies the form of these rules. Each US RTO/ISO-managed market has developed a large number of business practice manuals whose purpose is to convey its rules to its stakeholders. However, general guiding principles are not readily apparent from these dense difficult-to-read documents.

The basic contention of the current study is that the complexity of the rules governing current US RTO/ISO-managed wholesale power market operations primarily arises from the conceptual problems described in Section 3.2. The key conceptual problems are inappropriate focus on energy rather than power-paths as the essential transacted product; problematic attempt to determine separate prices for the attributes of power-paths, treated as separate products; and an inappropriate use of spot-market pricing principles for forward-market operations.

A primary indicator of these conceptual problems is seen in the settlement rules for these markets. The business practice manuals devoted to settlements consist largely of lengthy descriptions of complex OOM make-whole payments and payment adjustment rules.

15.8 SC MARKETS REDUCE GAMING OPPORTUNITIES

A *structural market advantage* is an advantage conferred upon a market participant through an instituted market rule; an example would be an eligibility rule that prevents entry by a competing resource based on irrelevant technological requirements. A *strategic market advantage* is an opportunity available to a market participant to gain advantage for themselves through strategic behavior, such as the reporting of an offer or bid that does not reflect the participant's true reservation values.[9]

The Linked Swing-Contract Market Design has been deliberately formulated to reduce structural market advantage. The design emphasizes open access for all entities that are able to offer reserve (dispatchable power-paths) or procure reserve

[9] *Market power* as defined by economists is a special case of strategic market advantage pertinent for competitive commodity markets: namely, the ability of a market participant to engage in strategic behavior that causes a commodity price to deviate from its competitive equilibrium level in a way that is advantageous to this participant. As discussed in Section 12.9, market power can be exercised in two basic ways: economic capacity withholding, and physical capacity withholding.

(delivered power-paths) during future operating periods. The two-part pricing design of reserve offers permits full coverage of avoidable costs, whatever form these costs take, with no need to resort to OOM make-whole payments. The ISO cost allocation rule, described in Section 6.7, is designed to ensure fair allocations of net reserve procurement costs and transmission service costs across market participants on the basis of their contributions to these costs.

In addition, as stressed in Section 15.7, the conceptual coherency of the Linked Swing-Contract Market Design reduces the complexity and opaqueness of its operating rules. This, in turn, facilitates oversight by independent market monitors tasked with preventing the exercise of strategic market advantage.

For example, in SC markets, the offer price payments received by cleared resources should be sufficient to cover their avoidable fixed costs, including their opportunity costs;[10] and the performance payments received by cleared resources should be sufficient to cover their variable costs. This cost coverage is necessary in order for these resources to remain in the reserve provision business. Competition either from existing participants or via threat of entry by new participants should drive these payments into alignment with true costs, since higher-cost reserve offers risk not being cleared. However, if competition is not sufficient to ensure this alignment, independent market monitors will need to monitor these costs. While this monitoring could be difficult, the goal of this monitoring is clear.

Last but not least, the Linked Swing-Contract Market Design has been deliberately formulated to ensure the time consistency of market settlements; see Section 10.3. An offer-price payment obligation is incurred at the time a reserve offer is cleared for a future operating period T. However, this payment can be amortized to help ensure appropriate steps are being taken to guarantee availability during period T. Performance payment obligations are only incurred for this cleared reserve offer ex post, after actual period-T performance has been verified. Similarly, payment obligations for price-sensitive reserve bids cleared for period T are only incurred ex post, after actual period-T reserve delivery. Consequently, market participants have no opportunity to engage in cross-market gaming of settlements through strategic offers and bids.

In contrast, as discussed in Section 15.7, the rules governing operations in current US RTO/ISO-managed wholesale power markets are exceedingly complicated and difficult to understand. This is particularly the case for settlement rules.

DAM/RTM LMPs, derived as energy spot prices, are not guaranteed to cover generator unit commitment costs. In addition, as discussed in Section 3.2.4, DAM/RTM LMPs are derived as dual variables for SCED optimizations, conditional on generator unit commitments. This can result in *lost commitment opportunity*

---

[10]Recall from Chapter 4 (footnote 3) that the standard economic definition of *opportunity cost* for a resource is the net earnings that the resource could have obtained in its next-best alternative use. The inclusion of opportunity cost in avoidable fixed cost is critical. Long-run inefficiency problems that can result from the use of "cost-based pricing," i.e. pricing based on audited costs that exclude opportunity costs, are carefully considered in [134].

(LCO) in the following sense: once SCED-optimal LMPs are determined, committed generators might find that deviating from their unit commitments would permit them to increase their net earnings; see [45, Section 2] and [165].

Consequently, much effort has been spent in US RTO/ISO-managed markets to devise OOM make-whole settlement rules to ensure more complete cost compensation for committed and cleared resources. The short-term goal is to reduce incentives to deviate from commitment and dispatch instructions. The longer-term goal is to prevent inefficient market entry, retention, and exit.

However, the need to implement and enforce these OOM rules increases RTO/ISO overhead costs. Moreover, the complexity of these OOM rules provides increased incentives and opportunities for the exercise of strategic market advantage.

For example, the OOM make-whole payments made to ensure that participants in DAM/RTM operations are compensated for their unit commitment costs can be gamed by these participants through strategic use of self-commitment and self-scheduling. Also, if an RTO/ISO institutes LCO compensation in order to prevent DAM/RTM-committed generators from regretting their inability to re-optimize their unit commitments, the generators can use their DAM/RTM energy and ancillary service supply offers to game this subsequent compensation.

In addition, participants in US RTO/ISO-managed markets have another important avenue for the exercise of strategic market advantage: namely, cross-market arbitrage. For example, as documented by Birge et al. [14], evidence suggests that some MISO participants use strategic DAM supply offers to induce artificial congestion in the DAM in order to increase the value of their positions in MISO's financial transmission right (FTR) market, resulting in overall net gains.

15.9 SC MARKETS HAVE SMALLER-SIZED OPTIMIZATIONS

For concreteness, this section compares the size of an SC DAM optimization to the size of a standard US RTO/ISO-managed DAM optimization. Size comparisons for markets with shorter look-ahead horizons would take a similar form. In addition, each dispatchable resource m participating in the SC DAM is assumed to submit a reserve offer consisting of a single contract SC_m.

As discussed in Chapter 13, US RTO/ISO-managed DAMs undertake two types of optimization (either separately or in integrated form) to determine unit commitment, economic dispatch, ancillary service, and pricing solutions for next-day operations: namely, SCUC and SCED. In contrast, the ISO managing an SC DAM undertakes a single comprehensive optimization to determine which reserve offers and price-sensitive reserve bids to clear for next-day operations.

The SCUC optimization for an RTO/ISO-managed DAM is typically formulated as a mixed integer linear programming (MILP) optimization. As seen in Chapters 7–9, the contract-clearing optimization for an SC DAM can also be expressed as an MILP optimization for various types of reserve offers and bids.

The size of a MILP optimization can be approximately measured by the number of binary decision variables and the number of constraints included in its formulation. This size measure will be used below to obtain a rough comparison between the size of an ISO DAM SCUC optimization and the size of an SC DAM contract-clearing optimization.

Suppose the current day is D and the planning horizon for each DAM consists of all 24 hours H of day D+1. Consider, first, the relative number of binary decision variables. For the DAM SCUC MILP optimization, the ISO has 24 binary decision variables for each participating dispatchable resource m: namely, a 0/1 unit commitment indicator for each hour H of day D+1. In contrast, for the SC DAM MILP optimization, the ISO has one binary decision variable for each participating dispatchable resource m: namely, a 0/1 contract-clearing indicator that covers all 24 hours of day D+1.

Next consider the relative number of constraints. For the standard DAM SCUC MILP optimization, the RTO/ISO must explicitly include the unit commitment restrictions (e.g. start-up/shut-down, minimum down/up time) for a participating dispatchable resource m among the constraints for this optimization. In contrast, for the SC DAM MILP optimization, each participating dispatchable resource m is responsible for ensuring the physical feasibility of its offered swing contract SC_m, which requires in particular that all power-path availability offered in SC_m must satisfy m's unit commitment restrictions. Thus, m's unit commitment restrictions are implicitly imposed through the form of SC_m. This implicit imposition can result in a reduction in the needed number of explicitly expressed MILP constraints.

Thus, measured in terms of binary decision variables and number of system constraints, the size of an SC DAM optimization is smaller than the size of a standard DAM SCUC optimization, all else equal. Moreover, this size advantage substantially increases as the number of participating dispatchable resources increases.

15.10 ADDITIONAL ADVANTAGES OF SC MARKETS

15.10.1 Ensure a Level Playing Field for Resource Participation

SC markets are open-access ISO-managed forward reserve markets that enable all dispatchable resources to compete fairly for the provision of reserve (dispatchable power-paths) in return for appropriate compensation. All resources able to offer reserve into an SC market for a future operating period are permitted to do so, without regard for irrelevant underlying differences in their technologies. Reserve offers (SC portfolios) permit diverse types of resources to seek appropriate compensation for the swing (flexibility) in their offered reserve, whatever form this swing might take.

15.10.2 Permit Co-Optimization of Diverse Reserve

An ISO managing an SC market M(T) for a future operating period T undertakes a contract-clearing optimization to determine cleared reserve offers and price-sensitive reserve bids for period T. This optimization permits the efficient advance procurement of diverse flexible reserve to meet net load balancing needs for T.

15.10.3 Appropriately Remunerate Diversity and Flexibility

The swing contracts constituting the reserve offers submitted into SC markets permit resources to seek full appropriate remuneration for the diversity and flexibility of their offered reserve. The two-part pricing feature of swing contracts permits cleared resources to receive appropriate compensation for reserve availability as well as reserve delivery.

15.10.4 Encourage Accurate Forecasting and Dispatch Following

Standardized rewards and/or penalties can be incorporated into the prices charged to LSEs for power-path deliveries under the terms of their cleared price-sensitive and/or fixed reserve bids to encourage submission of bids that accurately reflect the power-usage preferences and needs of customers. Similarly, standardized rewards and/or penalties can be incorporated into the performance payment methods included among the contractual terms of reserve offers to encourage reserve suppliers to follow real-time dispatch signals with required accuracy.

15.10.5 Ensure Settlement Time-Consistency

The two-part pricing feature of the reserve offers (SC portfolios) submitted into SC markets ensures settlement time-consistency. In particular, in broad support of the goals of FERC Order 755 [54], no performance payments are made in advance of verified actual performance.

CHAPTER 16

GRADUAL TRANSITION TO LINKED SWING-CONTRACT MARKETS

16.1 CHAPTER PREVIEW

Institutional stickiness results when participants fail to deviate from existing institutional arrangements, even when this deviation would result in a better overall outcome [172]. On the other hand, it is perfectly understandable that risk-averse participants are reluctant to shift to untested new arrangements.

The chances of moving from the design of current US RTO/ISO-managed wholesale power markets to the Linked Swing-Contract Market Design would be enhanced if a means could be found to introduce the latter design in small transitional steps. The key roadblock is not operational feasibility but rather the narrow manner in which resource offers are specified and compensated in current markets.

As carefully explained in Chapter 4, a swing contract SC_m offered by a dispatchable resource m into an ISO-managed swing-contract market M(T) for a future operating period T is a complete blueprint for a set $\mathbb{PP}_m$ of power-paths, each of which m could feasibly deliver at a specified grid location during period T in response to dispatch signals. The blueprint characterizes the degree of swing (flexibility) in these offered power-paths by specifying their jointly feasible attribute ranges. The offer price for SC_m permits m to ensure coverage of all avoidable fixed cost it would have to incur to guarantee the availability of the power-paths in $\mathbb{PP}_m$ for possible dispatch during period T. The performance payment method ϕ_m included among the terms of SC_m permits m to ensure recovery ex post of any variable cost it incurs for actual delivery of a power-path during period T in response to dispatch signals.

Suppose a dispatchable resource m is a participant on day D in a US RTO/ISO-managed day-ahead market (DAM), to be cleared using a standard SCUC/SCED optimization. As detailed in [120], the latter optimization is based on a DC optimal power flow formulation; hence, voltage magnitudes are fixed at base voltage levels, and reactive power levels are set to zero.

A New Swing-Contract Design for Wholesale Power Markets, First Edition, Leigh Tesfatsion.

Suppose *m* is given permission to submit a swing contract SC_m into this DAM that offers the availability of dispatchable power-paths for day D+1. All other dispatchable resource participants *j* submit their offers using standard DAM supply-offer forms. Finally, suppose no DAM participant attempts to game the DAM for strategic market advantage.

As seen in Chapter 7, the attribute ranges for *m*'s offered power-paths can be fully incorporated into the DAM SCUC/SCED system constraints. Moreover, the offer price and performance payment method for SC_m can be fully incorporated into the DAM SCUC/SCED objective function. However, optimization of the resulting extended DAM SCUC/SCED formulation would not necessarily result in an accurate merit order for commitment and dispatch for three basic reasons:

- First, the extended objective function includes (at most) a partial reporting of avoidable fixed cost for any resource *j* other than *m*: for example, as in ERCOT, a reporting of *j*'s cost for start-up energy and for the energy expended to inject power at a minimum sustainable level. Omitted types of avoidable fixed cost include equipment purchases; transmission access fees; and opportunity cost, such as the lost commitment opportunity (LCO) cost discussed in Section 15.8.
- Second, the extended objective function includes only a partial reporting of variable cost for any resource *j* other than *m*, for example, the fuel cost associated with different levels of energy production. Examples of omitted variable cost include transmission service charges, wear-and-tear depreciation of equipment, and employee wages and salaries.
- Third, the extended system constraints, based on a DC optimal power flow formulation, do not permit any voltage-support services offered by SC_m to be properly valued in terms of their contribution to system reliability [151, 152].

Consequently, although it is straightforward to incorporate an offered swing contract SC_m within a standard DAM SCUC/SCED optimization, it is by no means assured that SC_m will be correctly positioned in the resulting merit-order stack. In order for this incorporation to result in an accurate merit-order commitment and dispatch for next-day operations, the SCUC/SCED objective function would have to represent fully and accurately the total benefit and total avoidable cost associated with each feasible configuration of committed and cleared resources.[1]

---

[1]Recall from Section 2.2 that *short-run* efficiency is the nonwastage of *existing* resources. An inaccurate DAM merit order can result in a wasteful use of existing resources for next-day operations, hence, in short-run market inefficiency. In contrast, DAM settlements are a redistribution of existing income across the DAM participants. Thus, inaccurate DAM settlements due, for example, to the LMP computational fragility problems discussed in Section 3.2.4, do not affect short-run market efficiency. Over time, however, inaccurate DAM settlements can lead to inefficient entry, retention, and exit of DAM resources, hence, to longer-run market inefficiency.

Sections 16.2–16.4 show how a standard analytical mixed integer linear programming (MILP) formulation for a DAM SCUC/SCED optimization can be modified in a minimally disruptive manner to provide an even playing field for participation by dispatchable resources offering simple types of swing contracts. This modified analytical MILP DAM formulation, referred to as the *Transitional DAM*, could thus constitute a first transitional step toward a full swing-contract DAM design.

Wherever possible, the nomenclature used for the Transitional DAM is the same as the nomenclature used in earlier chapters for DAMs in standard or swing-contract form. All needed new notation is introduced and defined in Sections 16.2–16.4 at the point of first use.

16.2 A DAM FORMULATION PERMITTING GRADUAL TRANSITION

This section describes, in general terms, an ISO-managed DAM that permits gradual transition to an SC DAM design. Hereafter this DAM will be referred to as the *Transitional DAM*.

The objective function for the Transitional DAM incorporates avoidable fixed costs and variable costs for a collection $\mathbb{G}$ of dispatchable generators. One dispatchable generator in $\mathbb{G}$, denoted by m, is permitted to submit a supply offer in swing-contract form. All other dispatchable generators in $\mathbb{G}$ submit supply offers taking the standard ERCOT form explained in Section 5.4 with one departure: namely, generators are permitted to report shut-down cost attributes as well as start-up cost, minimum-energy cost, and variable cost attributes.

The Transitional DAM is assumed to take place on day D. All load-serving entity (LSE) demand bids submitted into the Transitional DAM on day D are assumed to be fixed (non-price-sensitive) demands for next-day power usage. This next day D+1 is partitioned into consecutive hours $k \in \mathbb{K} = \{1, \ldots, NK\}$, where $NK = 24$. The start-time t^s for hour 1 of day D+1 and the end-time t^e for hour NK of day D+1 are specific calendar times expressed at the granularity of minutes.

Dispatchable generator m submits a firm[2] swing contract into the DAM that offers swing (flexibility) in both power amplitude and ramp rate. The specific form of this swing contract is as follows:

$$\mathrm{SC}_m = [\alpha_m, \mathbb{PP}_m, \phi_m] \tag{16.1}$$

[2]Recall from Chapter 4 that a swing contract SC_m submitted into an ISO-managed market is referred to as *firm* if (i) its exercise set $\mathbb{T}_m^{\mathrm{ex}}$ consists of a single exercise time that coincides with the close of the market and (ii) a decision by the ISO to clear the contract at the close of the market automatically triggers the exercise of the contract. Following the convention in earlier chapters, the exercise set is omitted from the representation of the firm swing-contract (16.1).

where

$$\alpha_m = \text{Offer price}$$

$$\mathbb{PP}_m = (b_m,\ t^s,\ \mathbb{P}_m,\ \mathbb{RR}_m,\ \text{UT}_m,\ \text{DT}_m,\ t^e)$$

$$b_m = \text{Delivery location for } m$$

$$t^s = \text{Start-time for } m\text{'s power delivery}$$

$$\mathbb{P}_m = [P_m^{\text{min}},\ P_m^{\text{max}}] = \text{Range of feasible sustainable power levels } p_m$$

$$\mathbb{RR}_m = [-R_m^{\text{D}},\ R_m^{\text{U}}] = \text{Range of feasible down/up ramp-rates } r_m$$

$$\text{UT}_m = \text{Minimum up-time for } m$$

$$\text{DT}_m = \text{Minimum down-time for } m$$

$$t^e = \text{End-time for } m\text{'s power delivery}$$

$$\phi_m = \text{Performance payment method for } m\text{'s power-path delivery.}$$

The offer price α_m (\$) requests compensation for the avoidable fixed cost that m would have to incur to ensure the availability of the offered power-path collection $\mathbb{PP}_m$ for day D+1 operations, should m be cleared. The delivery location b_m denotes the grid location at which m can inject power. The minimum feasible sustainable power level P_m^{min} (MW) satisfies $P_m^{\text{min}} \geq 0$. The down/up limits $-R_m^{\text{D}}$ (MW/h) and R_m^{U} (MW/h) for m's feasible ramp-rates r_m (MW/h) satisfy $-R_m^{\text{D}} \leq 0 \leq R_m^{\text{U}}$; these down/up limits are applicable for start-up, normal and shut-down conditions. The minimum up-time UT_m and minimum down-time DT_m for m are each measured in hours. Finally, the performance payment method ϕ_m requests compensation for any variable cost that m incurs for the dispatched delivery of an offered power-path during day D+1.

The ISO's decision variables include unit commitments and dispatch levels for the dispatchable generators $g \in \mathbb{G}$. The ISO's objective is to select values for its decision variables that achieve maximum possible expected total net benefit for the Transitional DAM participants. However, since all LSE demand bids take a fixed-demand form, maximization of expected total net benefit is equivalently expressed as the minimization of expected total avoidable cost.

The ISO's cost-minimization problem is formulated as a MILP SCUC/SCED optimization, conditional on initial state conditions and subject to system constraints. The system constraints include

- transmission line power flow constraints;
- power balance constraints (with slack variables);
- dispatchable generator capacity constraints;
- dispatchable generator ramping constraints (start-up, normal, and shut-down);

- dispatchable generator minimum up-time/down-time constraints;
- dispatchable generator hot-start constraints;
- dispatchable generator start-up/shut-down cost constraints;
- system-wide down/up reserve requirement constraints;
- zonal down/up reserve requirement constraints;
- voltage angle constraints;
- production cost approximation constraints.

These system constraints are based on a standard DC optimal power flow (DC-OPF) approximation. Thus, the resistance for each transmission line is assumed to be negligible compared to the reactance, permitting the resistance for each transmission line to be set to 0. Also, the voltage magnitude at each bus is assumed to be equal to a common base voltage magnitude. Finally, the voltage angle difference $\Delta\theta(\ell)$ across any line ℓ is assumed to be sufficiently small to permit use of the following approximations: $\cos(\Delta\theta(\ell)) \approx 1$ in size and $\sin(\Delta\theta(\ell)) \approx \Delta\theta(\ell)$ in size.

16.3 COST FUNCTION PRELIMINARIES FOR THE TRANSITIONAL DAM

Each of the 24 hours k comprising day D+1, measured at the granularity of minutes, is represented as $k = [k^{\mathrm{s}}, k^{\mathrm{e}})$ with start-time k^{s} and end-time k^{e}. The length of k, measured in hours, is denoted by $\Delta t = 1$ hour.

The range of feasible sustainable power levels p (MW) for each dispatchable generator $g \in \mathbb{G}$ is given by an interval $\mathbb{P}_g = [P_g^{\mathrm{min}}, P_g^{\mathrm{max}}]$ with $0 \leq P_g^{\mathrm{min}}$. The ISO dispatches generation by means of dispatch set points. Specifically, the ISO's dispatch instruction conveyed to a generator g committed for hour k consists of a single dispatch set point $p_g(k) \in \mathbb{P}_g$ signaled to g at the start-time k^{s} for hour k, which indicates the power level that g is to provide at k^{s}.

The *avoidable production cost function* (\$/h) for each dispatchable generator $g \in \mathbb{G}$ for each hour k is assumed to be an increasing strictly convex function of g's power injection p. Specifically, this avoidable production cost function is assumed to take the following quadratic form: For all $p \in \mathbb{P}_g$,

$$\mathrm{C}_{g,k}(p) = a_g(k) + b_g(k)p + c_g(k)p^2 \tag{16.2}$$

where the cost coefficient $a_g(k) \geq 0$ has units \$/h, the cost coefficient $b_g(k) \geq 0$ has units \$/MWh, and the cost coefficient $c_g(k) > 0$ has units \$/[MW]^{2}h.

Generator g's *marginal production cost function* (\$/MWh) for hour k is thus a strictly increasing function taking the following form: For all $p \in \mathbb{P}_g$,

$$\mathrm{MC}_{g,k}(p) = \frac{\partial \mathrm{C}_{g,k}(p)}{\partial p} = [b_g(k) + 2c_g(k)p\ ] \tag{16.3}$$

Generator g's *variable production cost function* (\$/h) for hour k is given by the integration of the marginal production cost function (16.3) over intervals $[P_g^{\min}, p]$:

$$\mathrm{VC}_{g,k}(p) = \left[\int_{P_g^{\min}}^{p} [b_g(k) + 2c_g(k)q]\, dq\right], \quad \forall p \in \mathbb{P}_g \tag{16.4}$$

Generator g's *avoidable fixed production cost* (\$/h) for hour k is given by

$$\mathrm{AvoidFCProd}_{g,k} = \mathrm{C}_{g,k}(P_g^{\min}) = a_g(k) + b_g(k)P_g^{\min} + c_g(k)[P_g^{\min}]^2 \tag{16.5}$$

Note that g's avoidable production cost function (\$/h) for hour k, defined in (16.2), can be expressed as follows: For all $p \in \mathbb{P}_g$,

$$\mathrm{C}_{g,k}(p) = \mathrm{VC}_{g,k}(p) + \mathrm{AvoidFCProd}_{g,k} \tag{16.6}$$

Finally, g's *total avoidable cost* (\$) for all 24 hours k of day D+1 is the summation of the following four costs:

- **(c1)** g's total variable production cost (\$) incurred during D+1;
- **(c2)** g's total avoidable fixed production cost (\$) incurred during D+1;
- **(c3)** any start-up cost c_g^{U} (\$) incurred by g during D+1;
- **(c4)** any shut-down cost c_g^{D} (\$) incurred by g during D+1.

Explicit expressions for these four costs will next be derived.

The summation of costs (c1) and (c2), denoted by c_g^{P}, constitutes g's *total avoidable production cost* for day D+1; compare (16.6). The cost c_g^{P} is approximated in the Transitional DAM as follows:

$$c_g^{\mathrm{P}} = \sum_{k\in\mathbb{K}} c_g^{\mathrm{P}}(k) = \sum_{k\in\mathbb{K}} [\mathrm{C}_{g,k}(p_g(k)) \cdot u_g(k)] \cdot \Delta t \tag{16.7}$$

In (16.7), $u_g(k)$ denotes the unit commitment indicator function for hour k indicating whether g has been committed ($u_g(k) = 1$) or not committed ($u_g(k) = 0$) for hour k. Recall that $p_g(k)$ denotes g's dispatch set point at the start-time for hour k, assuming g has been committed for hour k.

The total avoidable *fixed* production cost (c2) for g is approximated in the Transitional DAM as follows:

$$\mathrm{TotAvoidFCProd}_g = \left[\sum_{k\in\mathbb{K}} [\mathrm{AvoidFCProd}_{g,k} \cdot u_g(k)]\right] \cdot \Delta t \tag{16.8}$$

Note that (16.8) is the portion of g's avoidable fixed cost that occurs contemporaneously with g's production during day D+1.

As will be shown in Section 16.4, g's start-up cost (c3) during day D+1 can be calculated as a function of the following five factors:

- g's down-time/up-time status $\hat{u}_g(0)$ at the start of hour 1 on day D+1, determined by the ISO's unit commitment decisions for g prior to day D;
- The ISO's unit commitment decision $u_g(k)$ for g in the day-D Transitional DAM for each hour k of day D+1;
- CSH_g = g's cold-start hours parameter (number of hours);
- CSC_g = g's cold-start cost parameter ($);
- HSC_g = g's hot-start cost parameter ($), satisfying $\text{HSC}_g \leq \text{CSC}_g$.

The three listed parameters have the following meaning. If g at the start of some hour k has been off-line for at least CSH_g consecutive hours *immediately prior* to k, then g is in a *cold-start state* and any start-up of g at the start of k incurs the cold-start cost CSC_g. Otherwise, g is in a *hot-start state* at the start of k and any start-up of g at the start of k incurs the hot-start cost HSC_g.[3]

In addition, as will be shown in Section 16.4, g's shut-down cost (c4) during day D+1 can be calculated as a function of the following two factors:

- the ISO's unit commitment decision $u_g(k)$ for g in the day-D Transitional DAM for each hour k of day D+1;
- g's shut-down cost parameter SDC_g ($).

The shut-down cost parameter determines the cost that g incurs between hours $k-1$ and k if g is committed for hour $k-1$ but not for hour k.

Thus, the start-up cost (c3) and shut-down cost (c4) are avoidable fixed costs for each $g \in \mathbb{G}$; they depend on g's hourly unit commitments during day D+1, but they do not depend on g's actual power dispatch levels during day D+1. It follows that the *total avoidable fixed cost* for each $g \in \mathbb{G}$ is given by

$$\text{TotAvoidFC}_g = \text{TotAvoidFCProd}_g + c_g^{\text{U}} + c_g^{\text{D}} \tag{16.9}$$

Recall from Section 16.2 that the offer price α_m ($) requested by dispatchable generator m in his swing-contract offer SC_m expressed in (16.1) is assumed to be equal to m's total avoidable fixed cost. By definition, this is the cost that m would have to incur to ensure the *availability* of m's offered collection $\mathbb{PP}_m$ of dispatchable power-paths for possible delivery during day D+1, independently of whether or not m is subsequently dispatched by the ISO to deliver one of these offered power-paths. Given the above cost specifications for all dispatchable generators $g \in \mathbb{G}$, including m, this offer price is given by

$$\alpha_m = \text{TotAvoidFC}_m \tag{16.10}$$

---

[3]This modeling of start-up cost is a simplified version of the "stairwise startup function" proposed by Carrión and Arroyo [30, Section II].

The performance payment method ϕ_m included by m in its swing-contract offer SC_m requests compensation for any variable cost that m incurs during day D+1, assuming the ISO clears SC_m. For the illustration at hand, this variable cost consists of any variable production cost that m incurs for the dispatched delivery of a power-path in $\mathbb{PP}_m$ during day D+1. To enable the ISO to incorporate m's variable cost in the objective function for the Transitional DAM, it is assumed that m's performance payment method ϕ_m includes m's avoidable production cost function $C_{m,k}(p)$ for each hour $k \in \mathbb{K}$.

Thus, dispatchable generator m submits its avoidable fixed cost and variable cost information to the ISO through its swing contract offer SC_m, along with its physical attributes; see (16.1). This permits the ISO to include these costs and attributes in the Transitional DAM SCUC/SCED optimization.

Every other dispatchable generator $g \in \mathbb{G}/\{m\}$ is assumed to transmit directly to the ISO all needed cost and physical attribute information for this optimization. The needed cost information consists of g's avoidable production cost function $C_{g,k}(p)$ for each hour $k \in \mathbb{K}$, g's cold-start hours parameter CSH_g, g's cold-start cost parameter CSC_g, g's hot-start cost parameter HSC_g satisfying $HSC_g \leq CSC_g$, and g's shut-down cost parameter SDC_g. The needed physical attribute information includes

- g's range $[-R_g^D, R_g^U]$ of feasible down/up ramp rates r_g (MW/h) for start-up, normal, and shut-down conditions, with limits $-R_g^D \leq 0 \leq R_g^U$;
- g's minimum up-time UT_g (measured in hours);
- g's minimum down-time DT_g (measured in hours).

The SCUC/SCED optimization for the Transitional DAM, presented below in Section 16.4 in full analytical form, uses a piecewise-linear approximation[4] for the avoidable production cost function $C_{g,k}(p)$ given in (16.2) for each dispatchable generator $g \in \mathbb{G}$ during each hour $k \in \mathbb{K}$. This piecewise-linear approximation permits this SCUC/SCED optimization to be solved as a standard MILP problem.

The piecewise-linear approximation for $C_{g,k}(p)$ is constructed by connecting finitely many power-cost points $\{(P_i, C_i) \mid i = 0, \ldots, NS_g(k)\}$ in the power-cost plane, where these points are then incorporated into the SCUC/SCED optimization as exogenous inputs. These power-cost points must satisfy

$$P_g^{\min} = P_0 < P_1 < P_2 \cdots < P_{NS_g(k)} = P_g^{\max} \tag{16.11}$$

with $C_i = C_{g,k}(P_i)$ for each $i = 1, \ldots, NS_g(k)$. A simple method for generating power-cost points satisfying these restrictions is as follows. Determine a uniform width for each power segment as follows:

$$PSW_g(k) = \frac{[P_g^{\max} - P_g^{\min}]}{NS_g(k)} \tag{16.12}$$

---

[4]See Section 8.3 for a detailed explanation of this approximation method.

Next, set the power point P_i equal to $P_0 + i \cdot \text{PSW}_g(k)$ for each $i = 1, \dots, NS_g(k)$. Finally, set the cost point C_i equal to $C_{g,k}(P_i)$ for each $i = 1, \dots, NS_g(k)$.

16.4 MILP SCUC/SCED OPTIMIZATION FOR THE TRANSITIONAL DAM

This section presents a complete analytical MILP formulation for the Transitional DAM SCUC/SCED optimization taking place on day D. This analytical formulation is a modified version of the SCUC/SCED formulation developed by Tesfatsion and Battula [187]. The latter formulation, in turn, extends a SCUC/SCED formulation originally developed by Carrión and Arroyo [30].

The ISO's objective is to select unit commitments and generation dispatch levels to minimize total avoidable cost $\widehat{\text{C}}(\text{D+1})$ for day D+1, conditional on initial state conditions and subject to system constraints. The cost $\widehat{\text{C}}(\text{D+1})$ is "expected" cost in the sense that it is calculated based on net load forecasts for day D+1. That is, certainty equivalence is used to approximate expectations.

The cost $\widehat{\text{C}}(\text{D+1})$ is a summation of total avoidable production cost c^{P}, total start-up cost c^{U}, total shut-down cost c^{D}, and imbalance cost IC. As will be clarified below, IC is a penalty (\$) assessed for the appearance of nonzero slack variables in the system power balance constraints. For each possible choice of unit commitments and dispatch levels, IC represents the ISO's assessment of additional cost that would have to be incurred for resource procurement in subsequent market or nonmarket processes to ensure net load balancing during day D+1.

Initial State Conditions. For each $g \in \mathbb{G}$:

$\hat{u}_g(0)$ = Down-time/up-time status of g at start of hour 0 (number of hours);
$p_g(0)$ = Power output (MW) of g at the start of hour 0.

Remarks: Hour 0 is the hour immediately preceding the start of day D+1. If the value of $\hat{u}_g(0)$ is negative (positive), it indicates the number of consecutive hours prior to *and including* hour 0 that g has been OFF (ON). Note that $\hat{u}_g(0)$ cannot be zero, by definition.

Derived Initial State Conditions and Parameter Values. For each $g \in \mathbb{G}$, $k \in \mathbb{K}$:

$u_g(0)$ = The OFF/ON (0/1) commitment status for g during hour 0
ITF_g = Number of hours (integer) that g must be offline *initially*
ITO_g = Number of hours (integer) that g must be online *initially*
SRU_g = Scaled ramp-up limit (MW) for g at the start of hour k
SRD_g = Scaled ramp-down limit (MW) for g at the start of hour k
SUT_g = Scaled minimum up-time (integer) for g
SDT_g = Scaled minimum down-time (integer) for g.

Calculation of Derived Initial State Conditions and Parameter Values. For each $g \in \mathbb{G}$ and $k \in \mathbb{K}$:

- If $\hat{u}_g(0) < 0$, $u_g(0) = 0$; otherwise, $u_g(0) = 1$
- Set $x = \text{round}([\text{DT}_g + \hat{u}_g(0)]/\Delta t)$.
- If $\hat{u}_g(0) < 0$, $\text{ITF}_g = \min\{\ NK,\ \max\{0,\ x\}\ \}$; otherwise, $\text{ITF}_g = 0$
- Set $y = \text{round}([\text{UT}_g - \hat{u}_g(0)]/\Delta t)$.
- If $\hat{u}_g(0) > 0$, $\text{ITO}_g = \min\{\ NK,\ \max\{0, y\}\ \}$; otherwise, $\text{ITO}_g = 0$
- $\text{SRU}_g = \min\ \{P_g^{\max}, R_g^U \cdot \Delta t\}$
- $\text{SRD}_g = \min\ \{P_g^{\max},\ R_g^D \cdot \Delta t\}$
- $\text{SUT}_g = \text{round}\ (\text{UT}_g/\Delta t)$.
- $\text{SDT}_g = \text{round}\ (\text{DT}_g/\Delta t)$.

Remarks: DT_g and UT_g denote the minimum down-time and minimum up-time for g, each measured in hours. Generator g's feasible ramp-rate range $[-R_g^D, R_g^U]$ for its down/up ramp-rates r_g(MW/h) is applicable for start-up, normal, and shut-down conditions. The function round(n, d) performs a round-off operation. The first argument n is the number to be rounded, and the second argument d is the number of decimal places to be included in the rounded number. If the second argument d is missing, this indicates $d = 0$. If the value of the decimal in the $(d + 1)$st decimal place is greater than 5, n is rounded up; otherwise, n is rounded down. The output is the resulting rounded-off number. The length of each hour k, measured in hours, is denoted by $\Delta t = 1$ hour.

ISO Objective:
Select decision variables to minimize total avoidable cost $\widehat{\text{C}}(\text{D+1})$ for day D+1, conditional on initial state conditions and subject to system constraints, where

$$\widehat{\text{C}}(\text{D+1}) = \sum_{k\in\mathbb{K}} \sum_{g\in\mathbb{G}} [c_g^{\text{P}}(k) + c_g^{\text{U}}(k) + c_g^{\text{D}}(k)] + \text{IC}(\text{D+1}) \tag{16.13}$$

ISO Decision Variables:
ISO binary-valued decision variables: For all $g \in \mathbb{G}$,

$$u_g(k) \in \{0, 1\}, \quad \forall k \in \mathbb{K}$$
$$u_m(k) = c_m, \quad \forall k \in \mathbb{K}/\{NK\}$$
$$c_m \in \{0, 1\}$$

Remarks: The binary-valued decision variable c_m is the contract-clearing indicator for dispatchable generator m's swing-contract offer (16.1). This swing contract is either cleared ($c_m = 1$) or not cleared ($c_m = 0$) for the first 23 hours of day D+1. The binary-valued decision variables $u_g(k)$ for $g \neq m$ are standard DAM binary unit commitment variables indicating whether g is committed ($u_g(k) = 1$) or not committed ($u_g(k) = 0$) for power dispatch during each hour k. The ISO makes one additional binary-valued decision: namely, a 0/1 setting for $u_m(NK)$ determining whether m is to be shut-down ($u_m(NK) = 0$) or not shut down ($u_m(NK) = 1$) during the final hour NK of day D+1.

ISO continuously valued decision variables:
For all $b \in \mathbb{B}$, $g \in \mathbb{G}$, $i \in NS_g(k)$, and $k \in \mathbb{K}$,

$$\delta_{i,g}(k)$$

$$\theta_b(k)$$

Remarks: The decision variables $\delta_{i,g}(k)$, $i = 1, \ldots, NS_g(k)$, appear in the system constraints (16.40)–(16.45) that determine the piece-wise linear approximation for g's avoidable production cost function $C_{g,k}(p)$ for hour k. The decision variable $\theta_b(k)$ is the voltage angle (radians) for bus b during hour k.

Derived Solution Values:
Solution values derived as functions of ISO decisions and system constraints: For all $b \in \mathbb{B}$, $g \in \mathbb{G}$, $k \in \mathbb{K}$, and $\ell \in \mathbb{L}$,

$$\text{IC(D+1)}, c_g^{\text{U}}(k),\ c_g^{\text{D}}(k),\ c_g^{\text{P}}(k)$$

$$hs_g(k)$$

$$p_g(k),\ \underline{p}_g(k),\ \overline{p}_g(k)$$

$$w_\ell(k)$$

$$\beta_b(k),\ \beta_b^-(k),\ \beta_b^+(k)$$

$$\theta_1(k)$$

Remarks: Explicit expressions for the power imbalance cost IC(D+1) (\$), start-up cost $c_g^{\text{U}}(k)$ (\$), shut-down cost $c_g^{\text{D}}(k)$ (\$), and production cost $c_g^{\text{P}}(k)$ (\$) as functions of ISO decisions and system constraints are given below in equations (16.17), (16.33), (16.34), and (16.40). The binary-valued (1/0) variable $hs_g(k)$ derived from equations (16.30)–(16.32) indicates whether or not g is in a hot-start state at the start of hour k. The power level $p_g(k)$ (MW) expressed by (16.41) is the ISO's dispatch set point for g at the start of hour k if g is committed for hour k. The power levels $\underline{p}_g(k)$ (MW) and $\overline{p}_g(k)$ (MW) are run-time lower and upper limits on $p_g(k)$ determined as functions of ISO decisions and the system constraints (16.18)–(16.23). The power level $w_\ell(k)$ (MW) expressed by (16.14) is the power flow on transmission line ℓ at the start of hour k. The value $\beta_b(k)$ (MW) is determined as the slack-variable value for the power balance constraint (16.16) for bus b at the start of hour k. The derived values $\beta_b^-(k)$ (MW) and $\beta_b^+(k)$ (MW) are the decomposition of $\beta_b(k)$ into negative and positive parts. Finally, the voltage angle $\theta_1(k)$ (radians) at angle reference bus 1 for hour k is determined by (16.39).

System Constraints:
Transmission line power flow constraints: For all $k \in \mathbb{K}$ and $\ell \in \mathbb{L}$,

$$w_\ell(k) = S_0 B(\ell)[\theta_{O(\ell)}(k) - \theta_{E(\ell)}(k)] \tag{16.14}$$

$$-F^{\max}(\ell) \leq w_\ell(k) \leq F^{\max}(\ell) \tag{16.15}$$

Remarks: In (16.14), the parameter value S_0(MW) is a positive setting for base power. The parameter value $B(\ell)$ denotes the inverse of the positively valued reactance (pu) for transmission line ℓ. The term $O(\ell)$ denotes the originating bus for transmission line ℓ, and the term $E(\ell)$ denotes the end-bus for transmission line ℓ. The parameter value $F^{\max}(\ell)$ in (16.15) is a positive value used to determine the lower and upper limits on the power flow $w_\ell(k)$ for line ℓ at the start of each hour k. All of these parameter values and terms are exogenous inputs for the SCUC/SCED optimization. The bracketed term in (16.14) is the *unit-free size* of the voltage angle difference for line ℓ at the start of hour k.

Power balance constraints (with slack variables): For all $b \in \mathbb{B}$ and $k \in \mathbb{K}$,

$$\sum_{g\in\mathbb{G}(b)} p_g(k) + \left[\sum_{\ell\in\mathbb{L}_{E(b)}} w_\ell(k) - \sum_{\ell\in\mathbb{L}_{O(b)}} w_\ell(k)\right] = \widehat{NL}_b^{\mathrm{f}}(k) + \beta_b(k) \tag{16.16}$$

Remarks: In constraint (16.16), the term $\widehat{NL}_b^{\mathrm{f}}(k)$ (MW) denotes the forecast for net must-service load at bus b at the start of hour k. The term $\mathbb{L}_{O(b)}$ denotes the subset of transmission lines originating at bus b, and the term $\mathbb{L}_{E(b)}$ denotes the subset of transmission lines ending at bus b. These three terms are exogenously given inputs for the SCUC/SCED optimization.

Slack variable constraints: For all $b \in \mathbb{B}$ and $k \in \mathbb{K}$,

$$\beta_b^-(k) = \max\{0, -\beta_b(k)\}$$

$$\beta_b^+(k) = \max\{0, \ \beta_b(k)\}$$

Imbalance cost:

$$\text{IC(D+1)} = \sum_{k\in\mathbb{K}} \sum_{b\in\mathbb{B}} [\Lambda^- \beta_b^-(k) + \Lambda^+ \beta_b^+(k)]\Delta t \tag{16.17}$$

Remark: In constraint (16.17), the terms Λ^-(\$/MWh) and Λ^+(\$/MWh) are nonnegative penalty weights imposed on negative and positive slack variable values $\beta_b(k)$(MW) in the power balance constraint (16.16). The term $\Delta t = 1$ hour is the length of each k in hours. These three terms are exogenously given inputs for the SCUC/SCED optimization.

Dispatchable generator capacity constraints: For all $g \in \mathbb{G}$ and $k \in \mathbb{K}$,

$$\underline{p}_g(k) \le p_g(k) \le \overline{p}_g(k) \tag{16.18}$$

$$\overline{p}_g(k) \le P_g^{\max} \cdot u_g(k) \tag{16.19}$$

$$\underline{p}_g(k) \ge P_g^{\min} \cdot u_g(k) \tag{16.20}$$

Dispatchable generator ramping constraints for start-up, normal, and shut-down conditions:

$$\overline{p}_g(k) \leq p_g(k-1) + \mathrm{SRU}_g \cdot u_g(k) + P_g^{\max}[1 - u_g(k)] ,$$
$$\forall g \in \mathbb{G},\ \forall k \in \mathbb{K} \tag{16.21}$$

$$\overline{p}_g(k-1) \leq \mathrm{SRD}_g[u_g(k-1) - u_g(k)] + P_g^{\max} u_g(k) ,$$
$$\forall g \in \mathbb{G},\ \forall k \in \mathbb{K} \tag{16.22}$$

$$p_g(k-1) \leq \underline{p}_g(k) + \mathrm{SRD}_g \cdot u_g(k-1) + P_g^{\max}[1 - u_g(k-1)] ,$$
$$\forall g \in \mathbb{G},\ \forall k \in \mathbb{K} \tag{16.23}$$

Remarks: Given system constraints (16.18)–(16.20) and the definitions for SRU_g and SRD_g, the following implications hold for these three ramping constraints: (i) In the *start-up case* $u_g(k-1) = 0$ and $u_g(k) = 1$, the first ramping constraint (16.21) imposes an important start-up ramping restriction on g's power usage during the successive hours $k - 1$ and k; the second and third ramping constraints (16.22) and (16.23) are guaranteed to hold. (ii) In the *normal up case* $u_g(k-1) = u_g(k) = 1$, the first and third ramping constraints (16.21) and (16.23) impose important down/up ramping restrictions on g's power usage during the successive hours $k - 1$ and k; the second ramping constraint (16.22) is guaranteed to hold. (iii) In the *shut-down case* $u_g(k-1) = 1$ and $u_g(k) = 0$, the second ramping constraint (16.22) imposes an important shut-down ramping restriction on g's power usage during the successive hours $k - 1$ and k that implies the third ramping restriction (16.23) also holds; the first ramping constraint (16.21) is guaranteed to hold. (iv) In the *normal down case* $u_g(k-1) = u_g(k) = 0$, all three ramping constraints (16.21)–(16.23) are guaranteed to hold.

Dispatchable generator minimum up-time constraints:

$$\sum_{k=1}^{\mathrm{ITO}_g} [1 - u_g(k)] = 0 \quad \text{for all } g \in \mathbb{G} \text{ with } \mathrm{ITO}_g \geq 1 \tag{16.24}$$

$$\sum_{n=k}^{k+\mathrm{SUT}_g-1} u_g(n) \geq \mathrm{SUT}_g[u_g(k) - u_g(k-1)],$$
$$\forall g \in \mathbb{G},\ \forall k = \mathrm{ITO}_g + 1, \ldots, NK - \mathrm{SUT}_g + 1 \tag{16.25}$$

$$\sum_{n=k}^{NK} (u_g(n) - [u_g(k) - u_g(k-1)]) \geq 0,$$
$$\forall g \in \mathbb{G},\ \forall k = NK - \mathrm{SUT}_g + 2, \ldots, NK \tag{16.26}$$

Remarks: To derive the minimum up-time constraints (16.24)–(16.26), consider the following. If $\mathrm{ITO}_g \geq 1$ for generator g, then by definition of ITO_g it must hold that $u_g(k) = 1$ for all hours k satisfying $1 \leq k \leq \text{ITO}_g$. For $\text{ITO}_g + 1 \leq k$, suppose a start-up event occurs for generator g in hour k; i.e. suppose $u_g(k-1) = 0$ and $u_g(k) = 1$, implying generator g is turned off in hour $k-1$ and on in hour k. Then, by definition of SUT_g, generator g must remain on for $\text{SUT}_g - 1$ additional hours, or until the end of the final modeled hour NK if $NK \leq k + \text{SUT}_g - 1$. The above minimum up-time constraints express these requirements in concise form.

Dispatchable generator minimum down-time constraints:

$$\sum_{k=1}^{\text{ITF}_g} u_g(k) = 0 \quad \forall g \in \mathbb{G} \text{ with } \text{ITF}_g \geq 1 \tag{16.27}$$

$$\sum_{n=k}^{k+\text{SDT}_g-1} [1 - u_g(n)] \geq \text{SDT}_g[u_g(k-1) - u_g(k)],$$
$$\forall g \in \mathbb{G},\ \forall k = \text{ITF}_g + 1, \ldots, NK - \text{SDT}_g + 1 \tag{16.28}$$

$$\sum_{n=k}^{NK}[1 - u_g(n) - [u_g(k-1) - u_g(k)]] \geq 0\,,$$
$$\forall g \in \mathbb{G},\ \forall k = NK - \text{SDT}_g + 2, \ldots, NK \tag{16.29}$$

Remarks: The derivation of minimum down-time constraints (16.27)–(16.29) is similar to the derivation of the minimum up-time constraints, except that one considers shut-down events with $u_g(k-1) = 1$ and $u_g(k) = 0$ rather than start-up events.

Dispatchable generator hot-start constraints:

$$hs_g(k) = 1, \quad \forall g \in \mathbb{G},\ 1 \leq k \leq \text{CSH}_g\ :\ k - \text{CSH}_g \leq \hat{u}_g(0) \tag{16.30}$$

$$hs_g(k) \leq \sum_{n=1}^{k-1} u_g(n), \quad \forall g \in \mathbb{G}, 1 \leq k \leq \text{CSH}_g\ :\ k - \text{CSH}_g > \hat{u}_g(0) \tag{16.31}$$

$$hs_g(k) \leq \sum_{n=k-\text{CSH}_g}^{k-1} u_g(n), \quad \forall g \in \mathbb{G},\ \forall k = \text{CSH}_g + 1, \ldots, NK \tag{16.32}$$

Remarks: As previously explained, a positive (negative) value for $\hat{u}_g(0)$ for some dispatchable generator $g \in \mathbb{G}$ indicates the number of consecutive hours prior to *and including* hour 0 that g has been turned ON (OFF). Note that $\hat{u}_g(0)$ cannot be zero, by definition. Also, $u_g(0)$ denotes the initial ON/OFF (1/0) status for generator g. If $\hat{u}_g(0) > 0$, then $u_g(0) = 1$; otherwise, $u_g(0) = 0$. Constraint (16.30) ensures that, if k does not exceed CSH_g, then generator g is in a hot-start state ($hs_g(k) = 1$) as long as g was ON either during hour 0 or during an hour prior to hour 0 that is sufficiently close to hour 0. Constraint (16.31) ensures that, if k does not exceed CSH_g, and g was *not* ON during hour

0 or during an hour prior to hour 0 that is sufficiently close to hour 0, then g is in a cold-start state ($hs_g(k) = 0$) unless g was ON during some hour between 1 and $k - 1$. Finally, constraint (16.32) ensures that, for k larger than CSH_g, g will be in a cold-start state ($hs_g(k) = 0$) if g was not committed during any of the CSH_g hours immediately preceding hour k. For reasons explained in the remarks following the next set of constraints (i.e., the generator start-up cost constraints), if generator g has a positive cold-start cost CSC_g, the cost-minimizing ISO will set $hs_g(k) = 1$ unless the generator hot-start constraints (16.30)–(16.32) force the ISO to set $hs_g(k) = 0$.

Dispatchable generator start-up cost constraints: For all $g \in \mathbb{G}$ and $k \in \mathbb{K}$,

$$c_g^{\mathrm{U}}(k) = \max\{0, U_g(k)\} \tag{16.33}$$

$$U_g(k) = \mathrm{CSC}_g - [\mathrm{CSC}_g - \mathrm{HSC}_g]hs_g(k) - \mathrm{CSC}_g[1 - [u_g(k) - u_g(k-1)]]$$

Remarks: The cold-start cost and hot-start cost parameters CSC_g and HSC_g are required to satisfy $\mathrm{CSC}_g \geq \mathrm{HSC}_g$. Consequently, (16.33) implies: (i) $c_g^{\mathrm{U}}(k) = \mathrm{CSC}_g$ if g starts up at the start-time for k (i.e. $u_g(k-1) = 0$ and $u_g(k) = 1$) while in a cold-start state ($hs_g(k) = 0$) and (ii) $c_g^{\mathrm{U}}(k) = \mathrm{HSC}_g$ if g starts up at the start-time for k while in a hot-start state ($hs_g(k) = 1$). Otherwise, $c_g^{\mathrm{U}}(k) = 0$. Thus, assuming $\mathrm{CSC}_g > 0$, in order to minimize total avoidable cost $\widehat{\mathrm{C}}(D+1)$ the ISO will strive to avoid starting up generator g in a cold-start state, all else equal. More generally, the cost-minimizing ISO will try to ensure $hs_g(k) = 1$ for each generator g it commits at the start of hour k.

Dispatchable generator shut-down cost constraints: For all $g \in \mathbb{G}$ and $k \in \mathbb{K}$,

$$c_g^{\mathrm{D}}(k) = \max\{0, D_g(k)\} \tag{16.34}$$

$$D_g(k) = \mathrm{SDC}_g[u_g(k-1) - u_g(k)]$$

Remark: Recall that SDC_g($) denotes the shut-down cost parameter for g.

System-wide down/up reserve requirement constraints: For all $k \in \mathbb{K}$,

$$\sum_{g \in \mathbb{G}} \underline{p}_g(k) \leq [1 - \mathrm{RRD}] \cdot \widehat{NL}^{\mathrm{f}}(k) \tag{16.35}$$

$$\sum_{g \in \mathbb{G}} \overline{p}_g(k) \geq [1 + \mathrm{RRU}] \cdot \widehat{NL}^{\mathrm{f}}(k) \tag{16.36}$$

Remarks: In constraints (16.35) and (16.36), the term $\widehat{NL}^{\mathrm{f}}(k)$ (MW) denotes the forecast for system-wide net must-service load at the start of hour k. The terms RRD (decimal percent) and RRU (decimal percent) are ISO-set system-wide non-negative reserve requirements for down-power and up-power, respectively. All of these terms are incorporated into the SCUC/SCED optimization as exogenously given inputs.

Zonal down/up reserve requirement constraints: For all $k \in \mathbb{K}$ and $z \in \mathbb{Z}$,

$$\sum_{g \in \mathbb{G}(z)} \underline{p}_g(k) \leq [1 - \text{RRD}(z)] \cdot \widehat{NL}_z^{\text{f}}(k) \tag{16.37}$$

$$\sum_{g \in \mathbb{G}(z)} \overline{p}_g(k) \geq [1 + \text{RRU}(z)] \cdot \widehat{NL}_z^{\text{f}}(k) \tag{16.38}$$

Remarks: In constraints (16.37) and (16.38) the term $\widehat{NL}_z^{\text{f}}(k)$ (MW) denotes the forecast for net must-service load in zone z at the start of hour k. The term $\mathbb{G}(z)$ denotes the subset of all dispatchable generators $g \in \mathbb{G}$ that are located within zone z. The terms RRD(z) (decimal percent) and RRU(z) (decimal percent) are ISO-set zonal nonnegative reserve requirements for down-power and up-power, respectively, in zone z. All of these terms are incorporated into the SCUC/SCED optimization as exogenously given inputs.

Voltage angle constraints: For all $b \in \mathbb{B}/\{1\}$ and $k \in \mathbb{K}$,

$$\theta_b(k) \in [-\pi, \pi]$$

$$\theta_1(k) = 0 \tag{16.39}$$

Avoidable production cost approximation constraints: For all $g \in \mathbb{G}$ and $k \in \mathbb{K}$,

$$c_g^{\text{p}}(k) = C_0 \Delta t \cdot u_g(k) + \sum_{i=1}^{NS_g(k)} (\text{MC}_i \cdot \delta_{i,g}(k))\Delta t \tag{16.40}$$

$$p_g(k) = P_0 \cdot u_g(k) + \sum_{i=1}^{NS_g(k)} \delta_{i,g}(k) \tag{16.41}$$

$$\delta_{i,g}(k) \leq P_{i,g} - P_{i-1,g} \quad \forall i = 1, \ldots, NS_g(k) - 1 \tag{16.42}$$

$$\delta_{NS_g(k),g}(k) \leq P_g^{\max} - P_{NS_g(k)-1,g} \tag{16.43}$$

$$\delta_{i,g}(k) \geq 0, \quad \forall i = 1, \ldots, NS_g(k) \tag{16.44}$$

$$\text{MC}_i = \frac{C_i - C_{i-1}}{P_i - P_{i-1}}, \quad \forall i = 1, \ldots, NS_g(k) \tag{16.45}$$

Remarks: As explained in Section 16.3, the power-cost points $\{(P_i, C_i) \mid i = 0, \ldots, NS_g(k)\}$ appearing in (16.40)–(16.45) are constructed from the avoidable production cost function provided by g to the ISO at the start of the Transitional DAM; these power-cost points are incorporated into the SCUC/SCED optimization as exogenously given inputs. If g is not committed in hour k, i.e. if $u_g(k) = 0$, the system constraints (16.18)–(16.20) require $p_g(k) = 0$; in this case, constraints (16.41)–(16.44) require $\delta_{i,g}(k) = 0$ for each $i = 1, \ldots, NS_g(k)$.

CHAPTER 17

SWING-CONTRACT SUPPORT FOR INTEGRATED TRANSMISSION AND DISTRIBUTION SYSTEMS

17.1 CHAPTER PREVIEW

The inclusion of variable energy resources (VERs) in RTO/ISO-managed wholesale power market operations is now growing rapidly, along with more active demand-side participation. These changes are amplifying the uncertainty and volatility of net load, including the possibility of sudden strong ramping events.

Consequently, RTO/ISOs are finding it difficult to ensure continual net load balance by traditional means: specifically, by the dispatch of unencumbered capacity from relatively inexpensive large-scale thermal generators with relatively slow ramp rates, supplemented at times by relatively expensive peaker generators with quick-start capability and fast ramp rates. Rather, RTO/ISOs need reserve that permits flexible dispatch of down/up power at relatively low cost during all operating hours.

Storage capabilities at the transmission level, while steadily improving, are not currently able to meet this RTO/ISO need. Power system researchers are thus also exploring whether this need could be met by harnessing net-load balancing services from resources located within distribution systems.[1]

A *grid-edge resource (GER)* is defined to be any entity capable of power usage and/or power output that has a *direct* point of connection to a distribution grid. Common examples of GERs include households, commercial businesses, and industrial plants. However, wind turbines, solar photovoltaic (PV) arrays, storage units, and other types of electrical facilities also constitute GERs if they are directly connected to a distribution grid. Finally, a *grid-edge resource aggregator (GERA)* is any entity

[1]For reviews of this research literature from diverse viewpoints, at different technical levels, see [15, 19, 40, 88, 103–106, 117, 150, 178, 182].

A New Swing-Contract Design for Wholesale Power Markets, First Edition, Leigh Tesfatsion.

that manages power supply, power usage, and/or ancillary service provision for a collection of GERs.

This chapter considers possible ways in which swing contracts could facilitate the harnessing of flexible dependable net-load balancing services from GERs in a manner consistent with system efficiency and reliability objectives. Four major premises of this chapter are as follows:

- First, power researchers need to consider with care the operations of *integrated transmission and distribution (ITD)* systems over time.
- Second, power researchers need to develop scalable approaches that permit the procurement of flexible net-load balancing services from ITD resources as the number of these resources continues to increase.
- Third, *independent distribution system operators (IDSOs)*, functioning as linkage entities at transmission and distribution (T–D) system interfaces, should be permitted to participate in distribution systems as GERAs and in transmission systems as providers of net-load balancing services harnessed from GERs in return for appropriate compensation.
- Fourth, IDSO participation in transmission systems as providers of net-load balancing services can be facilitated by appropriately designed swing contracts.

In keeping with these four premises, the remaining sections of this chapter propose a scalable framework for an ITD system that permits IDSOs to function as linkage entities at T–D system interfaces.[2] This framework consists of a hierarchical structuring of localized two-way communication links between entities at different system levels.

More precisely, the proposed framework models an RTO/ISO-managed transmission system linked to multiple IDSO-managed distribution systems. Each IDSO has a fiduciary responsibility to maintain the efficiency and reliability of its distribution system over time. As illustrated in Figure 17.1, each IDSO is in two-way communication with one or more GERAs tasked with handling power usage, power supply, and/or ancillary service provision for a collection of GERs. Finally, each GER manages the operation of one or more behind-the-meter electrical devices.

A key envisioned use of this framework is the support of IDSO-managed bid-based *transactive energy system (TES)* designs. A *TES design* is a collection of economic and control mechanisms permitting the dynamic balancing of power demands and supplies across an entire electrical infrastructure, using value as the key operational parameter [2, 68, 107, 116]. A *bid-based TES design* is a TES design for which valuations are based explicitly on purchase and sale reservation values.[3]

Section 17.2 proposes an IDSO-managed bid-based TES design that can be supported by the framework depicted in Figure 17.1. This design, referred to as the

[2]The developments reported in the remaining sections of this chapter are based on Battula et al. [8], Nguyen et al. [139], and Tesfatsion and Battula [186].

[3]A *purchase reservation value* for a quantity amount q at a particular time t is defined to be a buyer's maximum willingness to pay for q at time t. A *sale reservation value* for a quantity amount q at a particular time t defined to be the minimum payment that a seller is willing to receive for the sale of q at time t.

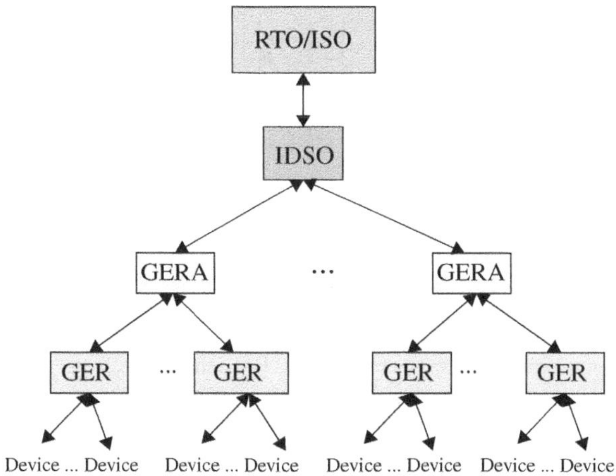

Figure 17.1 Scalable communication framework that supports the participation of IDSOs as linkage entities at transmission and distribution interfaces.

Five-Step TES Design, is a generalized version of the well-known PowerMatcher TES design developed by Kok [99, 149] to balance the power transactions of numerous power consuming and producing devices operating within an electrical infrastructure. The critical role of *Distribution Utilities (DUs)* within the Five-Step TES Design is considered in Section 17.3.

To illustrate the capabilities of the Five-Step TES Design in more concrete terms, Sections 17.4 and 17.5 focus on a special case in which each GER is a household with a smart (price-sensitive) heating, ventilation, and air conditioning (HVAC) system and a mix of conventional appliances. Each household repeatedly derives a refreshed price-sensitive bid function for its HVAC system, optimally determined as a function of its comfort-cost preferences and its current thermal dynamic state. This latest refreshed bid function is communicated to the IDSO, which in turn uses the aggregation of these bid functions to communicate prices back to the households in accordance with its system objectives.

Section 17.6 considers how an IDSO, operating at a T–D interface, could use the Five-Step TES Design to create additional revenue streams for GERs. For example, the IDSO could use the design to harness ancillary services from a collection of GERs, which it then submits into an ISO/RTO-managed wholesale power market by means of an appropriately designed swing-contract reserve offer. Any availability and performance payments received for these ancillary services could then be passed back to this collection of GERs.

17.2 TRANSACTIVE ENERGY SYSTEM DESIGN FOR ITD SYSTEMS

This section proposes an IDSO-managed bid-based TES design for a distribution system connected to a transmission system. The IDSO operates as a linkage entity at the T–D system interface; see Figure 17.2.

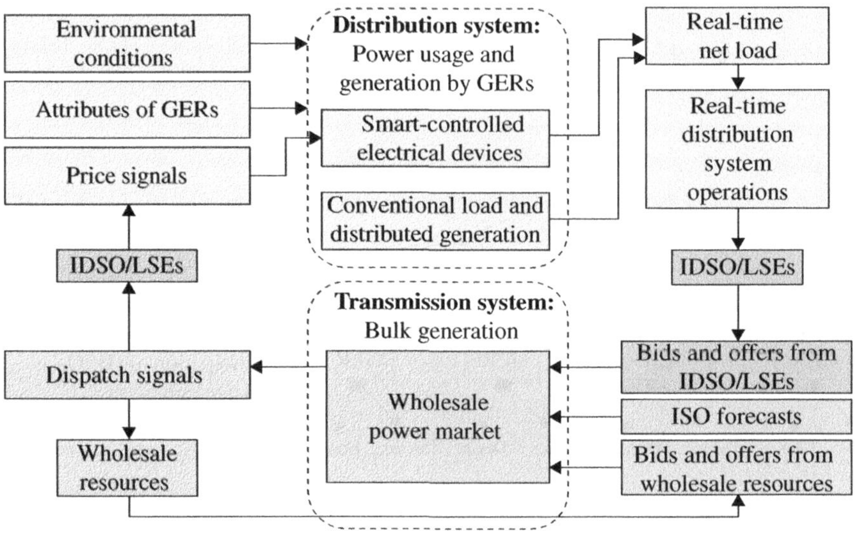

Figure 17.2 Illustration of an ITD system in which IDSOs function as T–D linkage entities.

As illustrated in Figure 17.3, the proposed design is supported by a communication network of *local intelligent software agents (LISAs)*. The IDSO constitutes the top node in the network. The LISAs operating at the edge of the distribution grid, referred to as *edge LISAs*, constitute the lowest layer in the network. Each edge LISA is associated with a particular GER that owns an array of electrical devices. Some of these devices have conventional controllers responding only to local state conditions, and some have smart controllers able to respond to signaled power prices. Finally, the network can also include one or more *interior LISAs* operating at interior layers of the network.

Each edge LISA constructs bid functions[4] for its associated GER that reflect the GER's current willingness and ability to buy or sell power as a function of the power price, conditional on local state conditions. Thus, an edge LISA functions as a device aggregator for its associated GER. Each edge LISA communicates its constructed bid functions to a LISA in the next-higher layer of the network, either an interior LISA or the IDSO itself.

[4]In standard auction theory, the auction participants are assumed to consist of buyers that submit "bids" for items they wish to buy and sellers that submit "offers" for items they wish to sell. However, there is no consideration of participation by *prosumers*, i.e. entities such as dealers that function as buyers or sellers depending on available prices and local state conditions. As will be seen in subsequent sections of this study, various types of GERs can participate in TES designs as prosumers. Moreover, as illustrated in Figure 17.4, the aggregation of pure demand and supply functions can result in aggregate functions taking a prosumer form. Consequently, for simplicity of exposition, the term *bid function* is hereafter used in this study to refer to any price-sensitive buy or sell request submitted by a participant in a TES design.

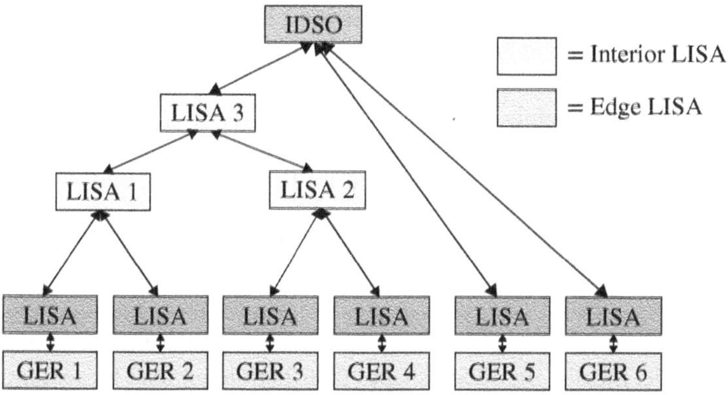

Figure 17.3 An illustrative LISA communication network for an IDSO-managed distribution system.

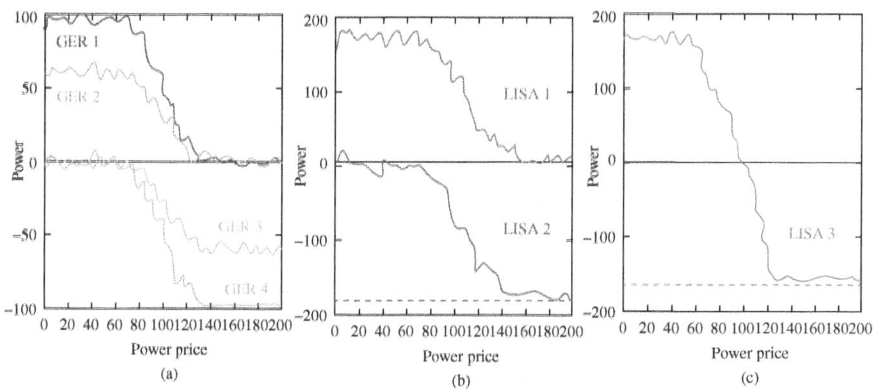

Figure 17.4 Illustrative bid aggregation process for the LISA communication network in Figure 17.3. Positive power levels denote power demands and negative power levels denote power supplies.

Each interior LISA aggregates all bids it receives from lower-layer LISAs and communicates this aggregate bid to a LISA in a next-higher layer. Thus, interior LISAs function in the network as GERAs.

Figure 17.4 illustrates this bid aggregation process for the LISA communication network depicted in Figure 17.3 given the simplifying assumption that each interior LISA aggregates all received bid functions into a single aggregate bid function. Figure 17.4a depicts possible bid functions for GERs 1–4. The depicted bid functions for GERs 1 and 2 are power demands, whereas the depicted bid functions for GERs 3 and 4 are power supplies.

Figure 17.4b shows how the interior LISA 1, operating as an aggregator for GERs 1 and 2, produces an aggregate bid function by summing their power requirements at each price.[5] Similarly, a second interior LISA 2 aggregates the bid functions for GERs 3 and 4. The aggregate bid function constructed by LISA 1 is essentially a pure demand function, whereas the aggregate bid function constructed by LISA 2 is essentially a pure supply function.

Finally, Figure 17.4c shows how the higher level interior LISA 3 then combines the aggregate bid functions constructed by interior LISAs 1 and 2 into a single aggregate bid function taking a prosumer form: for a sufficiently high price, power is supplied; and for a sufficiently low price, power is demanded.

The IDSO at the top of the LISA communication network receives bid functions from LISAs in lower layers. The IDSO uses aggregated forms of these bid functions to shape distribution system net load in accordance with its system efficiency and reliability objectives. Specifically, the IDSO shapes distribution net load by sending appropriately selected price signals back down the network to edge LISAs with associated GERs, hence to the price-sensitive controllers for the GERs' smart devices. This in turn triggers power responses from these smart devices.

However, to ensure the IDSO can maintain system efficiency and reliability in this manner, careful attention must be paid to the timing of the signals propagated up and down the LISA communication network. In the remaining sections of this chapter, attention will be focused on a network consisting solely of an IDSO and a collection of edge LISAs that manage the smart devices for particular GERs. It will be assumed that the IDSO attempts to achieve its system objectives by iteratively implementing the following *Five-Step TES Design* for this network, characterized by five critical time-rates:

Step 1: For each GER R, the price-sensitive controller for each smart device d(R) that belongs to R collects data on the state of R and d(R) at a *data check rate* that it uses to form a state-conditioned device bid function $\text{Bid}_d(\text{R})$ for power usage, power supply, or ancillary service provision.

Step 2: The edge LISA for each GER R combines the device bid functions $\text{Bid}_d(\text{R})$ into a state-conditioned aggregate bid function Bid(R) that it communicates to the IDSO at a *bid refresh rate*.

Step 3: The IDSO combines the GER bid functions Bid(R) into a vector of aggregate bid functions at an *aggregate bid refresh rate*.

Step 4: The IDSO uses these aggregate bid functions to determine price signals that it communicate back to the edge LISAs at a *price signal rate*.

Step 5: For each GER R, the price-sensitive controller for each smart device d(R) that belongs to R inserts its latest received price signal into its latest refreshed state-conditioned bid function $\text{Bid}_d(\text{R})$ at a *power control rate*, which triggers a power response from d(R).

---

[5]Note that the power range in Figure 17.4a is from −100 to +100, whereas the power range in Figure 17.4b,c is from −200 to +200.

This Five-Step TES Design generalizes the well-known PowerMatcher bid-based TES design developed by Kok [99] in four important regards. First, the manager of the Five-Step TES Design is an IDSO with a fiduciary responsibility for the efficiency and reliability of distribution system operations. Second, the IDSO functions as a linkage entity at the T–D interface of an ITD system. Third, the IDSO is able to submit bulk-power demands and supplies into an RTO/ISO-managed wholesale power market in order to service the power demands and supplies of GERs. Fourth, the IDSO is able to offer ancillary services into this wholesale power market, harnessed from GERs, thus creating additional revenue streams for the GERs.

As will be shown in Chapter 18, the absolute and relative settings of the five time-rates characterizing the Five-Step TES Design are critically important design elements. These time-rates must be set appropriately in order for the IDSO to be able to align private goals and constraints with its system objectives.

17.3 ROLE OF DISTRIBUTION UTILITIES

The ITD framework outlined in Section 17.2 envisions distribution systems managed by IDSOs, i.e. by independent entities tasked with fiduciary responsibility for ensuring the efficiency and reliability of distribution system operations.

The proposed role of an IDSO within a distribution system is thus analogous to the role of an RTO/ISO within a transmission system, with one key difference: IDSOs need to be able to manage the participation of a large and rapidly increasing number of GERs. Consequently, while RTO/ISOs have been able to implement centralized market processes, IDSOs need scalable decentralized management tools such as bid-based TES designs.

On the other hand, with regard to grid operations, the open access arguments used to support the separation of grid ownership from grid management at the transmission level apply equally well at the distribution level. Thus, in analogy to transmission system operations, it would seem reasonable and important to maintain *Distribution Utilities (DUs)* as the owners and operators of the physical assets that comprise the distribution grid while permitting the IDSO to oversee the daily management of grid access and use.

The duties of the DUs should include proper maintenance of all physical grid assets. In addition, DUs might need to take direct control actions at various times to ensure distribution system reliability. For example, in some circumstances it might be necessary to implement direct voltage control at the level of phase loads to ensure voltage magnitudes and voltage imbalance across phases stay within operationally desirable bounds.

These transmission service duties should be carried out by DUs in return for appropriate compensatory payments received from the IDSO. In turn, the IDSO should collect these payments from the GERs populating the distribution grid in direct proportion to their contribution to transmission service costs.

In short, DUs are envisioned to be entities hired by the IDSO to work on behalf of GER interests, as expressed by the IDSO's system objectives. DUs could take many forms. For example, DUs could be nonprofit entities (e.g. municipal utilities, electric cooperatives). Alternatively, DUs could be publicly or privately owned for-profit companies (e.g. investor-owned utilities) as long as their ability to exercise market power is curtailed due to (i) competition among existing DUs; (ii) competitive pressure from potential DU entry (contestability); and/or (iii) regulatory oversight of DU operations and compensation levels.

Finally, a plausible way that a current distribution system could gradually transition to an IDSO-managed form would be for an existing DU to slowly evolve into an IDSO through suitable changes in its organizational structure and rules of operation.

17.4 AN IDSO-MANAGED BID-BASED TES DESIGN FOR HOUSEHOLDS

To date, bid-based TES design research has largely focused on the achievement of prespecified system objectives. However, as stressed by Battula et al. [8], system objectives must be aligned with local customer goals and constraints if voluntary customer participation is to be sustained.

Thus, Battula et al. [8] consider the feasibility and desirability of undertaking bid-based TES design development from a customer-centric starting point. Proof-of-concept evidence is provided in support of the following five assertions:

i. The optimal general bid-function form for customers with *thermostatically controlled loads (TCLs)* can be derived from dynamic programming principles, based solely on general customer welfare and thermal dynamic attributes.

ii. Quantitative representations can be determined for optimal customer bid functions, given quantitative representations for basic customer welfare and thermal dynamic attributes.

iii. These basic customer attributes can be expressed in parametric form.

iv. Parameters appearing in these forms can be clustered into subsets expressing customer welfare and thermal dynamic attributes at a higher level of aggregation.

v. This clustering can be used to define a collection of representative customer types to facilitate the development and performance testing of bid-based TES designs aligned with local customer goals and constraints.

In particular, in support of assertion (i), Battula et al. [8] derive the optimal general bid-function formulation for a household with a smart (price-sensitive) HVAC system. The remainder of this section briefly summarizes this derivation.

Consider a household h that consists of a resident occupying a house at a particular location subject to external weather conditions; see Figure 17.5. The household has a smart electric HVAC system running in cooling mode with ON/OFF power settings. This HVAC system comprises a basic HVAC unit operating in parallel with a

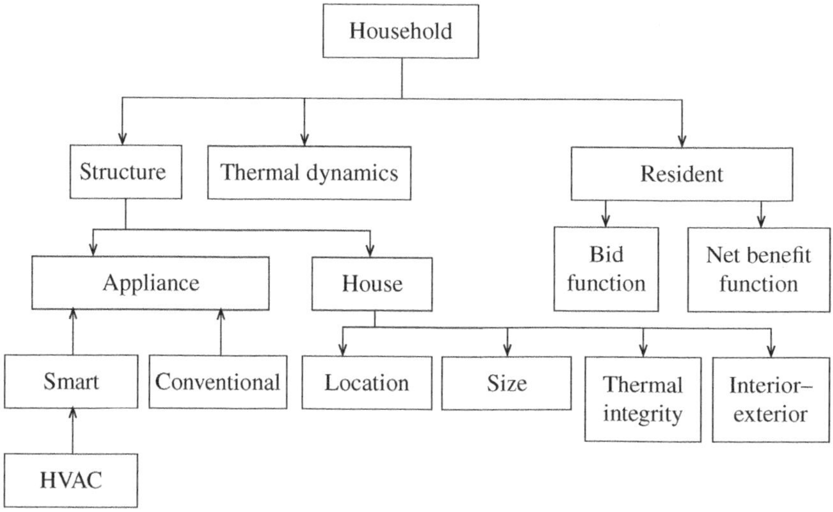

Figure 17.5 Hierarchy of basic attributes characterizing household h. Down-pointing arrows denote "has a" relationships and up-pointing arrows denote "is a" relationships.

one-speed fan for air circulation. The household's conventional appliances consist of lights, clothes-washer, refrigerator, dryer, freezer, range, and microwave.[6]

The goal of household h is to maximize its *net benefit* over time, measured as thermal comfort minus cost. In pursuit of this goal, household h participates in an IDSO-managed Five-Step TES Design consisting of an IDSO together with a collection of edge LISAs; see Section 17.2. The edge LISA for household h manages the daily power operation of h's HVAC system; hereafter, this edge LISA will simply be referred to as h's *HVAC controller*.

Household h's HVAC controller sends bids to the IDSO that express h's demands for HVAC power usage as a function of required price payment and h's supplies of ancillary service (HVAC power absorption) as a function of offered price compensation. In return, the IDSO sends price signals to h's HVAC controller that determine ON/OFF power control actions for h's HVAC system.

Let the time interval during which an ON/OFF power setting is maintained for household h's HVAC system be called a *control-step*. Each control-step has the same duration, determined by the power control rate setting for the Five-Step TES Design. The time-line for h can thus be divided into control-steps $n = [n^s, n^e)$ of equal duration. At the start-time n^s for each control-step n, a control signal is transmitted to h's HVAC controller either to retain or switch its current HVAC ON/OFF control setting. This control setting is then maintained for the remainder of control-step n.

[6]The derivation of an optimal bid-function formulation for a household h given in Battla et al. [8, Section VII] can easily be modified to handle an arbitrary mix of conventional appliances and an HVAC system running in heating mode instead of cooling mode.

The goal of household h at the start-time n^S for each control-step n is to maximize its net benefit (comfort minus cost) over the next N control-steps, where N denotes the household's *look-ahead horizon*. As will next be shown, household h can be in one of only two possible control-relevant states at the start-time n^S for n.

Let $\widehat{G}(n{:}ON)$ and $\widehat{G}(n{:}OFF)$ denote the maximum possible comfort the household forecasts it could achieve over control steps $n, n+1, \ldots, n+N-1$ if its HVAC system at time n^S were set to ON or OFF, respectively, and the ON/OFF HVAC controls for the remaining $N-1$ control steps $n+1, \ldots, n+N-1$ were subsequently set at optimal levels taking as given this time-n^S decision. Household h's two control-relevant states at time n^S are then as follows:

$$X_n^S \ (\textit{May run for ancillary service provision}): \quad \widehat{G}(n, ON) \leq \widehat{G}(n, OFF) \quad (17.1)$$

$$X_n^U \ (\textit{May run for power usage}): \quad \widehat{G}(n, ON) > \widehat{G}(n, OFF) \quad (17.2)$$

If household h is in state X_n^S at start-time n^S, it will not be willing to pay a positive price for HVAC power usage during n, no matter how small. However, household h could be induced to switch (or leave) its HVAC system ON if the *price received* for this HVAC power absorption (as ancillary service supply) is *sufficiently high*. Let this sufficiently high cut-off price be denoted by $-\Pi^*(X_n^S) \geq 0$. Conversely, if household h is in state X_n^U at start-time n^S, it will be willing to pay a positive price for HVAC power usage during n as long as this *price charged* is *sufficiently low*. Let this sufficiently low positive cut-off price be denoted by $\Pi^*(X_n^U) > 0$.

Consequently, household h's optimal bid function for control-step n has the general rectilinear form depicted in Figure 17.6, where P_n^* denotes the ON power consumption of household h's HVAC system during control-step n. Note the optimal bid form in the ancillary service provision state X_n^S constitutes a *supply* function for ancillary service (HVAC power absorption) as a function of price *received*. Conversely, the

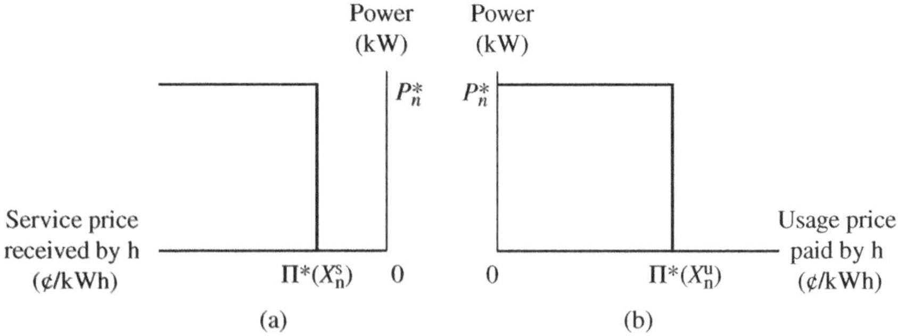

Figure 17.6 Household h's optimal state-dependent bid forms for (a) ancillary service provision and (b) power usage during a control-step n. A negative price denotes a price received by household h for provision of ancillary service (power absorption). A positive price denotes a price paid by household h for power usage.

optimal bid form in the power usage state X_n^{u} constitutes a *demand* function for HVAC power usage as a function of price *paid.*

As detailed in Battula et al. [8], the derivation of the bid-function form depicted in Figure 17.6 differs from household bid-function depictions and derivations in previous bid-based TES design research in four important respects:

1. The depicted bid-function form is *optimal* for any electrical appliance owned by household *h* that satisfies the following two conditions: (i) the appliance has a smart controller that turns the device either ON or OFF at the start of successive control-steps *n*; and (ii) the ON power usage of the appliance during any control-step *n* remains at an essentially constant level P_n^*.
2. Household h is shown to have only two possible control-relevant states at the start of a control-step *n*, an ancillary service state and a power usage state; and its optimal bid-function formulation is derived for each of these states.
3. Household h can communicate to the IDSO its entire state-conditioned bid function at the start of each control-step *n* by means of two scalar data points: namely, a cut-off price (either positive or negative) and a forecasted ON power level.
4. Permitting household h to communicate ancillary service bid functions to the IDSO when in an ancillary service state permits the IDSO to secure ancillary services from h in return for appropriate compensation.

Explicit quantitative representations are derived in Battula et al. [8] for household h's state-conditioned bid functions in Figure 17.6a,b as functions of basic parameters characterizing household h's comfort-cost preferences and thermal dynamic state. Correlated basic attribute parameters for household h are then clustered into subsets and used to construct a collection of representative household types.

17.5 IDSOS AS GRID-EDGE RESOURCE AGGREGATORS

The key question motivating the work reported in Battula et al. [8] and Nguyen et al. [139] is whether an IDSO-managed bid-based TES design can help an IDSO to maintain the efficiency and reliability of distribution system operations.

Preliminary test cases supportive of this supposition are reported under the following maintained assumptions.

- The distribution system is managed by an IDSO with fiduciary responsibility for the efficiency and reliability of distribution system operations;
- The distribution system is populated by households with diverse welfare and thermal dynamic attributes;
- The distribution system operates under a Five-Step TES Design;

- The IDSO uses this design to receive and aggregate household bids;
- The IDSO uses this design to communicate bid-conditional price signals back to households in accordance with its system objectives;
- These system objectives include peak-load reduction; target load matching; voltage imbalance control; and voltage limit control;
- To maintain its independent status, the IDSO allocates back to households all net costs incurred due to system operations.

An important aspect of these test cases is that the IDSO must sometimes use negative as well as positive price signals to achieve its system objectives. The use of negative prices indicates that the IDSO is relying on the compensated extraction of ancillary services (power absorption) from households.

These test cases will be discussed in greater detail in Chapter 18.

17.6 SWING-CONTRACT SUPPORT FOR IDSO PARTICIPATION IN WHOLESALE POWER MARKETS

Previous sections of this chapter have conceptually formulated an IDSO-managed bid-based TES design for an ITD system that permits an IDSO to manage a distribution system in accordance with its fiduciary responsibilities. Here management refers to the ability of an IDSO to service the power demands, power supplies, and ancillary service offers of distribution system participants.

However, to evaluate with care the performance of such designs, a computational platform is needed that permits the modeling of an ITD system operating during successive days under variously specified market rules and operating conditions. Moreover, in keeping with the purpose of the current study, the platform should permit the study of swing-contract support for IDSO participation in wholesale power markets.

The following chapter first reports on the development of such a computational platform: namely, the ITD TES Platform V2.0. It next summarizes illustrative test cases conducted by means of this platform that demonstrate how the IDSO-managed bid-based TES design presented in Sections 17.4 and 17.5 can help an IDSO achieve peak-load reduction and target load-matching objectives. Finally, it points to ongoing work addressing the ability of swing contracts to facilitate the participation of IDSOs in wholesale power markets.

CHAPTER 18

DESIGN EVALUATION VIA THE ITD TES PLATFORM

18.1 CHAPTER PREVIEW

As stressed in earlier chapters, a key anticipated benefit from transiting to a Linked Swing-Contract Market Design is a more level playing field for resource participation in centrally managed wholesale power market operations. As the number of wholesale power market participants grows, opportunities to exploit market rules for strategic own advantage should decrease, thus reducing the need for costly monitoring and enforcement.

Chapter 17 focuses on the encouragement of increased participation by the rapidly growing number of distribution-side resources. The chapter conjectures that the use of swing contracts by independent distribution system operators (IDSOs) could greatly facilitate this increased participation.

Specifically, Chapter 17 conjectures that IDSOs, operating as linkage entities at the transmission and distribution (T–D) interfaces of integrated transmission and distribution (ITD) systems, could use swing contracts to offer ancillary services into wholesale power markets. These ancillary services could be harnessed from grid-edge resources (GERs) in return for appropriate compensation by means of an IDSO-managed bid-based transactive energy system (TES) design.

The current chapter provides a more detailed discussion of work at Iowa State University (ISU) that has focused on this conjecture.[1] Section 18.2 presents a classification of *design readiness levels (DRLs)* 1–9 for proposed institutional designs. Section 18.3 explains the basic features of the ITD TES Platform V2.0, a computational platform that has specifically been developed to permit the performance evaluation of IDSO-managed bid-based TES designs at DRLs 4–6. Illustrative test cases are described in Section 18.4, and findings from these test cases are presented in Section 18.5. Code and data for the ITD TES Platform V2.0 and reported test cases are available at the ITD TES Platform Repository [92].

[1]The developments reported in this chapter are based on: Battula [6], Battula et al. [7], Battula et al. [8], Nguyen et al. [139], Tesfatsion [181], Tesfatsion and Battula [187], and Thomas and Tesfatsion [189].

A New Swing-Contract Design for Wholesale Power Markets, First Edition, Leigh Tesfatsion.

18.2 DESIGN READINESS LEVELS

Changes in the design of centrally managed wholesale power markets should be guided by research that is strongly supported by empirical evidence. However, reaching a point where a proposed new design is ready for real-world implementation will typically require a series of modeling efforts at different scales and with different degrees of empirical verisimilitude. Moving too soon to design implementation on the basis of over-simplified models entails major risk of adverse unintended consequences.

Consider, for example, the nine *DRLs* proposed in Table 18.1 for institutional design research.[2] All nine DRL levels are essential for ensuring conceptual design ideas are brought to real-world fruition. Explicit recognition and acceptance of this tiered approach to model valuation could encourage design researchers to become more supportive of each other's varied contributions.

One important issue for design researchers is how to ensure completion of the design arc from DRL 1 to DRL 9. Due to relatively limited data and computational

TABLE 18.1 Design readiness level (DRL) classification

Development level	DRL number	DRL description
Conceptual idea	DRL 1	Conceptual formulation of a design with desired attributes
Analytic formulation	DRL 2	Analytic characterization of a design with desired attributes
Modeling with low empirical fidelity	DRL 3	Analysis of design performance using a highly simplified model
Small-scale modeling with moderate empirical fidelity	DRL 4	Design performance tests using a small-scale model embodying several salient real-world aspects
Small-scale modeling with high empirical fidelity	DRL 5	Design performance tests using a small-scale model embodying many salient real-world aspects
Prototype small-scale modeling	DRL 6	Design performance tests using a small-scale model reflecting expected field conditions apart from scale
Prototype large-scale modeling	DRL 7	Design performance tests using a large-scale model reflecting expected field conditions
Field study	DRL 8	Performance tests of design in expected final form under expected field conditions
Real-world implementation	DRL 9	Implementation of design in final form under a full range of operating conditions

[2]These DRLs mimic, in rough form, the *technology readiness levels (TRLs)* devised by the US Department of Energy [43, p. 22] to rank the readiness of proposed new technologies for commercial application. The material in this section is adapted from Tesfatsion [181, Section 5].

capabilities, design researchers at universities have tended to work at DRLs 1–3. In contrast, design researchers within industry, government, and regulatory agencies typically work at DRLs 7–9. The interim DRLs 4–6 thus constitute a "valley of death" that hinders the careful step-by-step development and testing of design proposals from conceptual formulation to real-world implementation. Fortunately, continued improvements in software and hardware are permitting design researchers to develop computational platforms[3] able to bridge this valley of death.

Another important issue for design researchers is the proper role of stakeholders in the design development process. As emphasized by Roth and Wilson [154, p. 139], designs are intended for real-world deployment; thus, "designers have to be good listeners, prepared to learn from everyone."

Consequently, it is important to stress that the DRL classification provided in Table 18.1 does not necessarily represent a one-way road map from initial concept to completed application. Rather, DRLs 1–9 could constitute a single concept-to-application iteration in an ongoing *iterative participatory modeling (IPM)* process.

As explained with care in the materials provided at [184, IPM section], an IPM process is an open-ended collaborative learning process in which researchers join with stakeholders in a repeated looping through a multi-stage process involving conceptual formulation, model development, and real-world testing. The objective is to help stakeholders manage complex problems over time through a continuous learning process rather than to attempt the delivery of a definitive problem solution.

The RTO/ISOs for all seven US centrally managed wholesale power markets work hard to involve current market participants in all decisions affecting the operation of these markets. However, difficulties could arise when proposed changes in market design involve the inclusion of *additional* market participants, such as IDSO managers of distribution-side resources.

Even if such changes can be justified on *system* efficiency and reliability grounds, they are bound to result in winners and losers among current market participants. Thus, unless the changes are mandated by a higher level organization such as the U.S. Federal Energy Regulatory Commission (FERC) or the Power Utility Commission of Texas (PUCT), their implementation might require a temporary transitional phase during which the gains realized by some market participants are used to compensate the losses experienced by others.

18.3 AN ITD TES PLATFORM PERMITTING TES DESIGN EVALUATION

ISU researchers have been working since 2006 to develop open-source agent-based computational platforms that permit power system operations to be studied from both

[3]In the current study, the term *computational platform* is used to refer to a software framework together with a library of software classes that permit the plug-and-play development and study of a family of computational models.

an engineering and an economic point of view. For discussions of this work, as well as related work by other researchers, see [182, 183].

Recent ISU efforts [8, 139, 187] have focused on the development of the ITD TES Platform [92]. This platform permits the study of ITD systems operating under IDSO-managed bid-based TES designs. The latest platform version (V2.0) supports:

- modeling of ITD systems as open-ended dynamic systems;
- detailed modeling of T–D economic and control operations;
- physically based modeling of GERs and electrical devices;
- evaluation of IDSO-managed bid-based TES designs at DRLs 4–6;
- modeling of ITD decision-makers as strategic learning agents;
- evaluation of ITD system efficiency and reliability;
- evaluation of financial incentives and financial viability for each ITD participant as determined by their own local goals and constraints.

A partial agent hierarchy for the ITD TES Platform V2.0 is given in Figure 18.1. The transmission component for the ITD TES Platform V2.0 is implemented by AMES (*A*gent-based *M*odeling of *E*lectricity *S*ystems) [185]. AMES V5.0 is an open source agent-based computational platform (Java/Python) that permits users to model core institutional and operational aspects of current US RTO/ISO-managed wholesale power markets. For example, AMES V5.0 has been used to implement the ERCOT Test System [7], a specialized computational platform that models the operations of

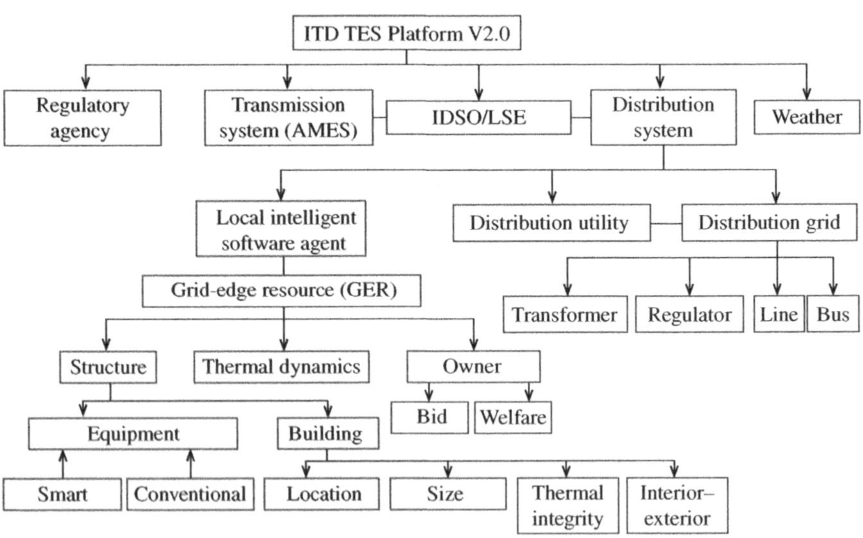

Figure 18.1 Partial agent hierarchy for the ITD TES Platform V2.0. Down-pointing arrows denote "has a" relationships and up-pointing arrows denote "is a" relationships.

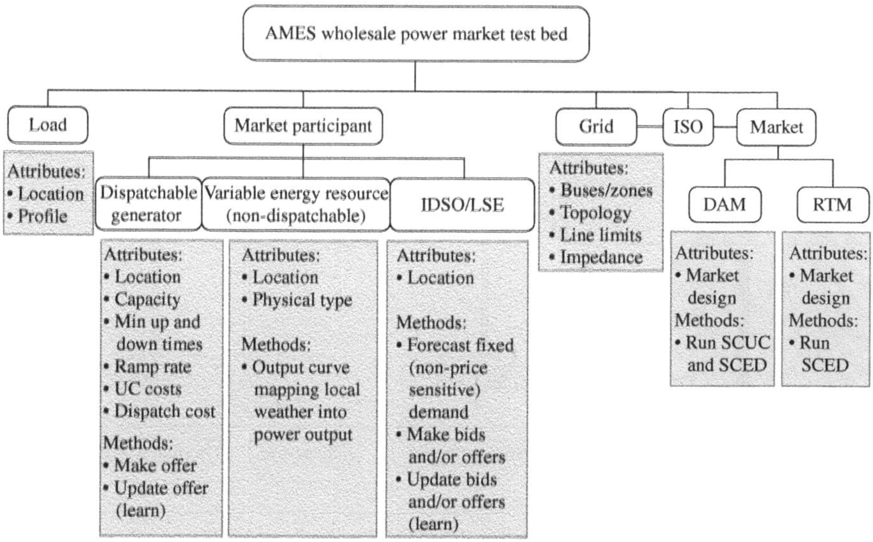

Figure 18.2 Partial agent hierarchy for AMES V5.0. The IDSO agent is an externally modeled agent at a T–D interface that can participate in AMES market operations.

ERCOT's day-ahead and real-time markets operating over a synthetically constructed ERCOT transmission grid during successive days.

A partial agent hierarchy for AMES V5.0 is given in Figure 18.2. As indicated, the decision-making agents include dispatchable generators, non-dispatchable variable energy resources, load-serving entities (LSEs), an IDSO, and an ISO. The *J*ava *Re*inforcement *L*earning *M*odule (*JReLM*) included within AMES V5.0 permits any decision-making agent to be equipped with reinforcement learning capabilities.

The complete analytical formulation for the Security-Constrained Unit Commitment (SCUC) and Security-Constrained Economic Dispatch (SCED) optimizations implemented in AMES V5.0 is carefully presented and explained in Tesfatsion and Battula [187]. This formulation permits unit commitment costs and dispatch costs for market participants to be incorporated into the objective function whether the supply offers of these participants are in standard or swing-contract form. For example, the Transitional DAM SCUC/SCED formulation presented in Chapter 16 of the current study is a special case of the AMES V5.0 SCUC/SCED formulation.

The distribution component for the ITD TES Platform V2.0 is implemented in part by GridLAB-D [66], an open-source computational platform (C/C++) modeling basic distribution system operations. GridLAB-D permits the modeling of GERs (business, commercial, and household entities), electrical devices, unbalanced distribution grid operations, and various types of voltage controls. In addition, the distribution component includes newly developed Python agents modeling an IDSO, GER owners/residents, and local intelligent software agents (LISAs) for the control of smart electrical devices.

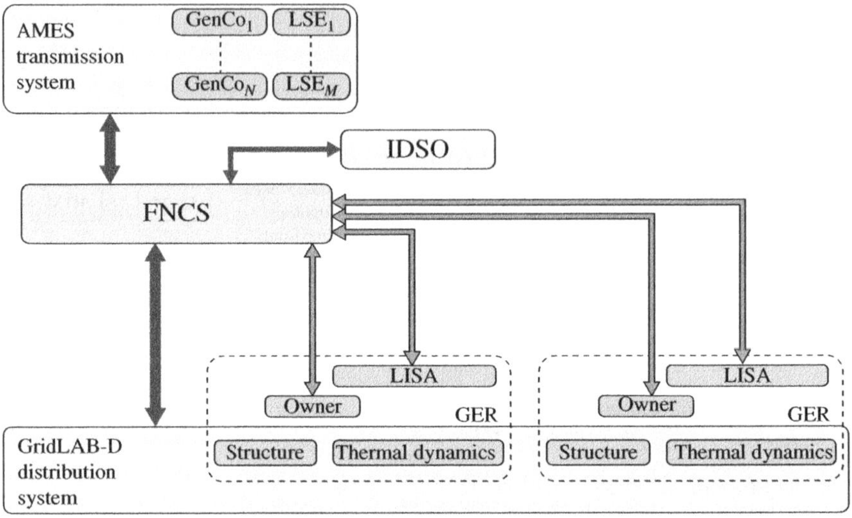

Figure 18.3 Key software components for the ITD TES Platform V2.0.

Finally, the *framework for network co-simulation (FNCS)* [35] is used to handle data exchange among agents. For example, FNCS enables the IDSO and LISA agents to engage in two-way communications. FNCS is a TCP/IP-based middleware developed by the Pacific Northwest National Laboratory (PNNL).

The key software components for the ITD TES Platform V2.0 are schematically depicted in Figure 18.3.

18.4 ILLUSTRATIVE TEST CASES: OVERVIEW

This section briefly describes several illustrative test cases conducted by means of the ITD TES Platform V2.0 to evaluate the performance of the IDSO-managed bid-based TES design for households presented in Sections 17.4 and 17.5.[4] Hereafter, these test cases are referred to as the *ITD Household Test Cases*. The general ITD feedback loop for each ITD Household Test Case is schematically depicted in Figure 18.4.

The transmission system for each test case is the standard 5-bus transmission grid developed by Lally [108]. The distribution system for each test case is the standard IEEE 123-bus distribution system [87] with three important modifications. First, all GERs take the form of households; 927 households are distributed across the 123 buses in proportion to the original loads, which are then omitted. Second, the distribution grid is connected to the transmission system at a substation; and all wholesale power is supplied to the distribution system through this T–D interface. Third, the distribution system is managed by an IDSO operating at this substation.

[4] A detailed description of these test cases, including code implementation, is given in [8].

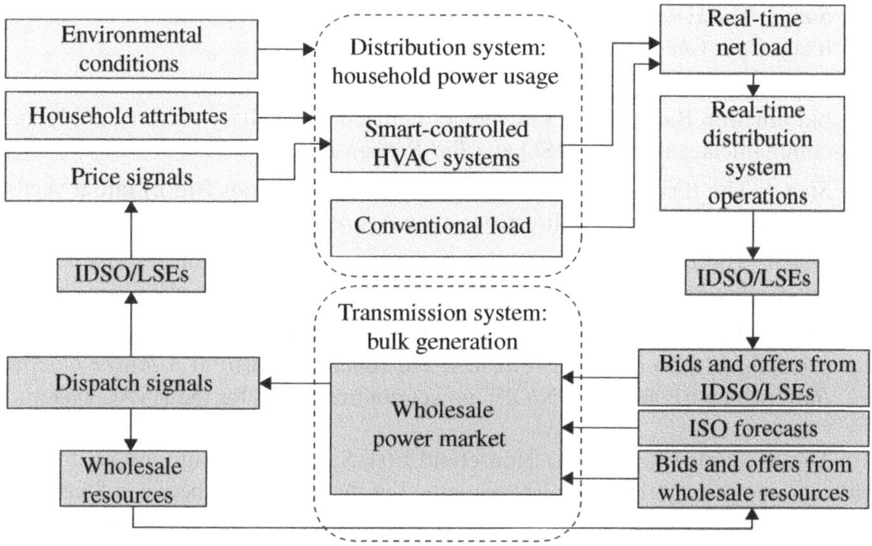

Figure 18.4 Feedback loop for the ITD Household Test Cases.

Each household is formulated in accordance with the household model described in Sections 17.4 and 17.5. Thus, each household consists of a resident, a house, a smart (price-sensitive) electric heating, ventilation, and air-conditioning (HVAC) system, and a collection of conventional (non-price-sensitive) appliances.

The price-sensitive power usage of each household's HVAC system is controlled by a LISA, hereafter simply referred to as an *HVAC controller*. Each household formulates state-conditioned bid functions expressing its power demands and its offered supplies of ancillary service (power absorption), as depicted in Figure 17.6. Household thermal dynamics are expressed in terms of time-varying temperatures for inside air and inside mass. Household welfare (net benefit) is expressed as resident thermal comfort minus net cost.[5]

Each household is implemented by means of the GridLAB-D (GLD) House Object [67].[6] The length of each control-step $n \geq 0$ is set to $\Delta\tau = 300$ seconds. Weather forcing terms consist of outside temperature, solar flux, and humidity data for hot summer days in Des Moines, Iowa. The location attributes specified for each household are also for Des Moines, Iowa.

The power demands and ancillary service supplies of each household h are managed by means of the following IDSO-managed *Household Five-Step TES Design*, characterized by five time-rates:

[5] A household's *net cost* is the cost it is charged for power usage net of any payments it receives for the provision of ancillary service (power absorption).

[6] See [8, Appendix E] for GLD House Object implementation details.

- *Step 1*: The HVAC controller for each household h collects data on the state of h at a *Data Check Rate*.
- *Step 2*: The HVAC controller for each household h forms a state-conditioned bid function Bid(h) for HVAC power demand or ancillary service supply and communicates it to the IDSO at a *Bid Refresh Rate*.
- *Step 3*: The IDSO combines the household bid functions Bid(h) into a vector AggBid of aggregate bid functions at an *Aggregate Bid Refresh Rate*.
- *Step 4*: The IDSO uses AggBid to determine and communicate price signals back to household HVAC controllers at a *Price Signal Rate*.
- *Step 5*: The HVAC controller for each household h inserts its latest received price signal into its latest refreshed bid function Bid(h) at a *Power Control Rate*, which triggers an ON/OFF power control action for the HVAC system.

The five time-rates for the Household Five-Step TES Design are set to $1/\Delta t$ with a common time-step $\Delta t = 300$ seconds. Let the time-delay between Step j and Step $j + 1$ in any given iteration of the five steps be denoted by ϵ_j for $j = 1, \ldots, 5$, where "Step 6" is equated with "Step 1" in the subsequent iteration. The time delays ϵ_j are commonly set for the ITD Household Test Cases so that their summation does not exceed Δt. Finally, let $t_j = t_{j-1} + \epsilon_j$ for $j = 1, \ldots, 5$. The iterated staggered implementation of Steps 1–5 for the Household Five-Step TES Design is then as depicted in Figure 18.5.

Finally, households are classified into representative types based on *structure quality*: "Low," "Medium," or "High." Each of these three structure quality types is determined by correlated parameter settings for the appliance and house structural

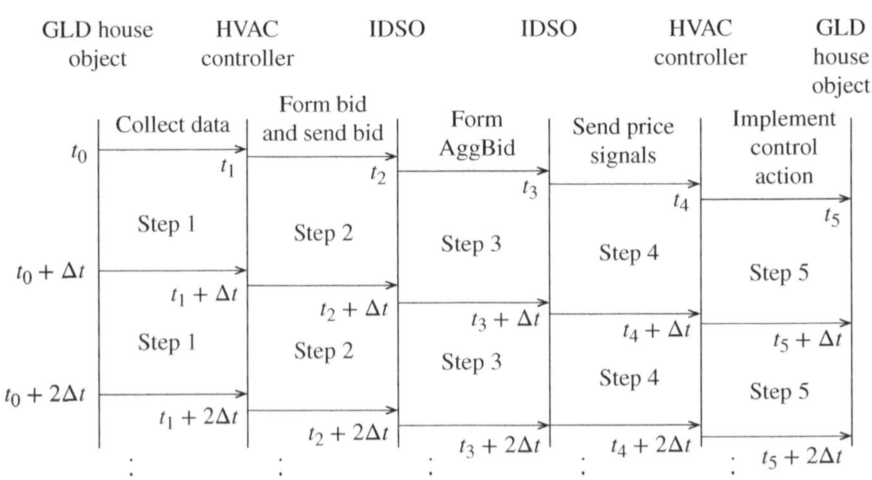

Figure 18.5 Staggered implementation of the five steps comprising the Household Five-Step TES Design for the ITD Household Test Cases.

attributes depicted in Figure 17.5, using the method developed in [8, Section VIII]. For example, a household with a "Low" structure quality type has an HVAC system with a low performance rating and a house that is small in size with poor thermal insulation.[7] The treatment factor for each of the ITD Household Test Cases reported below in Section 18.5 is the mix of household structure quality types.

18.5 ILLUSTRATIVE TEST CASES: REPORT

The Household ITD TES Cases reported in this section explore the performance capabilities of the Household Five-Step TES Design described in Section 18.4. Specifically, these test cases explore the extent to which the IDSO is able to use this design to achieve various system objectives. Two types of objectives are considered: peak-load reduction and target load matching.

18.5.1 IDSO Peak-Load Reduction Capabilities

The basic issue considered in the peak-load reduction test cases is as follows: Can the IDSO use retail price signals to achieve target peak-load reductions? The key finding is that the IDSO can indeed achieve target peak-load reductions through appropriate retail price signals; however, the mix of household structural quality types strongly affects the needed form of these retail price signals.

The IDSO on day D forecasts household peak load for day D+1. The IDSO uses this forecast to determine a target peak load for day D+1, given by forecasted peak load reduced by a designated percentage. During day D+1, the IDSO then uses latest refreshed household bids to send an appropriate sequence of retail price signals to households to maintain total household load at or below this target peak-load level.

Figures 18.6–18.8 report the ability of the IDSO to achieve a 0.5 MW peak-load reduction by means of retail prices π^{Ret} communicated to households with optimally formulated bid functions. These retail prices take either a flat-rate form (10¢/kWh) or a peak-load pricing form. All households have the same structure quality type, either all Low, all Medium, or all High.

Figure 18.9 reports the peak-load retail prices π^{Ret} required to achieve a targeted 0.5 MW peak-load reduction when households have all Low, all Medium, or all High structure quality types. The substantial differences in retail prices observed for these three test cases indicate that household structure quality should be given careful consideration in any peak-load reduction effort.

18.5.2 IDSO Load-Matching Capabilities

The basic issue explored in the load-matching test cases is as follows: Can the IDSO use retail price signals to match total household load to a target load profile? The key

[7] The specific correlated parameter values used to characterize each household structure quality type are given in [8, Appendix F].

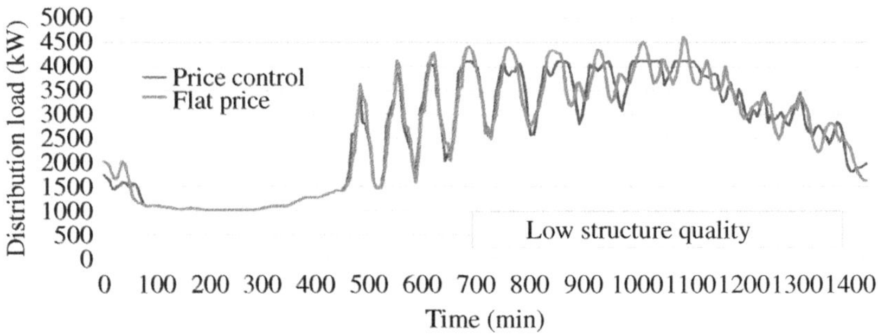

Figure 18.6 IDSO's ability to achieve a 0.5 MW peak-load reduction using flat-rate vs. peak-load pricing when all households have low structure quality.

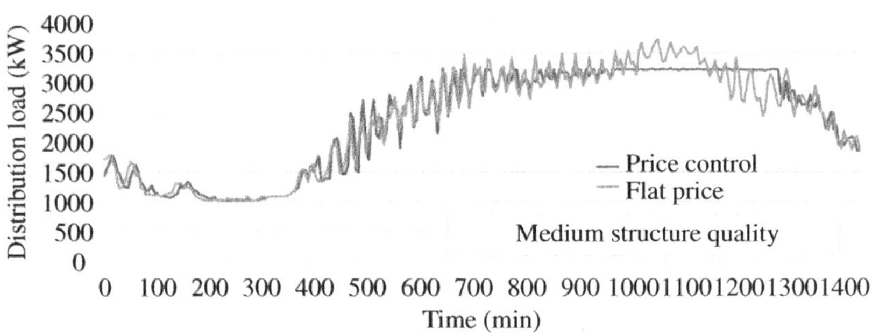

Figure 18.7 IDSO's ability to achieve a 0.5 MW peak-load reduction using flat-rate vs. peak-load pricing when all households have medium structure quality.

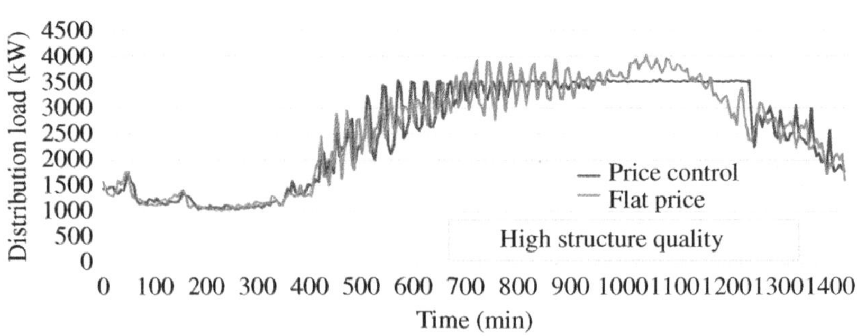

Figure 18.8 IDSO's ability to achieve a 0.5 MW peak-load reduction using flat-rate vs. peak-load pricing when all households have high structure quality.

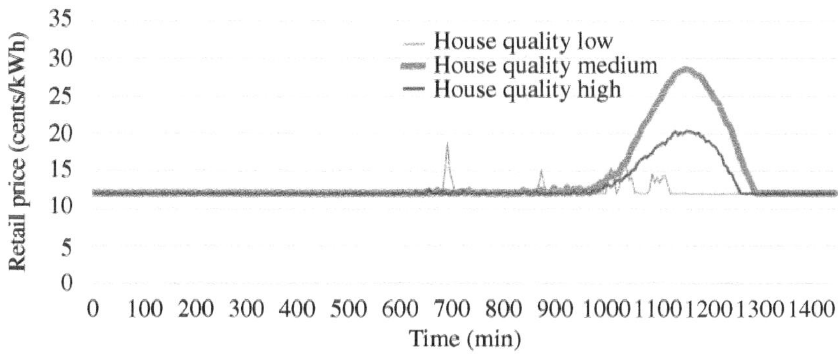

Figure 18.9 Retail price signals used by the IDSO to achieve a 0.5 MW peak load reduction when all households have the same structure quality type (all low, all medium, or all high).

finding is that the IDSO can indeed achieve target load matching; however, the IDSO at times might have to rely on household ancillary service provision to ensure this matching.

As for the peak-load reduction test cases, the IDSO manages a Household Five-Step TES design for a 123-bus distribution grid populated by 927 households. The IDSO is located at a substation that connects the distribution grid to a transmission grid, and all wholesale power supplied to the distribution system passes through this substation.

However, in contrast to the peak-load reduction test cases, the IDSO is now an active participant in an ISO-managed day-ahead market (DAM) operating over the transmission grid. On each day D, the IDSO submits a fixed demand bid into the DAM consisting of a forecasted 24-hour household load profile for day D+1. On day D+1, the IDSO attempts to ensure that actual household load does not deviate from the fixed demand bid it submitted into the DAM on day D.

Figure 18.10 reports load-matching outcomes for an illustrative case in which the distribution grid is populated with a random mix of households with Low, Medium, and High structure quality types.[8] As seen in Figure 18.10, the IDSO is successfully able to use retail price signals on day D+1 to match household load to the load profile it submitted to the day-D DAM as its fixed demand bid.

The retail price signals used by the IDSO to achieve the good load matching depicted in Figure 18.10 are shown in Figure 18.11. Note that all retail price signals are positive, indicating that all households are paying for demanded power usage.

Suppose, instead, that the IDSO submits into the day-D DAM the fixed demand bid (load profile) depicted in Figure 18.12. Once again, as seen in this figure, the IDSO is successfully able to use retail price signals on day D+1 to match total household load to this target load profile.

[8]The structure quality type of each household connected at each distribution grid bus is configured as Low, Medium, or High with probabilities (1/3, 1/3, 1/3).

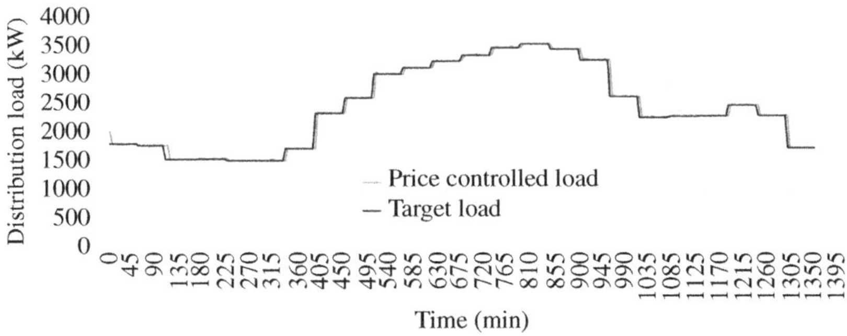

Figure 18.10 Ability of the IDSO to match total household load on day D+1 to a target load profile, given by the IDSO's fixed demand bid submitted into a day-ahead market on day D, with a mix of household structure quality types.

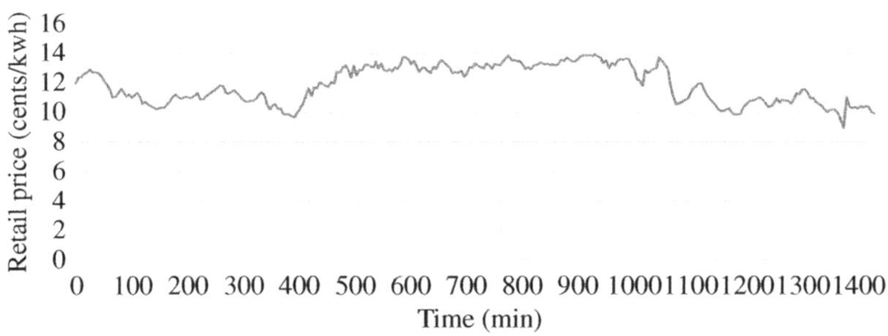

Figure 18.11 The retail price signals communicated by the IDSO to households on day D+1 to match total household load to the IDSO's day-D DAM fixed demand bid, depicted as the target load profile in Figure 18.10.

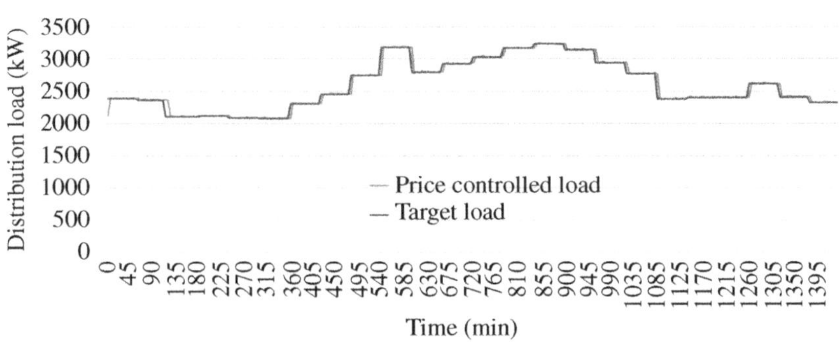

Figure 18.12 Ability of the IDSO to match total household load on day D+1 to a different target load profile, i.e. a different fixed demand bid submitted into the day-D DAM.

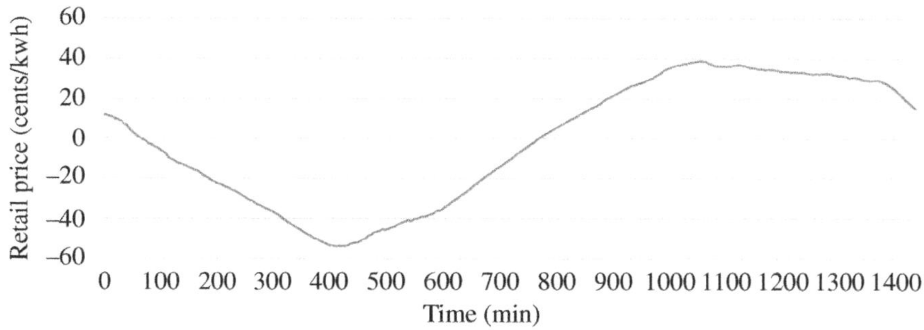

Figure 18.13 The positive and negative retail price signals communicated by the IDSO to households on day D+1 to match total household load to the target load profile depicted in Figure 18.12. The negative retail prices indicate the IDSO at times must depend on household ancillary service provision to achieve its load-matching goal.

The retail prices used by the IDSO to accomplish the good load matching depicted in Figure 18.12 are shown in Figure 18.13. The negative retail price signals seen in Figure 18.13 indicate that the IDSO must purchase ancillary services (power absorption) from households during some control-steps in order to match its target load profile.[9]

18.5.3 Household ITD Test Cases: Discussion

This section has reported test cases illustrating the performance of the IDSO-managed bid-based TES design for households presented in Sections 17.4 and 17.5. These test cases consider relatively simple types of IDSO system objectives: namely, peak-load reduction and target load matching.

The next step is to consider how an IDSO might leverage its direct point of connection to a transmission system to create additional value for distribution system participants. Additional value can be created in three basic ways. First, local demands for power usage can be met either by local generation or by wholesale generation. Second, local generation can be used to meet either local power demand or wholesale power demand. Third, local ancillary service offers can be used to meet either local reliability needs or transmission system reliability needs.

Ideally, these decisions should be determined on the basis of their overall impact on customer net benefit at both the distribution and transmission levels. However, the global centralized optimizations that would be needed to determine these net benefit effects are far beyond current computational capabilities. Consequently, system operators must instead rely on robust approximation methods.

[9] In this simple illustration, the IDSO communicates a single common retail price to each household at each price-signal step. However, efficient resource management would actually require the IDSO to send different retail price signals to households at each price-signal step depending on whether they are in an ancillary service provision state or a power usage state; see Figure 17.6.

Swing contracts permit resources to submit offers for reserve (dispatchable power-paths) in flexible diverse forms. Consequently, appropriately designed swing contracts could facilitate the ability of an IDSO to submit local generation and ancillary service offers into a wholesale power market.

For example, consider the swing contract offering swing (flexibility) in power and ramp illustrated in Figure 5.4. The discussion in Section 5.5 conjectures that this type of swing contract could be used by an IDSO to offer ancillary services (down/up power) into an RTO/ISO-managed DAM operating over a transmission grid, where these services have been harnessed from the equipment and appliances of businesses and households populating a distribution system connected to the transmission grid. The IDSO could then distribute its ancillary service payments back to these businesses and households as an additional revenue stream.

The ability of swing contracts to support IDSO participation in RTO/ISO-managed wholesale power markets is an interesting challenging issue. Battula [6] is currently undertaking a careful conceptual and computational study of this issue.

CHAPTER 19

POTENTIAL FUTURE RESEARCH DIRECTIONS

THE LINKED SWING-CONTRACT MARKET DESIGN raises various challenging issues, both conceptual and practical, that would benefit from further study. Summary descriptions of these issues are given below. These issues will be addressed in future studies.

19.1 EFFECTIVE USE OF OPTION SWING CONTRACTS

How can *option* swing contracts be most effectively formulated and used in support of the Linked Swing-Contract Market Design?

Decommitment of previously committed generation units is an ongoing concern in current US RTO/ISO-managed wholesale power markets. As operating periods approach, units committed in earlier security-constrained unit commitment (SCUC) optimizations to hedge against possible contingencies during these periods (e.g. line outages due to severe weather events) might no longer be viewed as necessary. However, RTO/ISOs are rightly reluctant to decommit earlier committed units, as this could be negatively construed as breakage of contracts.

The use of option swing contracts in European, American, and/or Bermudan form would permit an ISO managing a swing-contract market M(T) to avoid having to make firm generator unit commitments for the future operating period T. As noted in Section 4.3, swing contracts in European option form would provide flexibility to the ISO with regard to the entailment (or not) of dispatch obligations. Swing contracts in American and Bermudan option form would provide further flexibility to the ISO with regard to the exercise time itself.

Exercise-time flexibility could be particularly important for longer term swing-contract markets. For example, it could help an ISO to ensure the long-run availability of reserve offering a range of ramp-rate and/or start-time capabilities, only some of which will be needed to meet actual real-time net load balancing needs.

A New Swing-Contract Design for Wholesale Power Markets, First Edition, Leigh Tesfatsion.

However, in order for swing contracts to be submitted into swing-contract markets in option form, their offer prices and performance payment methods must be formulated in a manner that is computationally tractable as well as conceptually sensible.

19.2 REPRESENTATION OF RESERVE BIDS

Chapter 9 of this study demonstrates how various types of price-sensitive reserve bids could be used by load-serving entities (LSEs) to represent the power-usage benefits of their customers in swing-contract markets.

However, the appropriate formulation of reserve bids remains an important challenging frontier for swing-contract market research. Customers withdraw power from the grid for use as an input to a multistage production chain. This power enables the provision of intermediate services such as air conditioning, cooking, and transportation, which in turn enable the provision of final goods such as thermal comfort, food, and vacation cruises. In consequence, customer demands for power-path deliveries can be idiosyncratic and complex in form.

Moreover, customers also care about the *reliability* of their power-path deliveries. Reserve bids submitted into swing-contract markets on behalf of customers should express these customers' willingness to pay for reserve as a general insurance product, not simply as a way to meet specific power usage needs.

How should the reserve bids submitted into swing-contract markets be formulated to express these broader customer preferences for reserve procurement, thus permitting a more accurate measure of customer benefits from this procurement? This problem is particularly critical for longer term swing-contract markets whose goal is adequate reserve procurement for relatively distant operating periods with highly uncertain power usage needs.

19.3 COMPENSATION FOR STORAGE SERVICES

An example of a swing contract permitting a battery resource to offer charge and discharge capabilities as an ancillary service, in return for availability and performance cost compensation, is provided in Section 5.6. How should swing contracts be formulated to permit storage devices to offer a fuller range of possible services in return for availability and performance cost compensation? As noted by Joskow [96, Section VI], these possible services include efficient time-shifting of power injections and withdrawals; peak-load balancing; peak-load shaving; frequency regulation; deferral of transmission and distribution grid investments; and the emergency provision of backup power.

19.4 COMPENSATION FOR RELIABILITY SERVICES

How can swing contracts be formulated to permit availability and performance cost compensation for inertia, and for voltage support via reactive power production and absorption [202]? Also, the reliability services provided by primary frequency response are not compensated in current US RTO/ISO-managed wholesale power markets [46]. How might swing contracts be formulated to permit availability and performance cost compensation for primary frequency response services?

19.5 REPRESENTATION OF POWER-PATHS

The current study represents power-paths as real-valued functions over designated operating periods. Step functions are used to approximate these power-paths.[1]

However, power-paths could also be represented in the frequency domain, as linear combinations of sine and cosine basis functions [188, Appendix]. Other classes of basis functions could be used as well. What advantages or disadvantages might these various basis-function representations entail, particularly with regard to the ability to express the costs and benefits associated with the production, transmission, and delivery of power-paths?

Alternatively, could perturbation theory be used to represent power-paths as perturbations of given base power-paths, where each perturbation is expressed as a function of finitely many real-valued attribute parameters? If so, the functional analysis problem of selecting power-paths from functional domains to maximize total net benefit could instead be expressed as a problem of selecting vectors from a finite-dimensional Euclidean space to maximize total net benefit.

19.6 IMPLEMENTATION OF CONTRACT-CLEARING OPTIMIZATIONS FOR SWING-CONTRACT MARKETS

Could the decomposition of net load uncertainty sets into *system pattern regions* [63, 203] permit the off-line determination of region-conditional optimal contract-clearing solutions for swing-contract markets, thus reducing computation time for the determination of system-wide solutions? Reduced computation time could permit important additional system reliability considerations to be incorporated into the optimization, such as reactive power production and absorption for voltage support.

For example, building on Zhou et al. [203], Zhong [202] proposes a multistage SCUC optimization approach to reduce the computational complexity of existing SCUC/SCED optimization formulations. This approach combines the offline characterization of optimal SCUC/SCED decisions conditional on critical ("system pattern") regions with a time-decomposition of the full SCUC optimization problem into three rolling-horizon sub-problems for slow-start thermal generators, fast-start generators, and grid-scale energy storage. The approach also incorporates voltage security constraints and reactive power dispatch into what is called a generalized DC (genDC) power flow representation for the SCUC/SCED system constraints.

[1] As explained in Chapter 7 (footnote 5), step functions are universal approximators for the class of all continuous real-valued functions defined over compact intervals of the real line.

19.7 GRADUAL TRANSITION TO A SWING-CONTRACT MARKET

As detailed in Chapter 16, swing-contract reserve offers can be incorporated into otherwise standard DAM/RTM processes. Permitting RTO/ISOs to experiment in a transparent manner with this type of incorporation, with solicitation of feedback from stakeholders, would permit careful gradual field testing of the Linked Swing-Contract Market Design.

On the other hand, key stakeholder voting blocs might have a vested interested in preventing any such transition. For example, encouraging participation of IDSOs as ancillary service providers might be viewed by some current stakeholders as a threat to their revenue streams.

What might be the best way to encourage RTO/ISOs to engage in careful gradual testing of the Linked Swing-Contract Market Design?

CHAPTER 20

CONCLUSION: THE DOTS KEEP CONNECTING

CURRENT US RTO/ISO-MANAGED WHOLESALE power markets maintain a conceptually problematic distinction between "reserve" and "energy" as transacted products. Spot-market prices, supplemented by out-of-market make-whole payments, are used to determine product settlements for forward-market operations. These design features result in system operations and settlement processes that are complicated and confusing for market participants.

In contrast, the Linked Swing-Contract Market Design proposed in this study is conceptually coherent. The design envisions a centrally managed wholesale power market as a linked collection of ISO-managed forward reserve markets. Reserve is characterized as diverse collections of dispatchable power-paths, where a power-path is a flow of power into or out of a grid at a specific grid location during a specific operating period.

Reserve offers submitted into a swing-contract market consist of swing-contract portfolios. Each swing contract is a two-part pricing contract that offers a collection of dispatchable power-paths for possible ISO dispatch during a designated future operating period. The two-part pricing permits cleared resources to ensure full separate coverage of their availability cost ex ante and their performance cost ex post. Reserve bids submitted into a swing-contract market are price-sensitive and/or fixed demands for power-path deliveries. The ISO clears reserve offers and price-sensitive reserve bids in order to maximize the expected total net benefit of market participants, conditional on initial state conditions and subject to system constraints. The ISO allocates net reserve procurement costs and transmission service costs across market participants on the basis of their contributions to these costs.

Previous chapters have asserted that substantial advantages could be obtained by the adoption of the Linked Swing-Contract Market Design for centrally managed wholesale power market operations. The proposal of a new market design, however coherent its foundations, does not guarantee its practical effectiveness. Nevertheless, a famous quote from Jobs [94] provides inspiration:

A New Swing-Contract Design for Wholesale Power Markets, First Edition, Leigh Tesfatsion.

> (Y)ou can't connect the dots looking forward; you can only connect them looking backward. So you have to trust that the dots will somehow connect in your future.

So far the dots have indeed continued to connect for the Linked Swing-Contract Market Design as each new challenge has arisen and been addressed.

APPENDIX A

APPENDICES

TABLE A.1 Quick-reference glossary of standard acronyms

Acronym	Description
AGC	Automatic generation control
CAISO	California Independent System Operator
CFD	Contract-for-difference
D	Generic symbol for a day
DAM	Day-ahead market
DIR	Dispatchable intermittent resource
ERCOT	Electric Reliability Council of Texas
FERC	Federal Energy Regulatory Commission
FTR	Financial transmission right
GenCo	Generation company
H	Generic symbol for an hour
ISO	Independent system operator
ISO-NE	Independent System Operator for New England
LCO	Lost commitment opportunity
LMP	Locational marginal price
LOLP	Loss of load probability
LSE	Load-serving entity
MCP	Market-clearing price (reserve)
MILP	Mixed integer linear programming
MISO	Midcontinent Independent System Operator
MOP	Multistage optimization problem
MWh	Megawatt hour
NERC	North American Electric Reliability Corporation
NYISO	Independent System Operator for New York
OOM	Out-of-market
OPF	Optimal power flow
PJM	PJM interconnection
QSE	Qualified scheduling entity (ERCOT)
RAC	Reliability assessment commitment
RMR	Reliability-must-run
RTM	Real-time market
RTO	Regional transmission organization

(continued)

A New Swing-Contract Design for Wholesale Power Markets, First Edition, Leigh Tesfatsion.

TABLE A.1 *(continued)*

Acronym	Description
RUC	Residual (or reliability) unit commitment
SCED	Security-constrained economic dispatch
SCUC	Security-constrained unit commitment
SPP	Southwest Power Pool
TOU	Time-of-use
TSO	Transmission system operator
VER	Variable energy resource
VOLL	Value of lost load

TABLE A.2 Quick-reference glossary of transmission system terms

Term	Description
Ancillary service	Service provided to help ensure system reliability
Availability cost	Avoidable fixed cost incurred to ensure reserve availability
Commitment	Scheduling of a resource for possible future dispatch
Dispatch	Signaling a resource to inject/withdraw power
Energy	Abbreviation for electric energy (MWh)
Fixed demand	Non-price-sensitive demand for a specified amount of must-service load
Fixed reserve bid	Non-price-sensitive bid submitted into a swing-contract market requesting the future delivery of a specified power-path
Generation	Production of power, either for local behind-the-meter use or for injection into a grid
Load	Used as a synonym for grid power withdrawal; technically, a grid device or grid component to which power is delivered
Locational marginal price	Energy price conditioned on delivery time and location
Merit-order dispatch	Dispatch in accordance with net benefit contribution
Must-service load	Non-dispatched load that must be balanced by power injection under normal grid operating conditions
Net load	Load minus non-dispatched power injection
Net must-service load	Must-service load minus non-dispatched power injection
Net reserve cost	Reserve procurement cost net of reserve revenue receipts
Non-dispatchable power	Power not under dispatchable control

(continued)

TABLE A.2 *(continued)*

Term	Description
Performance	Delivery of a power-path in response to dispatch signals
Performance cost	The variable cost incurred to deliver a power-path in response to dispatch signals
Power	Abbreviation for electric power (MW)
Power absorption	Incremental down/up changes in power withdrawal offered into a power system as an ancillary service
Power imbalance	Discrepancy between power injection and withdrawal on a grid
Power injection	Insertion of power into a grid
Power-path	Power injected and/or withdrawn at a single grid location during a specified time interval
Power-path delivery	A power-path produced during an operating period in response to dispatch instructions
Power usage	Use of power as an intermediate good to further some end
Power withdrawal	Extraction or loss of power from a grid
Price-sensitive demand	Demand schedule indicating maximum willingness to pay for each successive increment of power withdrawn from a grid
Price-sensitive reserve bid	Reserve bid accompanied by willingness-to-pay information
Reserve	Dispatchable power-paths
Reserve bid	A demand for reserve submitted into a swing-contract market
Reserve offer	A request to supply reserve that has been submitted by a dispatchable resource into a swing-contract market
Transmission service cost	Variable cost incurred for grid operation and maintenance

TABLE A.3 Quick-reference glossary of economic terms

Term	Description
Asset	Anything of durable value, whether physical or financial
Avoidable cost	A cost that can be avoided by not undertaking production
Avoidable fixed cost	Avoidable cost not dependent on specific level of production
Commodity	Asset with a standard unit of measurement that permits one unit to be substituted for another with no change in valuation
Competitive market	A commodity market whose buyers and sellers are price takers

(continued)

TABLE A.3 *(continued)*

Term	Description
Demand schedule (inverse)	A schedule expressing a buyer's purchase reservation value for each successive unit of a commodity
Efficiency	Non-wastage of resources
Firm contract	Non-contingent contract whose terms are binding on all parties
Forward market	Market for which product payment is contractually determined in advance of product delivery
Good	Exchangeable physical item whose consumption (or use) provides benefit to the consumer (or user)
Hedonic pricing	Pricing of a product by means of prices separately assigned to its external circumstances and/or its intrinsic physical attributes
Joint products	Products jointly produced from a given set of inputs
Law of one price	In the absence of trade frictions, each unit of a commodity available at a given time should sell at the same price
Net benefit	Benefit minus avoidable cost
Net buyer surplus	Difference between a buyer's maximum willingness to pay for a quantity purchase q and the buyer's actual payment for q
Net seller surplus	Difference between a seller's actual compensation for a quantity sale q and the seller's minimum acceptable compensation for q
Option contract	A contingent contract for which the holder has the right, but not the obligation, to exercise the contract
Pareto efficiency	No wastage of opportunity to increase benefit for some at no cost to others via a feasible reallocation of resources
Price taker	A buyer or seller that considers a market price to be given, i.e. not able to be influenced by their own market activities
Productive efficiency	No physical wastage of production inputs and/or outputs
Purchase reservation value	The maximum price a buyer is willing to pay to purchase a specific quantity amount q
Revenue insufficiency	Seller revenue is not sufficient to cover seller avoidable cost
Sale reservation value	The minimum price a seller is willing to accept in payment for the sale of a specific quantity amount q
Service	Action taken by an entity that provides benefit to another entity
Spot market	Market for which product delivery and product payment are determined at the same time
Supply schedule (inverse)	A schedule expressing a seller's sale reservation value for each successive unit of a commodity
Utility function	Function measuring the benefit gained by an end-use customer from the consumption (or use) of goods (or services)
Transaction cost	Avoidable fixed cost incurred to organize a production process
Variable cost	Avoidable cost that depends on the specific level of production

TABLE A.4 Nomenclature for a swing-contract market

Term	Description
Acronyms and generics	
GenPort	Generator portfolio of swing contracts (SCs)
H	Generic symbol for an hour
IRR	Inherent reserve range for an SC portfolio
ISOPort(T)	ISO portfolio of reserve offers and bids cleared for T
$k = [k^s, k^e)$	Time-step with start-time k^s and end-time k^e
LAH(T)	Look-ahead horizon between close of M(T) and start of T
LFM(T)	Long-term forward market (swing) held in advance of T
M(T)	SC market held in advance of T
m	Dispatchable resource participating in M(T)
$\mathbb{PP}$	Generic symbol for the power-path set offered by an SC
$\mathbf{p}$	Generic symbol for a power-path $\{p(t) \mid t \in \mathrm{T}\}$
n	Non-dispatchable generator participating in M(T)
p	Generic symbol for a power level (MW)
r	Generic symbol for a ramp-rate (MW/min or MW/Δt)
SC	Swing contract taking the general form SC $=(\alpha, \mathbb{T}^{ex}, \mathbb{PP}, \phi)$
SFM(T)	Short-term forward market (swing) held in advance of T
$\mathrm{T} = [t^s, t^e)$	Operating period with start-time t^s and end-time t^e
$\mathbb{T}^{ex}$	Set of SC exercise times t^{ex}
VSFM(T)	Very short-term forward market (swing) held in advance of T
α	Generic symbol for an SC offer price
Δt	Length of a time-step k measured in hours
ϕ	Generic symbol for an SC performance payment method
Sets and subsets	
$\mathbb{B} = \{1, \ldots, NB\}$	Index set for the buses b of a transmission grid
$\mathbb{B}(z) \subseteq \mathbb{B}$	Subset of buses constituting reserve zone z
$\mathbb{C}_j$	Collection of customers serviced by load-serving entity j
$\mathbb{K} = \{1, \ldots, NK\}$	Index set for the time-steps k forming a partition of T
$\mathbb{K}_m \subseteq \mathbb{K}$	Subset of time-steps forming a partition of $\mathrm{T}_m \subseteq \mathrm{T}$
$\mathbb{K}^0 = \{0, 1, \ldots, NK\}$	Augmentation of $\mathbb{K}$ to include time-step 0
$\mathbb{K}^0_m$	Augmentation of $\mathbb{K}_m$ to include time-step 0
$\mathbb{L} \subseteq \mathbb{B} \times \mathbb{B}$	Index set for the lines ℓ of a transmission grid
$\mathbb{L}_{O(b)} \subseteq \mathbb{L}$	Subset of transmission lines originating at bus b
$\mathbb{L}_{E(b)} \subseteq \mathbb{L}$	Subset of transmission lines ending at bus b
$\mathbb{LS}$	Index set for load-serving entities j
$\mathbb{LS}(b) \subseteq \mathbb{LS}$	Subset of load-serving entities serving customers at bus b
$\mathbb{LS}(z) \subseteq \mathbb{LS}$	Subset of load-serving entities serving customers in zone z
$\mathbb{M}$	Index set for dispatchable resources m
$\mathbb{M}(b) \subseteq \mathbb{M}$	Subset of dispatchable resources located at bus b
$\mathbb{M}(z) \subseteq \mathbb{M}$	Subset of dispatchable resources located in zone z
$\mathbb{NG}$	Index set for non-dispatchable generators n
$\mathbb{NG}(b) \subseteq \mathbb{NG}$	Subset of non-dispatchable generators located at bus b
$\mathbb{NG}(z) \subseteq \mathbb{NG}$	Subset of non-dispatchable generators located in zone z
$\mathbb{PP}_m$	Set of dispatchable power-paths $\mathbf{p}$ offered into M(T) by m
$\mathbb{P}_m$	Set of feasible sustainable power levels p (MW) for m
$\mathbb{RR}_m$	Set of feasible ramp-rates r (MW/min or MW/Δt) for m
$\mathbb{T}^{ex}_m$	Set of possible exercise times t^{ex} for swing contract SC_m
$\mathbb{Z} = \{1, \ldots, NZ\}$	Set of indices z for reserve zones $\mathbb{B}(z) \subseteq \mathbb{B}$

(continued)

TABLE A.4 *(continued)*

Term	Description
Parameters	
$B(\ell)$	Inverse of reactance (pu) for transmission line ℓ
b_j	Bus location for the customers serviced by LSE j
b_m	Bus location for dispatchable resource m
b_n	Bus location for non-dispatchable generator n
$E(\ell)$	End bus for transmission line ℓ
F_ℓ^{max}	Power capacity limit (MW) for transmission line ℓ
$\widehat{\mathrm{NB}}(\mathrm{T}) = \mathrm{B}(\mathrm{T}) - \widehat{\mathrm{C}}(\mathrm{T})$	Forecast (\$) for total net benefit for operating period T
$\widehat{NL}_b^f(k)$	Forecast (MW) for net must-service load at bus b at time-step k
$\widehat{NL}_z^f(k)$	Forecast (MW) for net must-service load in zone z at time-step k
$\widehat{NL}^f(k)$	Forecast (MW) for system-wide net must-service load at time-step k
$O(\ell)$	Originating bus for transmission line ℓ
$\hat{p}_j^f(k)$	Forecast (MW) for must-service load at b_j at time-step k
$\hat{p}_n(k)$	Forecast (MW) for the power injection of $n \in \mathbb{NG}$ at time-step k
$P_m^{\min}$	Minimum feasible sustainable power level (MW) for m
$P_m^{\max}$	Maximum feasible sustainable power level (MW) for m
R_m^D	Ramp-down limit (MW/Δt) for m
R_m^U	Ramp-up limit (MW/Δt) for m
$\mathrm{RRD}(z, k)$	Reserve required (decimal %) for down-power in z at k
$\mathrm{RRU}(z, k)$	Reserve required (decimal %) for up-power in z at k
$\mathrm{RRD}(k)$	Reserve required (decimal %) for down-power system wide at k
$\mathrm{RRU}(k)$	Reserve required (decimal %) for up-power system wide at k
S_o	Positive base power (MW)
$\mathrm{T}_m = [t_m^{\mathrm{s}}, t_m^{\mathrm{e}}) \subseteq \mathrm{T}$	Delivery period for a swing contract SC_m offered by m
Δk	Common length of each time-step k, in minutes
Δt	Common length of each time-step k, in hours (e.g. 3h, 1h/12, …)
Λ^-, Λ^+	Penalty weights (\$/MWh) for power imbalance
μ_m	Weight factor (unit-free scalar) for m's avoidable cost function
$\theta_1(k)$	Bus voltage angle (radians) for reference bus 1 at time-step k
Functions and variables	
$A_m(k)$	Availability offer indicator for m: 1 if $k \in \mathbb{K}_m$; 0 otherwise
$\mathrm{B}(\mathrm{T})$	Total benefit (\$) for operating period T
$\widehat{\mathrm{C}}(\mathrm{T})$	Forecasted total avoidable cost (\$) for operating period T
$\mathrm{C}_m^{\mathrm{E}}(\mathbf{p})$	Energy cost reimbursement (\$) required by m for $\mathbf{p} \in \mathbb{PP}_m$
$\mathrm{C}_m^{\mathrm{R}}(\mathbf{p})$	Ramp cost reimbursement (\$) required by m for $\mathbf{p} \in \mathbb{PP}_m$
$\underline{p}_m(k)$	Run-time minimum power level (MW) of m at time-step k
$\overline{p}_m(k)$	Run-time maximum power level (MW) of m at time-step k
SC_m	Swing contract offered by m
$v_m(k) = c_m A_m(k)$	1 if m is available for power delivery at time-step k; 0 otherwise
$w_\ell(k)$	Transmission line power flow (MW) for line ℓ at time-step k
α_m	Offer price (\$) for a swing contract offered by m
$\beta_b(k)$	Power balance slack variable (MW) for bus b at time-step k
ϕ_m	Performance payment method for a swing contract offered by m
$\pi_m^{\mathrm{E}}(k)$	Energy price (\$/MWh) requested by m at time-step k
$\pi_m^{\mathrm{R}}(k)$	Ramp price (\$/MW) requested by m at time-step k

(continued)

TABLE A.4 *(continued)*

Term	Description
ISO decision variables	
c_m	Contract clearing indicator: 1 if SC_m is cleared; 0 otherwise
$p_j^s(k)$	Price-sensitive power usage (MW) of customers in $\mathbb{C}_j$
$p_m(k)$	Power level (MW) of m at bus b_m at time-step k
$\theta_b(k)$	Voltage angle (radians) at bus $b \in \mathbb{B}/\{1\}$ at time-step k
Performance metrics	
$\mathrm{E}^{\mathrm{xp}}[\mathrm{PC}(k)]$	Expected ex-post performance cost ($) for time-step k
$\mathrm{E}^{\mathrm{xp}}[\mathrm{IC}(k)]$	Expected ex-post imbalance cost ($) for time-step k
$\mathrm{IRR}(z, k)$	Inherent reserve range (MW) for zone z at time-step k
$\mathrm{OC}(k)$	Offer cost ($) for time-step k

TABLE A.5 Nomenclature for a distribution system

Term	Description
Acronyms and generics	
b	A grid location where power is injected and/or withdrawn
DU	Distribution utility
IDSO	Independent distribution system operator
GER	Grid-edge resource
GERA	Grid-edge resource aggregator
h	Generic symbol for a household
HVAC	Heating, ventilation, and air conditioning
ITD	Integrated transmission and distribution
LISA	Local intelligent software agent
N	Look-ahead horizon for a household
n	Generic symbol for a control-step
n^{s}	Start-time for a control-step n
n^{e}	End-time for a control-step n
p	Generic designation for a power level
PV	Photovoltaic
T-D	Transmission and distribution
TCL	Thermostatically controlled load
TES	Transactive energy system
π^{Ret}	Generic designation for a retail price
Functions and variables	
$\widehat{\mathrm{G}}_{\mathrm{h}}(n, \mathrm{ON})$	Max future comfort (utils) for h if HVAC is set ON at n^{s}
$\widehat{\mathrm{G}}_{\mathrm{h}}(n, \mathrm{OFF})$	Max future comfort (utils) for h if HVAC is set OFF at n^{s}
P_n^*	Forecasted HVAC ON power usage (MW) for control-step n
X_n^{s}	"May run for ancillary service provision" state for h
X_n^{u}	"May run for power usage" state for h
$-\Pi^*(X_n^{\mathrm{s}})$	Cut-off price for ancillary service supplied by h during n
$\Pi^*(X_n^{\mathrm{u}})$	Cut-off price for power usage demanded by h during n

REFERENCES

[1] Abani, A.O., Hary, N., Saguan, M., and Rious, V. (2016). Risk aversion and generation adequacy in liberalized electricity markets: benefits of capacity markets. *13th International Conference on the European Energy Market (EEM)*, pp. 1–5. https://doi.org/10.1016/j.enpol.2017.10.008.

[2] Abrishambaf, O., Lezama, F., Faria, P., and Vale, Z. (2019). Towards transactive systems: an analysis on current trends. *Energy Strategy Reviews* 26 (100418). https://doi.org/10.1016/j.esr.2019.100418.

[3] Alderete, G.B. (2005). Alternative models to analyze market power and financial transmission rights in electricity markets. PhD thesis. Canada: Electrical and Computer Engineering, University of Waterloo.

[4] Ausubel, L.M. and Cramton, P. (2010). Using forward markets to improve electricity market design. *Utility Policy* 18: 195–200. https://doi.org/10.1016/j.jup.2010.05.004.

[5] Baldick, R. (2006). *Applied Optimization: Formulation and Algorithms for Engineering Systems.* Cambridge: Cambridge University Press.

[6] Battula, S. (2020). Transactive energy system design for integrated transmission and distribution systems. PhD thesis. Ames, IA: Iowa State University (in preparation).

[7] Battula, S., Tesfatsion, L., and McDermott, T.E. (2020) An ERCOT test system for market design studies. *Applied Energy* 275, October, to appear. DOI: 10.1016/j.apenergy.2020.115182

[8] Battula, S., Tesfatsion, L., and Wang, Z. (2020) A customer-centric approach to bid-based transactive energy system design. *IEEE Transactions on Smart Grid* 11(6): 4996–5008. DOI: 10.1109/TSG.2020.3008611

[9] Baumol, W.J., Panzar, J.C., and Willig, R.D. (1982). *Contestable Markets and the Theory of Industry Structure.* New York, NY: Harcourt Brace Jovanovich, Inc..

[10] Benjamin, R. (2010). A Further Inquiry into Financial Transmission Right Properties. *Report.* Round Table Group, Inc.

[11] Bertsimas, D., Litvinov, E., Sun, X.A. et al. (2013). Adaptive robust optimization for the security constrained unit commitment problem. *IEEE Transactions on Power Systems* 28 (1): 52–63.

[12] Bhagwat, P.C., de Vries, L.J., and Hobbs, B.F. (2016). Expert survey on capacity markets in the US: lessons for the EU. *Utility Policy* 38: 11–17. http://doi.org/10.1016/j.jup.2015.11.005.

[13] Bidwell, M. (2005). Reliability options: a market-oriented approach to long-term adequacy. *The Electricity Journal* 18 (5): 11–25.

[14] Birge, J., Hortaçsu, A., Mercadal, I., and Pavlin, M. (2017). Limits to arbitrage in electricity markets: a case study of MISO, *CEEPR WP 2017-003.* MIT Center for Energy and Environmental Policy Research (CEEPR) Working Paper Series.

[15] Birk, M., Chaves-Ávila, J.P., Gómez, T., and Tabors, R. (2017). TSO/DSO coordination in a context of distributed energy resource penetration, *CEEPR WP 2017-017*, MIT Center for Energy and Environmental Policy Research.

[16] Bjørndal, E., Bjørndal, M., Midthun, K., and Tomasgard, A. (2018). Stochastic electricity dispatch: a challenge for market design. *Energy* 150: 992–1005. DOI: 10.1016.j.energy.2018.02.055

[17] Bublitz, A., Keles, D., Zimmermann, F. et al. (2019). A survey on electricity market design: insights from theory and real-world implementations of capacity remuneration mechanisms. *Energy Economics* 80: 1059–1078. https://doi.org/10.1016/j.eneco.2019.01.030.

A New Swing-Contract Design for Wholesale Power Markets, First Edition, Leigh Tesfatsion.

[18] Bunn, D. (2004). Structural and behavioural foundations of competitive electricity prices. In: *Introduction: Modelling Prices in Competitive Electricity Markets* (ed. D. Bunn), 1–17. Wiley.

[19] Burger, S., Chaves-Ávila, J.P., Batlle, C., and Pérez-Arriaga, I.J. (2017). A review of the value of aggregators in electricity systems. *Renewable and Sustainable Energy Reviews* 77: 395–405.

[20] Bushnell, J.B. (2013). JP Morgan and market complexity. *Energy Economics Exchange Blog*. http://energyathaas.wordpress.com/2013/08/12/jp-morgan-and-market-complexity (accessed 12 July 2020).

[21] Bushnell, J.B., Harvey, S.M., and Hobbs, B.F. (2012). Opinion on pay for performance: FERC order 755. *Members of the Market Surveillance Committee of the California ISO*.

[22] Byers, C., Levin, T., and Botterud, A. (2018). Capacity market design and renewable energy: performance incentives, qualifying capacity, and demand curves. *Electricity Journal* 31: 65–74. https://doi.org/10.1016/j.tej.2018.01.006.

[23] CAISO (2004). Transmission Economic Assessment Methodology, California Independent System Operator (CAISO), June 2004.

[24] CAISO (2009). Market Optimization Details. *Technical Bulletin 2009-06-05*. California ISO, Revised November 19.

[25] CAISO (2019a). California Independent System Operator (CAISO) Homepage. http://www.caiso.com/ (accessed 12 July 2020).

[26] CAISO (2019b). Price performance in the CAISO's energy markets. White Paper, California Independent System Operator, April 3. http://www.caiso.com/Documents/WhitePaper-PricePerformanceAnalysis-Apr3-2019.pdf (accessed 12 July 2020).

[27] CAISO (2019c). Business Practice Manual for Market Instruments, California Independent System Operator (CAISO), Version 53, Last Revised: May 2.

[28] Caramanis, M.C., Bohn, R.E., and Schweppe, F.C. (1987). System security control and optimal pricing of electricity. *Electric Power Energy Systems* 9: 217–224.

[29] Carpentier, P., Chancelier, J.P., Cohen, G. et al. (2012). Dynamic consistency for stochastic optimal control problems. *Annals of Operations Research* 200 (1): 247–263.

[30] Carrión, M. and Arroyo, J. (2006). A computationally efficient mixed-integer linear formulation for the thermal unit commitment problem. *IEEE Transactions on Power Systems* 21 (3): 1371–1378.

[31] Chao, H.P. (2019). Incentives for efficient pricing mechanisms in markets with non-convexities. *Journal of Regulatory Economics* 56 (1): 33–58.

[32] Chao, H.P. and Wilson, R. (2002). Multi-dimensional procurement auctions for power reserves: robust-incentive compatible scoring and settlement rules. *Journal of Regulatory Economics* 22 (2): 161–183.

[33] Chao, H.P. and Wilson, R. (2004). Resource Adequacy and Market Power Mitigation Via Option Contacts. *EPRI Report*. Electric Power Research Institute (EPRI), Palo Alto, CA.

[34] Chipman, J.S., Hurwicz, L., Richter, M.K., and Sonnenschein, H.F. (eds.) (1971). *Preferences, Utility and Demand*. New York, NY: Harcourt Brace Jovanovich.

[35] Ciraci, S., Daily, J., Fuller, J. et al. (2014). FNCS: a framework for power system and communication networks co-simulation. *Symposium on Theory of Modeling and Simulation*. San Diego, CA: ACM.

[36] Conejo, A.J., Contreras, J., Lima, D.A., and Padilha-Feltrin, A. (2007). Z_{bus} transmission network cost allocation. *IEEE Transactions on Power Systems* 22 (1): 342–349.

[37] Cramton, P. (2017). Electricity market design. *Oxford Review of Economic Policy* 33 (4): 589–612.

[38] Cramton, P. and Ockenfels, A. (2012). Economics and design of capacity markets for the power Sector. *Zeitschrift für Energiewirtschaft* 36 (2): 113–134.

[39] Dahlke, S. (2012). Grid-scale energy storage for integrating renewable energy: updates on FERC Order 755 and DOE-funded demonstration projects, Environmental Studies Student Work. Paper 1. http://digitalcommons.csbsju.edu/environmental_studies_students/1 (accessed 12 July 2020).

[40] De Martini, P. and Kristov, L. (2015). Planning, Market Design, Operation and Oversight. Berkeley Lab, *Report No. 2, LBNL-100397*.

[41] Deaton, A. and Muellbauer, J. (1980). *Economics and Consumer Behavior*. Cambridge: Cambridge University Press.

[42] Deng, S.J. and Oren, S.S. (2006). Electricity derivatives and risk management. *Energy* 31: 940–953.

[43] DOE (2011). DOE Technology Readiness Assessment Guide. Office of Management, Department of Energy, DOE G 413.3-4A.

[44] Ela, E., Milligan, M., and Kirby, B. (2011). Operating Reserves and Variable Generation, National Renewable Energy Laboratory. *Technical Report NREL/TP-5500-51976*. National Renewable Energy Laboratory (NREL), Golden, Colorado, USA.

[45] Eldridge, B., O'Neill, R., and Hobbs, B.F. (2018). Pricing in day-ahead electricity markets with near-optimal unit commitment, Working Paper EPRG 1840. Energy Policy Research Group, Cambridge Judge Business School, University of Cambridge, UK.

[46] Ellison, J.F., Tesfatsion, L.S., Loose, V.W., and Byrne, R.H. (2012). A Survey of Operating Reserve Markets in U.S. ISO/RTO-Managed Electric Energy Regions. Sandia National Laboratories *Report (SAND2012-1000)*.

[47] ERCOT (2017). Nodal 101. Electric Reliability Council of Texas (ERCOT), Course Modules 1-6. http://www.ercot.com/mktrules/nprotocols/current (accessed 12 July 2020).

[48] ERCOT (2018). Market Information. Electric Reliability Council of Texas (ERCOT). http://www.ercot.com/mktinfo (accessed 12 July 2020).

[49] ERCOT (2019a). Nodal Protocols. Section 4: Day-Ahead Operations. Electric Reliability Council of Texas (ERCOT).

[50] ERCOT (2019b). Nodal Protocols. Section 6: Adjustment Period and Real-time Operations. Electric Reliability Council of Texas (ERCOT).

[51] ERCOT (2019d). Grid Information. Electric Reliability Council of Texas (ERCOT). http://www.ercot.com/gridinfo/ (accessed 12 July 2020).

[52] FERC (2003). Notice of White Paper. U.S. Federal Energy Regulatory Commission (FERC).

[53] FERC (2008). Assessment of Demand Response and Advanced Metering. Staff Report. U.S. Federal Energy Regulatory Commission (FERC).

[54] FERC (2011). Frequency Regulation Compensation in Organized Wholesale Power Markets. U.S. Federal Energy Regulatory Commission (FERC) Final Rule, Order No. 755, October 20.

[55] FERC (2015). Energy Primer: A Handbook of Energy Market Basics. U.S. Federal Energy Regulatory Commission (FERC).

[56] FERC (2016a). Offer Caps in Markets Operated by Regional Transmission Organizations and Independent System Operators., U.S. Federal Energy Regulatory Commission (FERC), Final Rule, Order No. 831.

[57] FERC (2016b). Fast-Start Pricing Operated in Markets Operated by Regional Transmission Organizations and Independent System Operators. U.S. Federal Energy Regulatory Commission (FERC), Notice of Proposed Rule-Making, Docket No. RM17-3-000.

[58] FERC (2018a). Electric Storage Participation in Markets Operated by Regional Transmission Organizations and Independent System Operators. U.S. Federal Energy Regulatory Commission (FERC). Final Rule, Order No. 841.

[59] FERC (2018b). Essential Reliability Services and the Evolving Bulk-Power System: Primary Frequency Response. U.S. Federal Energy Regulatory Commission (FERC), Final Rule, Order No. 842.

[60] FERC (2018c). Uplift Cost Allocation and Transparency in Markets Operated by Regional Transmission Organizations and Independent System Operators, U.S. Federal Energy Regulatory Commission (FERC), Final Rule, Order No. 844.

[61] FERC (2019). Map of Regional Transmission Organizations and Independent System Operators, U.S. Federal Energy Regulatory Commission (FERC).

[62] Fletcher, R. (2000). *Practical Methods of Optimization*, 2e. Chichester: Wiley.

[63] Geng, X. and Xie, L. (2017). Learning the LMP-load coupling from data: a support vector machine based approach. *IEEE Transactions on Power Systems* 32 (2): 1127–1138.

[64] Graves, F.C., Read, E.G., Hanser, P.Q., and Earle, R.L. (1998). One-part markets for electric power: ensuring the benefits of competition. In: *Power Systems Restructuring, The Springer International*

Series in Engineering and Computer Science (Power Electronics and Power Systems), Chapter 7 (ed. M. Ilic, F. Galiana, and L. Fink), 243–280. Boston, MA: Springer.

[65] Gribik, P.R., Hogan, W.W., and Pope, S.L. (2007). Market-Clearing Electricity Prices and Energy Uplift, Working Paper, John F. Kennedy School of Government, Harvard University, Cambridge, MA.

[66] GridLAB-D (2018). GridLAB-D: The Next Generation Simulation Software. http://www.gridlabd.org/ (accessed 12 July 2020).

[67] GridLAB-D (2019). GridLAB-D House Object Documentation. http://gridlab-d.shoutwiki.com/wiki/House (accessed 12 July 2020).

[68] GridWise Architecture Council (2015). GridWise Transactive Energy Framework Framework: Version 1.0. Pacific Northwest National Laboratory *Report No. PNNL-22946*.

[69] Gross, G. and Bompard, E. (2004). Optimal power flow application issues in the pool paradigm. *Electrical Power and Energy Systems* 26: 787–796.

[70] Haarbrücker, G. and Kuhn, D. (2009). Valuation of electricity swing options by multistage stochastic programming. *Automatica* 45: 889–899.

[71] Hand, M.M., Baldwin, S., DeMeo, E. et al. (eds.) (2012). Renewable Electricity Futures Study (4 vols.). National Renewable Energy Laboratory (NREL), *NREL/TP-6A20-52409*.

[72] Hansen, J., Somani, A., Sun, Y., and Zhang, Y. (2016). Auction design for power markets based on standardized contracts for energy and reserves. *Proceedings of the IEEE Power and Energy Society General Meeting*, Boston, MA (electronic).

[73] Hartman, D. (2017). Refreshing price formation policy in wholesale power markets. R Street Policy Study No. 106.

[74] Hausman, E., Hornby, R., and Smith, A. (2008). *Bilateral Contracting in Deregulated Electricity Markets.* Synapse Energy Economics, Inc.

[75] Helman, U., Hobbs, B.F., and O'Neill, R. (2008). The design of US wholesale energy and ancillary service auction markets: theory and practice. In: *Competitive Electricity Markets: Design, Implementation, and Performance*, Chapter 5 (ed. F.P. Sioshansi), 179–243. Elsevier Global Energy Policy and Economics Series.

[76] Heo, D.Y. and Tesfatsion, L.S. (2015a). Standardized contracts with swing for the market-supported procurement of energy and reserve: illustrative examples. Working Paper No. 13018, *Economics Working Paper Series*. Ames, IA: Iowa State University. Original release: November 2013; Revised: June 2015. http://www2.econ.iastate.edu/tesfatsi/StandardizedContracts.HeoTesfatsion.WP13018.pdf (accessed 12 July 2020).

[77] Heo, D.Y. and Tesfatsion, L.S. (2015b). Facilitating appropriate compensation of electric energy and reserve through standardized contracts with swing. *Journal of Energy Markets* 8 (4): 93–121. https://doi.org/10.21314/JEM.2015.135.

[78] Hinman, C. (2015). Pay for Performance Regulation (FERC Order 755) Updated with Year One Design Changes. California Independent System Operator (CAISO).

[79] Hogan, W.W. (1997). Reshaping the electricity industry. Unpublished working paper, Center for Business and Government, John F. Kennedy School of Government. Cambridge, MA: Harvard University.

[80] Hogan, W.W. (2005). On an "energy only" electricity market design for resource adequacy. Unpublished working paper, Center for Business and Government, John F. Kennedy School of Government. Cambridge, MA: Harvard University.

[81] Hogan, W.W. (2016). *Electricity Market Design, Workshop on Optimization and Equilibrium in Energy Economics, Institute for Pure and Applied Mathematics (IPAM).* Los Angeles, CA: University of Southern California.

[82] Hogan, M. (2017). Follow the missing money: ensuring reliability at least cost to consumers in the transition to a low-carbon power system. *Electricity Journal* 30: 55–61. https://doi.org/10.1016.j.tej.2016.12.006.

[83] Hogan, W.W. and Pope, S.L. (2017). Priorities for the Evolution of an Energy-Only Electricity Market Design in ERCOT. Report. FTI Consulting. https://sites.hks.harvard.edu/fs/whogan/Hogan_Pope_ERCOT_050917.pdf (accessed 12 July 2020).

[84] Hsieh, E. and Anderson, R. (2017). Grid flexibility: the quiet revolution. *Electricity Journal* 30: 1–8.

[85] Hua, B., Schiro, D.A., Zheng, T. et al. (2019). Pricing in multi-interval real-time markets. *IEEE Transactions on Power systems* 34 (4): 2696–2705. https://doi.org/10.1109/TPWRS.2019.2891541.

[86] Hurwicz, L. (1973). The design of mechanisms for resource allocation. *The American Economic Review* 63 (2): 1–30.

[87] IEEE (2014). IEEE 123 Node Test Feeder. IEEE Distribution System Analysis Subcommittee. http://sites.ieee.org/pes-testfeeders/resources/ (accessed 12 July 2020).

[88] Ilić, M.D. (2016). Toward a unified modeling and control for sustainable and resilient energy systems. *Foundations and Trends in Electric Energy Systems* 1 (1): 1–141.

[89] ISO-NE (2017). Annual Markets Report for 2016, Internal Market Monitor. ISO New England, Inc.

[90] ISO-NE (2018). Independent System Operator of New England. http://www.iso-ne.com/ (accessed 12 July 2020).

[91] ISO-NE (2019). Energy Security Improvements. ISO New England Discussion Paper, Version 1. https://www.iso-ne.com/static-assets/documents/2019/04/a00_iso_discussion_paper_energy_security_improvements.pdf (accessed 12 July 2020).

[92] ITDProject (2019). *GitHub Household Formulation Code/Data Repository*. Ames, IA: ISU. https://github.com/ITDProject/HouseholdFormulationRepository (accessed 12 July 2020).

[93] Jain, K. and Mahdian, M. (2007). *Cost-Sharing in Algorithmic Game Theory*. Cambridge: Cambridge University Press.

[94] Jobs, S. (2005). Stanford Commencement Address.

[95] Joskow, P.L. (2008). Capacity payments in imperfect electricity markets: need and design. *Utilities Policy* 16 (3): 159–170.

[96] Joskow, P. (2019). Challenges for wholesale electricity markets with intermittent renewable generation at scale: the U.S. experience. *Oxford Review of Economic Policy* 35 (2): 291–331.

[97] Kirschen, D.S. and Strbac, G. (2018). *Fundamentals of Power System Economics*, 2e. New York, NY: Wiley.

[98] Klemperer, P. (2004). *Auctions: Theory and Practice*. Princeton, NJ: Princeton University Press. http://www.nuff.ox.ac.uk/economics/papers/2004/W9/AuctionsTheoryPractice.pdf (accessed 12 July 2020).

[99] Kok, K. (2013). The PowerMatcher: Smart coordination for the smart electricity grid. *Siks Dissertation Series No. 2013-17*. The Netherlands: Dutch Research School for Information and Knowledge Systems, TNO. http://dare.ubvu.vu.nl/handle/1871/43567 (accessed 12 July 2020).

[100] Krantz, D.H., Luce, R.D., Suppes, P., and Tversky, A. (1971). *Foundations of Measurement*, vol. I. New York, NY: Academic Press.

[101] Kreps, D.M. (1990). *A Course in Microeconomic Theory*. Princeton, NJ: Princeton University Press.

[102] Krishnamurthy, D., Li, W., and Tesfatsion, L. (2016). An 8-zone test system based on ISO New England data: development and application. *IEEE Transactions on Power Systems* 31 (1): 234–246. Preprint: http://www2.econ.iastate.edu/tesfatsi/8ZoneISONETestSystem.RevisedAppendix.pdf (accessed 12 July 2020).

[103] Kristov, L. (2016). The electricity system in 2030: a history of the future. *GO15 TSO/DSO Workshop*, San Francisco, CA.

[104] Kristov, L. (2019). The bottom-up (r)evolution of the electric power system. *IEEE Power & Energy Magazine* 17 (2): 42–49.

[105] Kristov, L. and De Martini, P. (2014). 21st Century Electric Distribution System Operations. Unpublished report.

[106] Kristov, L., De Martini, P., and Taft, J.D. (2016). A tale of two visions. *IEEE Power & Energy Magazine* 14 (3): 63–69.

[107] Küster, K.K., Aoki, A.R., and Lambert-Torres, G. (2019). Transaction-based operation of electric distribution systems: a review. *International Transactions on Electrical Energy Systems*. https://doi.org/10.1002/2050–7038.12194.

[108] Lally, J. (2002). Financial transmission rights: auction example. Section 6 in Financial Transmission Rights Draft 01-10-02, m-06 ed., ISO New England, Inc.

[109] LANL (2015). Thinking inside the box. In: *1663: The Los Alamos Science and Technology Magazine*. Los Alamos, New Mexico, USA, Los Alamos National Laboratory. 18–24.

[110] Li, H., Sun, J., and Tesfatsion, L. (2008). Dynamic LMP response under alternative price-cap and price-sensitive demand scenarios. *Proceedings of the IEEE Power and Energy Society General Meeting*, Pittsburgh, PA, (electronic).

[111] Li, H., Sun, J., and Tesfatsion, L. (2009). Separation and volatility of locational marginal prices in restructured wholesale power markets. Working Paper No. 09009, *Economics Working Paper Series*. Ames, IA: Iowa State University.

[112] Li, H. and Tesfatsion, L. (2011). ISO net surplus collection and allocation in wholesale power markets under locational marginal pricing. *IEEE Transactions on Power Systems* 26 (2): 627–641.

[113] Li, W. and Tesfatsion, L. (2016). Market provision of flexible energy/reserve contracts: optimization formulation. *Proceedings of the IEEE Power and Energy Society General Meeting*, Boston, MA (electronic).

[114] Li, W. and Tesfatsion, L. (2017). An 8-zone ISO-NE test system with physically-based wind power. Working Paper No. 17017, *Economics Working Paper Series*. Ames, IA: ISU. http://www2.econ.iastate.edu/tesfatsi/EightZoneISONETestSystemWithWind.LiTesfatsion.pdf (accessed 13 July 2020).

[115] Li, W. and Tesfatsion, L. (2018). A swing-contract market design for flexible service provision in electric power systems. In: *Energy Markets and Responsive Grids: Modelling, Control, and Optimization*, IMA Volumes in Mathematics and its Applications, Chapter 5, vol. 162 (ed. S Meyn, T. Samad, I. Hiskens, and J. Stoustrup), 105–127. Springer.

[116] Li, S., Lian, J., Conejo, A., and Zhang, W. (2019). Transactive Energy System: Market-Based Coordination of Distributed Energy Resources. arXiv:1908.03641v1 [math.OC].

[117] Li, S., Zhang, W., Lian, J., and Kalsi, K. (2016). Market-based coordination of thermostatically controlled loads - Part I: A mechanism design formulation. *IEEE Transactions on Power Systems* 31 (2): 1170–1178.

[118] Lima, D.A., Padilha-Feltrin, A., and Contreras, J. (2009). An overview of network cost allocation methods. *Electric Power Systems Research* 79 (5): 750–758.

[119] Litvinov, E., Zhao, F., and Zheng, T. (2009). Alternative auction objectives and pricing schemes in short-term electricity markets. *Proceedings of the IEEE Power and Energy Society General Meeting*, Calgary, Canada.

[120] Liu, H., Tesfatsion, L., and Chowdhury, A.A. (2009). Derivation of locational marginal prices for restructured wholesale power markets. *Journal of Energy Markets* 2 (1): 3–27.

[121] Lorca, A., Sun, X.A., Litvinov, E., and Zheng, T. (2016). Multistage adaptive robust optimization for the unit commitment problem. *Operations Research* 64 (1): 32–51.

[122] Ma, S., Wang, Z., and Tesfatsion, L. (2019). Swing contracts with dynamic reserves for flexible service management. *IEEE Transactions on Power Systems* 34 (5): 4024–4037. http://www2.econ.iastate.edu/tesfatsi/SwingContractsWithDynamicReserves.MaEtAl.pdf

[123] Ma, X., Chen, Y., and Wan, J. (2009a). Midwest ISO co-optimization based real-time dispatch and pricing of energy and ancillary services. *Proceedings of the IEEE Power and Energy Society General Meeting*. (electronic).

[124] Ma, X., Song, H., Hong, M. et al. (2009b). The security-constrained commitment and dispatch for Midwest ISO day-ahead co-optimized energy and ancillary service market. *Proceedings of the IEEE Power and Energy Society General Meeting* (electronic).

[125] Marshall, A. (1890). *Principles of Economics*. London: Macmillan and Co., Ltd.

[126] Milgrom, P. (2004). *Putting Auction Theory to Work*. Cambridge: Cambridge University Press.

[127] Milligan, M., Frew, B.A., Bloom, A. et al. (2016). Wholesale electricity market design with increasing levels of renewable generation: revenue sufficiency and long-run reliability. *The Electricity Journal* 29: 26–38.

[128] MISO (2011). Dispatchable Intermittent Resource Workshop II, Midcontinent Independent System Operator (MISO).

[129] MISO (2015). State of the Market Report 2014. Independent Market Monitor, Midcontinent Independent System Operator (MISO).

[130] MISO (2018a). Midcontinent Independent System Operator (MISO). https://www.misoenergy.org/ (accessed 13 July 2020).

[131] MISO (2018b). Energy and Operating Reserve Markets. Business Practices Manual BPM-002-r18, Midcontinent Independent System Operator (MISO). https://www.misoenergy.org/legal/business-practice-manuals/ (accessed 13 July 2020).
[132] Morales, J.M., Conejo, A., and Pérez-Ruiz, J. (2009). Economic valuation of reserves in power systems with high penetration of wind power. *IEEE Transactions on Power Systems* 24 (2): 900–910.
[133] Moye, R. and Meyn, S. (2018). Redesign of U.S. electricity capacity markets. In: *Energy Markets and Responsive Grids: Modelling, Control, and Optimization, The IMA Volumes in Mathematics and its Applications Series*, Chapter 4, vol. 162 (ed. S. Meyn, T. Samad, I. Hiskens, and J. Stoustrup), pp. 73–104. Springer.
[134] Munoz, F.D., Wogrin, S., Oren, S.S., and Hobbs, B.F. (2018). Economic inefficiencies of cost-based electricity market designs. *The Energy Journal* 39 (3): 51–68.
[135] NAS (2016). *Analytic Research Foundations for the Next-Generation Electric Grid*. Washington, DC: National Academies of Science, The National Academies Press.
[136] Navid, N. and Rosenwald, G. (2013). Ramp capability product design for MISO markets. Midcontinent Independent System Operator (MISO).
[137] Newbery, D. (2016). Missing money and missing markets: reliability, capacity auctions, and interconnectors. *Energy Policy* 94: 401–410. https://doi.org/10.1016/j.enpol.2015.10.028.
[138] Newell, DB, and Tiesinga, E. (eds.) (2019). *The International System of Units (SI)*, Special Publication 330. National Institute of Standards and Technology (NIST), U.S. Department of Commerce. https://www.nist.gov/pml/special-publication-330.
[139] Nguyen, H.T., Battula, S., Takkala, R.R. et al. (2019). An integrated transmission and distribution test system for evaluation of transactive energy designs. *Applied Energy* 240: 666–679.
[140] NYISO (2018). New York Independent System Operator (NYISO). http://www.nyiso.com/ (accessed 13 July 2020).
[141] Oren, S.S. (2005a). Generation adequacy via call options obligations: safe passage to the promised land. *The Electricity Journal* 18 (9): 28–42.
[142] Oren, S.S. (2005b). Ensuring generation adequacy in competitive electricity markets. In: *Electricity Deregulation: Choices and Challenges*, Chapter 10 (ed. M.J. Griffin and S.L. Puller). University of Chicago Press. 388–413
[143] Oren, S.S. and Gross, G. (2009). Economic Impact Assessment of Transmission Enhancement Projects.. Final Report, PSERC Publication 09-07. Power Systems Engineering Research Center (PSERC).
[144] Orfanogianni, T. and Gross, G. (2007). A general formulation for LMP evaluation. *IEEE Transactions on Power Systems* 22 (3): 1163–1173.
[145] Papavasiliou, A., Oren, S., and O'Neill, R. (2011). Reserve requirements for wind power integration: a scenario-based stochastic programming framework. *IEEE Transactions on Power Systems* 26 (4): 2197–2206.
[146] Pfeifenberger, J., Sees, K., and Schumacher, A. (2009). *A Comparison of PJM's RPM with Alternative Energy and Capacity Market Designs*. The Brattle Group.
[147] Pflug, G.-Ch. and Pichler, A. (2014). *Multistage Stochastic Optimization*. Springer.
[148] PJM (2018). PJM RTO. http://www.pjm.com/ (accessed 13 July 2020).
[149] PowerMatcher (2020). http://flexiblepower.github.io/technology/powermatcher/.
[150] Rahimi, F.A. and Mokhtari, S. (2018). Distribution management system for the grid of the future. *IEEE Electrification Magazine* 6 (2): 84–94.
[151] Rebours, Y.G., Kirschen, D.S., Trotignon, M., and Rossignol, S. (2007a). A survey of frequency and voltage control ancillary services - Part I: Technical features. *IEEE Transactions on Power Systems* 22 (1): 350–357.
[152] Rebours, Y.G., Kirschen, D.S., Trotignon, M., and Rossignol, S. (2007b). A survey of frequency and voltage control ancillary services - Part II: Economic features. *IEEE Transactions on Power Systems* 22 (1): 358–366.
[153] Ross, S.M. and Zhu, Z. (2008). On the structure of a swing contract's optimal value and optimal strategy. *Journal of Applied Probability* 45 (1): 1–15.

[154] Roth, A.E. and Wilson, R.B. (2019). How market design emerged from game theory: a mutual interview. *Journal of Economic Perspectives* 33 (3): 118–143.

[155] Royal Swedish Academy of Sciences (2007). Mechanism design theory. Scientific background on the Sveriges Riksbank Prize in Economic Sciences in Memory of Alfred Nobel 2007, Compiled by the Prize Committee.

[156] Ruiz, P.A., Philbrick, C.R., Zak, E. et al. (2009). Uncertainty management in the unit commitment problem. *IEEE Transactions on Power Systems* 24 (2): 642–651.

[157] Salazar, H. (2008). A critical appraisal of economic-driven transmission enhancement. M.S. Creative Component. Ames, IA: Department of Economics, Iowa State University.

[158] Sauma, E. and Oren, S. (2009). Do generation firms in restructured electricity markets have incentives to support social-welfare-improving transmission investments? *Energy Economics* 31: 676–689.

[159] Schweppe, F.C., Caramanis, M.C., Tabors, R.D., and Bohn, R.E. (1988). *Spot Pricing of Electricity*, Fourth Printing (2000). Boston, MA: Kluwer Academic Publishers.

[160] Schweppe, F.C., Tabors, R.D., Kirtley, J.L. Jr. et al. (1980). Homeostatic utility control. *IEEE Transactions on Power Systems* 99 (3): 1151–1163.

[161] Seliga, K., George, S., and DePillis, M. (2014). Energy market offer flexibility. ISO-NE Webex Broadcast: Customer Training Webinar. http://www.iso-ne.com/support/training/courses/energy_mkt_ancil_serv_top/ (accessed 13 July 2020).

[162] Sensfuß, F., Ragwitz, M., and Genoese, M. (2008). The merit-order effect: a detailed analysis of the price effect of renewable electricity generation on spot market prices in Germany. *Energy Policy* 36 (8): 3086–3094. https://doi.org/10.1016/j.enpol.2008.03.035.

[163] Shane, F., Loewenstein, G., and O'Donoghue, T. (2002). Time discounting and time preference: a critical review. *Journal of Economic Literature* 40 (2): 351–401. https://doi.org/10.1257/002205102320161311

[164] Shapley, L.S. (1953). A value for n-person games. In: *Contributions to the Theory of Games, Volume II, Annals of Mathematical Studies*, vol. 28 (ed. H.W. Kuhn and A.W. Tucker), 307–317. Princeton, NJ: Princeton University Press.

[165] Shiro, D. and White, M. (2015). Real-time price formation: energy market offer design. Technical Session #11. Sturbridge, MA: ISO New England.

[166] Sioshansi, R., O'Neill, R., and Oren, S.S. (2008). Economic consequences of alternative solution methods for centralized unit commitment in day-ahead electricity markets. *IEEE Transactions on Power Systems* 23 (2): 344–352.

[167] Somani, A. and Tesfatsion, L. (2008). An agent-based test bed study of wholesale power market performance measures. *IEEE Computational Intelligence Magazine* 3 (4): 56–72.

[168] Spence, D.B. (2018). Naïve energy markets. In: *Energy Markets and Responsive Grids: Modelling, Control, and Optimization, IMA Volumes in Mathematics and its Applications* Chapter 2, vol. 162 (ed. S. Meyn, T. Samad, I. Hiskens, and J. Stoustrup), 29–58. Springer.

[169] Spence, D. and Bush, D. (2009). Why does ERCOT have only one regulator? In: *Electricity Restructuring: The Texas Story*, Chapter 1 (ed. L.L. Kiesling and A.N. Kleit), 9–21. Washington, DC: The AEI Press.

[170] SPP (2018). Southwest Power Pool. https://www.spp.org/ (accessed 13 July 2020).

[171] Stein, A. (2016). Distributed reliability. *University of Colorado Law Review* 87: 888–962.

[172] Stein, A. (2017). Breaking energy path dependencies. *Brooklyn Law Review* 822: 559–604.

[173] Stoft, S. (2002). *Power System Economics: Designing Markets for Electricity*. New York: Wiley-Interscience.

[174] Strotz, R.H. (1956). Myopia and inconsistency in dynamic utility maximization. *The Review of Economic Studies* 23 (3): 165–180.

[175] Sun, J. and Tesfatsion, L. (2007). Open-source software for power industry research, teaching, and training: a DC-OPF illustration. *Proceedings of the IEEE Power and Energy Society General Meeting*, Tampa, Florida.

[176] Sun, J. and Tesfatsion, L. (2010). DC Optimal Power Flow Formulation and Testing Using QuadProgJ, Working Paper No. 06014, Department of Economics, Iowa State University Last Revised: March 2010. http://www2.econ.iastate.edu/tesfatsi/DC-OPF.JSLT.pdf (accessed 13 July 2020).

[177] Tackett, M.H. (2009). Experience with implementing simultaneous co-optimization in the Midwest ISO energy & operating reserve markets. *Proceedings, IEEE Power & Energy Society General Meeting* (electronic).

[178] Taft, J. (2017). Electric Grid Market-Control Structure. *Report PNNL-26753*. Richland, WA: Pacific Northwest National Lab (PNNL).

[179] Tesfatsion, L. (1986). Time inconsistency of benevolent government economies. *Journal of Public Economics* 31: 25–52.

[180] Tesfatsion, L. (2009). Auction basics for wholesale power markets: objectives and pricing rules. *Proceedings of IEEE PES General Meeting*, Calgary, Alberta, CA (26–30 July 2009) (electronic). http://www2.econ.iastate.edu/tesfatsi/AuctionBasics.IEEEPES2009.LT.pdf (accessed 13 July 2020).

[181] Tesfatsion, L. (2017). Modeling economic systems as locally-constructive sequential games. *Journal of Economic Methodology* 24 (4): 384–409.

[182] Tesfatsion, L.S. (2018). Electric power markets in transition: agent-based modeling tools for transactive energy support. In: *Handbook of Computational Economics 4: Heterogeneous Agent Models, Handbooks in Economics Series*, Chapter 13 (ed. C. Hommes and B. Le Baron), 715–766. Amsterdam, The Netherlands: North Holland (Elsevier). http://www2.econ.iastate.edu/tesfatsi/TESHandbookChapter.LTesfatsion.pdf

[183] Tesfatsion, L. (2020a). Agent-based research on restructured electricity markets. http://www2.econ.iastate.edu/tesfatsi/aelect.htm (accessed 13 July 2020).

[184] Tesfatsion, L. (2020b). Empirical validation and verification of agent-based models. http://www2.econ.iastate.edu/tesfatsi/empvalid.htm (accessed 13 July 2020).

[185] Tesfatsion, L. (2020c). AMES wholesale power market test bed. http://www2.econ.iastate.edu/tesfatsi/AMESMarketHome.htm (accessed 13 July 2020).

[186] Tesfatsion, L. and Battula, S. (2020). Notes on the GridLAB-D household equivalent thermal parameter model. Working Paper No. 19001. Ames, IA: Department of Economics, ISU. https://lib.dr.iastate.edu/econ_workingpapers/60 (accessed 13 July 2020).

[187] Tesfatsion, L. and Battula, S. (2020). Analytical SCUC/SCED optimization formulation for AMES V5.0. Working Paper No. 20014. Ames, IA: Department of Economics, ISU. https://lib.dr.iastate.edu/econ_workingpapers/109/

[188] Tesfatsion, L., Silva-Monroy, C.S., Loose, V.W. et al. (2013). New wholesale power market design using linked forward markets. Sandia National Laboratories Report (SAND2013-2789). http://www2.econ.iastate.edu/tesfatsi/MarketDesignSAND2013-2789.LTEtAl.pdf (accessed 13 July 2020).

[189] Thomas, A.G. and Tesfatsion, L. (2018). Braided cobwebs: cautionary tales for dynamic pricing in retail electric power markets. *IEEE Transactions on Power Systems* 6 (33): 6870–6882. DOI: 10.1109/TPWRS.2018.2832471.

[190] Tielens, P. and Van Hertem, D. (2016). The relevance of inertia in power systems. *Renewable and Sustainable Energy Reviews* 55: 999–1009.

[191] Tuohy, A., Meibom, P., Denny, E., and O'Malley, M. (2009). Unit commitment for systems with significant wind penetration. *IEEE Transactions on Power Systems* 24 (2): 592–601.

[192] Twomey, P., Green, R., Neuhoff, K., and Newbery, D. (2005). A review of the monitoring of market power: possible roles of TSOs in monitoring for market power issues in congested transmission systems, 05-002 WP, Center for Energy and Environmental Policy Research.

[193] Vrakopoulou, M., Margellos, K., Lygeros, J., and Andersson, G. (2013). A probabilistic framework for reserve scheduling and N-1 security assessment of systems high wind power penetration. *IEEE Transactions on Power Systems* 28 (4): 3885–3896.

[194] Weiss, A. (1990). *Efficiency Wages: Models of Unemployment, Layoffs, and Wage Dispersion*. Princeton, NJ: Princeton University Press.

[195] Weller, P.A. (1978). Consistent intertemporal decision making under uncertainty. *Review of Economic Studies* 45: 263–266. https://doi.org/10.2307/2297340.

[196] Wu, F., Varaiya, P., Spiller, P., and Oren, S. (1996). Folk theorems on transmission access: proofs and counterexamples. *Journal of Regulatory Economics* 10 (1): 5–23.

[197] Xu, L. and Tretheway, D. (2014). *Flexible Ramping Products Incorporating FMM and EIM*. California ISO. http://www.caiso.com/Documents/.

[198] Yan, J.H., Stern, G.A., Luh, P.B., and Zhao, F. (2008). Payment versus bid cost minimization in ISO markets. *IEEE Power & Energy Magazine* 6 (2): 24–36.

[199] Yu, N., Somani, A., and Tesfatsion, L. (2010). Financial risk management in restructured wholesale power markets: concepts and tools. *Proceedings of the IEEE Power and Energy Society General Meeting*. Minneapolis, MN (electronic).

[200] Yu, N., Tesfatsion, L., and Liu, C.C. (2012). Financial bilateral contract negotiation in wholesale power markets using Nash bargaining theory. *IEEE Transactions on Power Systems* 27 (1): 251–267.

[201] Zhao, S., Lin, X., Aliprantis, D. et al. (2016). Online multi-stage decisions for robust power-grid operations under high renewable uncertainty. *INFOCom-2016, IEEE Conference on Computer Communications*, April 1–9.

[202] Zhong, W. (2019). Demand uncertainties management in SCUC and voltage security enhancement for SCED. PhD dissertation. Case Western Reserve University.

[203] Zhou, Q., Tesfatsion, L., and Liu, C.-C. (2011). Short-term congestion forecasting in wholesale power markets. *IEEE Transactions on Power Systems* 26 (4): 2185–2196.

[204] Zou, Y., Lin, X., Aliprantis, D., and Chen, M. (2018). Robust multi-stage power grid operations with energy storage. *INFOCom-2018, IEEE Conference on Computer Communications*, pp. 2483–2491.

Index

A New Swing-Contract Design for Wholesale Power Markets, First Edition, Leigh Tesfatsion.

IEEE Press Series on Power Engineering

Series Editor: M. E. El-Hawary, Dalhousie University, Halifax, Nova Scotia, Canada

The mission of IEEE Press Series on Power Engineering is to publish leading-edge books that cover the broad spectrum of current and forward-looking technologies in this fast-moving area. The series attracts highly acclaimed authors from industry/academia to provide accessible coverage of current and emerging topics in power engineering and allied fields. Our target audience includes the power engineering professional who is interested in enhancing their knowledge and perspective in their areas of interest.

1. *Electric Power Systems: Design and Analysis, Revised Printing*
Mohamed E. El-Hawary

2. *Power System Stability*
Edward W. Kimbark

3. *Analysis of Faulted Power Systems*
Paul M. Anderson

4. *Inspection of Large Synchronous Machines: Checklists, Failure Identification, and Troubleshooting*
Isidor Kerszenbaum

5. *Electric Power Applications of Fuzzy Systems*
Mohamed E. El-Hawary

6. *Power System Protection*
Paul M. Anderson

7. *Subsynchronous Resonance in Power Systems*
Paul M. Anderson, B.L. Agrawal, and J.E. Van Ness

8. *Understanding Power Quality Problems: Voltage Sags and Interruptions*
Math H. Bollen

9. *Analysis of Electric Machinery*
Paul C. Krause, Oleg Wasynczuk, and S.D. Sudhoff

10. *Power System Control and Stability, Revised Printing*
Paul M. Anderson and A.A. Fouad

11. *Principles of Electric Machines with Power Electronic Applications, Second Edition*
Mohamed E. El-Hawary

12. *Pulse Width Modulation for Power Converters: Principles and Practice*
D. Grahame Holmes and Thomas A. Lipo

13. *Analysis of Electric Machinery and Drive Systems, Second Edition*
Paul C. Krause, Oleg Wasynczuk, and Scott D. Sudhoff

14. *Risk Assessment for Power Systems: Models, Methods, and Applications*
Wenyuan Li

15. *Optimization Principles: Practical Applications to the Operation of Markets of the Electric Power Industry*
Narayan S. Rau

16. *Electric Economics: Regulation and Deregulation*
Geoffrey Rothwell and Tomas Gomez

17. *Electric Power Systems: Analysis and Control*
Fabio Saccomanno

18. *Electrical Insulation for Rotating Machines: Design, Evaluation, Aging, Testing, and Repair, Second Edition*
Greg C. Stone, Ian Culbert, Edward A. Boulter, and Hussein Dhirani

19. *Signal Processing of Power Quality Disturbances*
Math H. J. Bollen and Irene Y. H. Gu

20. *Instantaneous Power Theory and Applications to Power Conditioning*
Hirofumi Akagi, Edson H. Watanabe, and Mauricio Aredes

21. *Maintaining Mission Critical Systems in a 24/7 Environment*
Peter M. Curtis

22. *Elements of Tidal-Electric Engineering*
Robert H. Clark

23. *Handbook of Large Turbo-Generator Operation and Maintenance, Second Edition*
Geoff Klempner and Isidor Kerszenbaum

24. *Introduction to Electrical Power Systems*
Mohamed E. El-Hawary

25. *Modeling and Control of Fuel Cells: Distributed Generation Applications*
M. Hashem Nehrir and Caisheng Wang

26. *Power Distribution System Reliability: Practical Methods and Applications*
Ali A. Chowdhury and Don O. Koval

27. *Introduction to FACTS Controllers: Theory, Modeling, and Applications*
Kalyan K. Sen and Mey Ling Sen

28. *Economic Market Design and Planning for Electric Power Systems*
James Momoh and Lamine Mili

29. *Operation and Control of Electric Energy Processing Systems*
James Momoh and Lamine Mili

30. *Restructured Electric Power Systems: Analysis of Electricity Markets with Equilibrium Models*
Xiao-Ping Zhang

31. *An Introduction to Wavelet Modulated Inverters*
S. A. Saleh and M. Azizur Rahman

32. *Control of Electric Machine Drive Systems*
Seung-Ki Sul

33. *Probabilistic Transmission System Planning*
Wenyuan Li

34. *Electricity Power Generation: The Changing Dimensions*
Digambar M. Tagare

35. *Electric Distribution Systems*
Abdelhay A. Sallam and Om P. Malik

36. *Practical Lighting Design with LEDs*
Ron Lenk and Carol Lenk

37. *High Voltage and Electrical Insulation Engineering*
Ravindra Arora and Wolfgang Mosch

38. *Maintaining Mission Critical Systems in a 24/7 Environment, Second Edition*
Peter M. Curtis

39. *Power Conversion and Control of Wind Energy Systems*
Bin Wu, Yongqiang Lang, Navid Zargari, and Samir Kouro

40. *Integration of Distributed Generation in the Power System*
Math H. J. Bollen and Fainan Hassan

41. *Doubly Fed Induction Machine: Modeling and Control for Wind Energy Generation*
Gonzalo Abad, Jesús López, Miguel Rodrigues, Luis Marroyo, and Grzegorz Iwanski

42. *High Voltage Protection for Telecommunications*
Steven W. Blume

43. *Smart Grid: Fundamentals of Design and Analysis*
James Momoh

44. *Electromechanical Motion Devices, Second Edition*
Paul Krause, Oleg Wasynczuk, and Steven Pekarek

45. *Electrical Energy Conversion and Transport: An Interactive Computer-Based Approach, Second Edition*
George G. Karady and Keith E. Holbert

46. *ARC Flash Hazard and Analysis and Mitigation*
J. C. Das

47. *Handbook of Electrical Power System Dynamics: Modeling, Stability, and Control*
Mircea Eremia and Mohammad Shahidehpour

48. *Analysis of Electric Machinery and Drive Systems, Third Edition*
Paul Krause, Oleg Wasynczuk, Scott Sudhoff, and Steven Pekarek

49. *Extruded Cables for High-Voltage Direct-Current Transmission: Advances in Research and Development*
Giovanni Mazzanti and Massimo Marzinotto

50. *Power Magnetic Devices: A Multi-Objective Design Approach*
S. D. Sudhoff

51. *Risk Assessment of Power Systems: Models, Methods, and Applications, Second Edition*
Wenyuan Li

52. *Practical Power System Operation*
Ebrahim Vaahedi

53. *The Selection Process of Biomass Materials for the Production of Bio-Fuels and Co-Firing*
Najib Altawell

54. *Electrical Insulation for Rotating Machines: Design, Evaluation, Aging, Testing, and Repair, Second Edition*
Greg C. Stone, Ian Culbert, Edward A. Boulter, and Hussein Dhirani

55. *Principles of Electrical Safety*
Peter E. Sutherland

56. *Advanced Power Electronics Converters: PWM Converters Processing AC Voltages*
Euzeli Cipriano dos Santos Jr. and Edison Roberto Cabral da Silva

57. *Optimization of Power System Operation, Second Edition*
Jizhong Zhu

58. *Power System Harmonics and Passive Filter Designs*
J. C. Das

59. *Digital Control of High-Frequency Switched-Mode Power Converters*
Luca Corradini, Dragan Maksimovic, Paolo Mattavelli, and Regan Zane

60. *Industrial Power Distribution, Second Edition*
Ralph E. Fehr, III

61. *HVDC Grids: For Offshore and Supergrid of the Future*
Dirk Van Hertem, Oriol Gomis-Bellmunt, and Jun Liang

62. *Advanced Solutions in Power Systems: HVDC, FACTS, and Artificial Intelligence*
Mircea Eremia, Chen-Ching Liu, and Abdel-Aty Edris

63. *Operation and Maintenance of Large Turbo-Generators*
Geoff Klempner and Isidor Kerszenbaum

64. *Electrical Energy Conversion and Transport: An Interactive Computer-Based Approach*
George G. Karady and Keith E. Holbert

65. *Modeling and High-Performance Control of Electric Machines*
John Chiasson

66. *Rating of Electric Power Cables in Unfavorable Thermal Environment*
George J. Anders

67. *Electric Power System Basics for the Nonelectrical Professional*
Steven W. Blume

68. *Modern Heuristic Optimization Techniques: Theory and Applications to Power Systems*
Kwang Y. Lee and Mohamed A. El-Sharkawi

69. *Real-Time Stability Assessment in Modern Power System Control Centers*
Savu C. Savulescu

70. *Optimization of Power System Operation*
Jizhong Zhu

71. *Insulators for Icing and Polluted Environments*
Masoud Farzaneh and William A. Chisholm

72. *PID and Predictive Control of Electric Devices and Power Converters Using MATLAB®/Simulink®*
Liuping Wang, Shan Chai, Dae Yoo, Lu Gan, and Ki Ng

73. *Power Grid Operation in a Market Environment: Economic Efficiency and Risk Mitigation*
Hong Chen

74. *Electric Power System Basics for the Nonelectrical Professional, Second Edition*
Steven W. Blume

75. *Energy Production Systems Engineering*
Thomas Howard Blair

76. *Model Predictive Control of Wind Energy Conversion Systems*
Venkata Yaramasu and Bin Wu

77. *Understanding Symmetrical Components for Power System Modeling*
J. C. Das

78. *High-Power Converters and AC Drives, Second Edition*
Bin Wu and Mehdi Narimani

79. *Current Signature Analysis for Condition Monitoring of Cage Induction Motors: Industrial Application and Case Histories*
William T. Thomson and Ian Culburt

80. *Introduction to Electric Power and Drive Systems*
Paul Krause, Oleg Wasynczuk, Timothy O'Connell, and Maher Hasan

81. *Instantaneous Power Theory and Applications to Power Conditioning, Second Edition*
Hirofumi Akagi, Edson Hirokazu Watanabe, and Mauricio Aredes

82. *Practical Lighting Design with LEDs, Second Edition*
Ron Lenk and Carol Lenk

83. *Introduction to AC Machine Design*
Thomas A. Lipo

84. *Advances in Electric Power and Energy Systems: Load and Price Forecasting*
Mohamed E. El-Hawary

85. *Electricity Markets: Theories and Applications*
Jeremy Lin and Fernando H. Magnago

86. *Multiphysics Simulation by Design for Electrical Machines, Power Electronics, and Drives*
Marius Rosu, Ping Zhou, Dingsheng Lin, Dan Ionel, Mircea Popescu, Frede Blaabjerg, Vandana Rallabandi, and David Staton

87. *Modular Multilevel Converters: Analysis, Control, and Applications*
Sixing Du, Apparao Dekka, Bin Wu, and Navid Zargari

88. *Electrical Railway Transportation Systems*
Morris Brenna, Federica Foiadelli, and Dario Zaninelli

89. *Energy Processing and Smart Grid*
James A. Momoh

90. *Handbook of Large Turbo-Generator Operation and Maintenance, Third Edition*
Geoff Klempner and Isidor Kerszenbaum

91. *Advanced Control of Doubly Fed Induction Generator for Wind Power Systems*
Dehong Xu, Frede Blaabjerg, Wenjie Chen, Nan Zhu

92. *Electric Distribution Systems, Second Edition*
Abdelhay A. Sallam and Om P. Malik

93. *Power Electronics in Renewable Energy Systems and Smart Grid: Technology and Applications*
Bimal K. Bose

94. *Distributed Fiber Optic Sensing and Dynamic Rating of Power Cables*
Sudhakar Cherukupalli and George J. Anders

95. *Power System and Control and Stability, Third Edition*
Vijay Vittal, James D. McCalley, Paul M. Anderson, A. A. Fouad

96. *Electromechanical Motion Devices: Rotating Magnetic Field-Based Analysis and Online Animations, Third Edition*
Paul Krause, Oleg Wasynczuk, Steven D. Pekarek, and Timothy O'Connell

97. *Applications of Modern Heuristic Optimization Methods in Power and Energy Systems*
Kwang Y. Lee and Zita A. Vale

98. *Handbook of Large Hydro Generators: Operation and Maintenance*
Glenn Mottershead, Stefano Bomben, Isidor Kerszenbaum, Geoff Klempner

99. *Advances in Electric Power and Energy: Static State Estimation*
Mohamed E. El-Hawary

100. *Arc Flash Hazard Analysis and Mitigation, Second Edition*
J.C. Das

101. *Distributed Energy Management of Electrical Power Systems*
Yinliang Xu, Wei Zhang, Wenxin Liu, Wen Yu

102. *A New Swing-Contract Design for Wholesale Power Markets*
Leigh Tesfatsion

Printed and bound by CPI Group (UK) Ltd, Croydon, CR0 4YY

17/09/2026

14964318-0002